MW01672613

Clinical Engineering: principles and practices

Clinical Engineering: principles and practices

edited by

JOHN G. WEBSTER and ALBERT M. COOK

Prentice-Hall, Inc. Englewood Cliffs, New Jersey 07632

Library of Congress Cataloging in Publication Data
Main entry under title:

Clinical engineering.

Includes bibliographies and index.
1. Hospitals — Engineering departments. 2. Hospitals — Furniture, equipment, etc. 3. Biomedical engineering. I. Webster, John G., date II. Cook, Albert M. [DNLM: 1. Biomedical engineering. 2. Technology, Medical. 3. Delivery of health care — United States. QT34.3 C641]
RA967.8.C45 658.2'7 78-25776
ISBN 0-13-137737-X

Editorial/production supervision by Phyllis Springmeyer and Steven Bobker
Interior design by Phyllis Hall
Cover design by Edsal Enterprises
Manufacturing buyer: Gordon Osbourne

© 1979 by Prentice-Hall, Inc., Englewood Cliffs, N.J. 07632

Printed in the United States of America

All rights reserved. No part of this book may be reproduced in any form or by any means without permission in writing from the publisher.

10 9 8 7 6 5 4 3 2 1

Prentice-Hall International, Inc., *London*
Prentice-Hall of Australia Pty. Limited, *Sydney*
Prentice-Hall of Canada, Ltd., *Toronto*
Prentice-Hall of India Private Limited, *New Delhi*
Prentice-Hall of Japan, Inc., *Tokyo*
Prentice-Hall of Southeast Asia Pte. Ltd., *Singapore*
Whitehall Books Limited, *Wellington, New Zealand*

Contents

2

10

11

12

13
Management and Supervision 373

Michael O. Brinkman

14
Legal, Professional, and Ethical Aspects 407

Richard J. Jendrucko and Richard E. Favoriti

15
Resources 435

Robert A. Peura and John G. Webster

Contributing Authors

Bruce H. Barkalow, Integral Biomedical Engineering, Inc., 601 University Ave., Suite 261, Sacramento, CA 95825.

William F. Betts, Biomedical Engineering, Arizona Medical Center, Tucson, AZ 85724.

Michael O. Brinkman, Hospital Maintenance Consultants, Box 100, Lebanon, WI 53047.

Robert J. Cangelosi, Bureau of Medical Devices—HFK 300, 8757 Georgia Avenue, Silver Springs, MD 20910.

David E. Clapp, Bioengineering Program, Texas A&M University, College Station, TX 77843.

Burt Dodson, Jr., Carolina Hospital and Health Services, P.O. Box 4587, Charlotte, NC 28204.

Dean D. Duncan, Department of Industrial Engineering, New Mexico State University, Las Cruces, NM 88003.

Contributing Authors

Richard E. Favoriti, Burditt and Calkins, 135 South LaSalle Street, Chicago, IL 60603.

James J. Freeman, Department of Electrical Engineering, University of Detroit, Detroit, MI 48221.

Emanuel Furst, Biomedical Engineering, Arizona Medical Center, Tucson, AZ 85724.

David Gordon, Clinical Engineering Services Group, Johns Hopkins Hospital, Baltimore, MD 21205.

Richard J. Jendrucko, Department of Engineering Science and Mechanics, University of Tennessee, Knoxville, TN 37916.

George H. Mack, University Hospitals, Tucson, AZ 85724.

Terence F. Moore, Midland Hospital Center, 4005 Orchard Drive, Midland, MI 48640.

Kenneth C. Mylrea, Department of Electrical Engineering, University of Arizona, Tucson, AZ 85721.

J. Barry Oakes, Applied Physics Laboratory, Johns Hopkins Road, Laurel, MD 20810.

Lee E. Ostrander, Center for Biomedical Engineering, Rensselaer Polytechnic Institute, Troy, NY 12181.

Robert A. Peura, Electrical Engineering Department, Worcester Polytechnic Institute, Worcester, MA 01609.

Alan K. Reeter, Biomedical Engineering, Arizona Medical Center, Tucson, AZ 85724.

Fred J. Savaglio, Shared Clinical Engineering Service, 3601 South 42nd Street, Suite C, Phoenix, AZ 85040.

Earl A. Simendinger, University Hospitals of Cleveland, 2101 Edelberg Rd., Cleveland, OH 44101.

James B. Smathers, Bioengineering Program, Texas A&M University, College Station, TX 77843.

David L. Stoner, Bioengineering Program, Texas A&M University, College Station, TX 77843.

Stephen E. Thompson, Center for Biomedical Engineering, Rensselaer Polytechnic Institute, Troy, NY 12181.

W. Sanford Topham, Department of Biomedical Engineering, Case Western Reserve University, Cleveland, OH 44106.

Preface

This book describes how to set up a clinical engineering program in a hospital. It covers both the regulations that require a clinical engineering program and the directions for implementing these regulations. It also provides useful information for the improvement of existing programs.

There is a special reason for this book. The new engineer who works in the medical device industry usually works with more experienced engineers. They generally act as a guide for him, and provide him with a valuable and effective internship. In contrast, the new engineer who works in a hospital may be the only engineer there. His coworkers are health-care personnel who have little knowledge of medical equipment and cannot offer him an experienced perspective. Thus this book seeks to fill this gap by providing information and guidance both on what he needs to accomplish and means for accomplishing it effectively.

The first five chapters provide basic background knowledge

necessary for productive work in a hospital. Chapter 1 describes the health-care system: the hospital, health maintenance organizations, and governmental planning organizations. Chapter 2 presents the hospital organizational structure: the administrator, the medical staff, etc. Chapter 3 analyzes health-care economics: what causes rising health-care costs and who pays for them. Chapter 4 sets forth the standards required by law and by accrediting agencies. Chapter 5 describes how information flows in a hospital: both physical samples and administrative orders.

Chapter 6 provides an overall view of the clinical engineering program, which is then detailed in the rest of the book. Chapter 7 presents the safety program, both electrical and nonelectrical. Chapter 8 describes the procedures for equipment acquisition, while Chapter 9 organizes the required documentation. Chapter 10 discusses both how to organize repair when necessary and how to minimize it by preventive maintenance. Because the clinical engineer is involved in installation and remodeling, Chapter 11 covers building facilities. Because he has a responsibility for educating health-care personnel as well as himself, Chapter 12 covers education. Since he must manage his program and may direct the work of others, Chapter 13 describes management and supervision. Because health care presents special risks, Chapter 14 covers legal, ethical, and professional aspects. Finally, since this book cannot cover all aspects, Chapter 15 lists the most important sources for acquiring more information.

This book is primarily written for a senior–graduate course in engineering and thus learning objectives are listed at the beginning of each chapter. The material emphasizes the organizational and managerial aspects of the clinical engineering program. Since the book's emphasis is not highly technical, it should also appeal to biomedical equipment technicians, hospital administrators, medical equipment designers, and sales personnel, as well as others who have an interest in proper operation of medical equipment.

This book is the outgrowth of workshops and surveys made by the editors (Webster, 1976; Webster and Cook, 1976; Cook, 1978), and we wish to thank all who provided us information that helped to define the need and scope of this book.

The editors and authors would welcome your suggestions for corrections and improvement of subsequent printings and editions.

John G. Webster
Dept. of Electrical & Computer Engineering
University of Wisconsin
1425 Johnson Drive
Madison, WI 53706

Albert M. Cook
Biomedical Engineering Program
California State University
6000 Jay Street
Sacramento, CA 95819

References

Webster, J. G. (1976). What every clinical engineer ought to know. In J. B. Oakes (ed.), Proceedings of the conference on clinical engineering education, Baltimore, MD, May 2-4, 1976.

Webster, J. G. and A. M. Cook (1976). Workshop on "What every clinical engineer ought to know." *Proc. Ann. Conf. Eng. Med. Biol.* 18:474.

Cook, A. M. (1978). Skills required for clinical engineering practice—results of a survey. *J. Clin. Eng.*, 3(3):279-283.

Clinical Engineering: principles and practices

1

Health-Care Delivery Systems

TERENCE F. MOORE and EARL A. SIMENDINGER

1.1 Learning Objectives

Upon completing this chapter you should be able to:

1. List the major elements of the health-care delivery system and describe their interrelationships.
2. Explain the major limitations of the current U.S. health-care delivery system.
3. Describe the major factors affecting demand for health services.
4. Describe the health-care planning process and the three major roles of government in health-care delivery.

1.2 Introduction

Technology has been an integral part of medicine for generations. Engineers and physical scientists have made contributions to biological and medical knowledge since antiquity, and physicians and biologists have contributed to the advancement of the physical sciences (Geddes, 1977). As systems for health-care delivery have evolved, engineers have continued to play a role and technology has been a major part of this evolution. Although the marriage of technology and health care has been stormy at times, there have been significant advances and changes in the health-care system brought about by technology. Despite past successes, many problems plague modern health-care systems. Some of these problems have their roots in technological change and some depend on technology for solution. This chapter places the health-care delivery system in perspective for the clinical engineer.

1.3 Historical Overview

Medical practice

During the 19th century, for geographic and historical reasons, American medical practice grew away from the European practice of medicine. The resulting methods were basically influenced by Europe, but they were different; the doctors' education and background were different; medical schools were different; the hospitals were differently organized, sponsored, and funded. And America invented group practice.

By the 1930s many of the problems related to getting and giving medical care were well established. Because there were fewer practitioners trained, there was less competition. Rapid urbanization after 1900 reduced the number of doctors willing to move out into the rural community to practice in impoverished areas. An increasing number of specialists changed the ratio of patients to family doctors drastically. Moreover, the cost of medical care was rising.

A major change in American medical practice occurred in 1910, when the Flexner report, *Medical Education in the U.S. and Canada*, appeared. It created a sensation at the time, and was harshly critical of the poor quality of medical education generally and the shockingly poor quality of teachers and practitioners. The Flexner report received much of the credit for the events that resulted after its publication. As a result of this report, the quality of medical education was significantly upgraded and many of the poorer-quality medical education programs were eliminated.

In 1965 the Coggershall report, *Planning for Medical Progress*, perhaps had as much impact as the Flexner report of 1910. Stressing the existence of a manpower crisis in the health services, the Coggershall report called for creation of 9 new university health science centers and 126 area health education centers to bring central health services to within one hour's driving time for over 95 percent of Americans. The Coggershall report also stated that there must be more attention paid to the medical needs of all society, with a consequent shift toward professionalism and away from entrepreneurial ends.

Since the turn of the century, disease patterns have changed considerably as treatments have changed. Many communicable diseases have almost been eliminated. Polio, measles, and mumps have all been conquered except for cases where preventive measures have not been administered. Even tuberculosis, with mass detection techniques and new drugs, is slowly being brought under control. The death rate has been reduced and the average length of hospital stay has been cut in half in the past 15 years.

Acute episodes of chronic diseases, however, have replaced acute diseases as the chief causes of death. In 1900, 40 of every 100,000 of the U.S. population died of diphtheria; by 1940 this figure was reduced to 1. There were 210 deaths per 100,000 population due to pneumonia in 1900; by 1940 there were 70. However, deaths from heart disease rose from 137 in 1900 to 292 in 1940, and the cancer death rate went from 64 to 210 in the same time period (Durbin and Springall, 1969). In the future, new cures and sociological changes will undoubtedly influence the patterns of disease, and consequently, the health-care delivery system. Technology has also brought about changes in the distribution of diseases. For example, reducing the time of transportation of heart attack victims to hospitals has reduced the fatality rate during transportation (Brown, 1972).

Technology in health care

Health-care delivery and medicine have been dependent on technology for many years. Geddes (1977) has traced the role of instruments in medical diagnosis. Clinical and basic science investigations have always depended on careful observation and measurement. As investigators began to exceed the limits of their natural senses they needed instruments that would help them to study physiological events. As early as the 14th century, instrumentation was employed to improve the observation of physiological processes. Instrumentation was included in the repertoire of the clinician more gradually. The development of the X-ray, the ECG, and chemical analysis equipment applicable to biological specimens have all had profound effects on medical practice. Some of these developments took place prior to the 1940s, but the major growth in diagnostic and therapeutic equipment came after World War II (Rushmer, 1972). The growth in medical devices since that time has been exponential, and has had profound effects on health-care delivery.

The influx of technology has resulted in new subspecialties in medicine, new allied health fields, and generally increasing costs for health care. This cost has been justified on the basis of better care provided through increasingly sophisticated instrumentation for diagnosis and therapy. It is the large amount of equipment around which medical care is centered that has led to the increasing role for engineers, particularly clinical engineers, in the health-care system.

Health manpower

Medical practice has changed from a group of general practitioners evenly distributed throughout the country to a group of specialists concentrated in the cities (Rushmer, 1972). These changes represent more than geographic alterations or increased specialization. The physician of the early 1900s delivered care in the home, made minimal use of hospitals and depended on families for nursing care. This medical practitioner was not expected to cure all ills, but rather to ease the patient's and his family's minds and provide some relief (Rushmer, 1972).

Technology played a large role in changing this situation. As more equipment became available and was concentrated in hospitals, the hospital became the center of medical care. Technology helped improve monitoring and reduce the risk of surgical procedures. It also helped create medical subspecialties which were possible only with increasingly sophisticated equipment.

With each advance in technology, new technicians and other allied health personnel were required. Specialized technicians became necessary for ECGs, blood-pump operation, heart catheterization, and other specific functions. There are now a large number of allied health professionals whose careers depend on technology. These advances in health care have, however, resulted in a reduction in the number of physicians in primary care. Thirty years ago, only 17 percent of physicians were in specialties while today 70 percent are (Rushmer, 1972).

A major problem facing our nation's health-care delivery system is that of providing the necessary physician manpower to deliver the health services that the public demands. There are about 330,000 M.D.'s today, but only approximately 215,000 actually treat patients. Other physician activities include administration, industry, and public health as well as full-time teaching and research. It is obvious, however, that increasing the number of physicians alone will not solve the problems of health-care delivery.

An additional problem is the maldistribution of physicians, who are reluctant to practice in rural or inner city areas. The burden is on the American people to seek help and find it when they are sick. If they live in the inner city, they may find their physicians have left and that help can be had only by going to the city hospital and waiting for hours. If they are sick while out of town or after hours, they will have difficulty finding physicians to see them. If they are elderly, disabled, or very ill and ask a physician to come to them, they will most likely not see one.

1.4 The Illness System — a Nonsystem

The word "system" is a euphemism that is often used to characterize the functions of health care. In reality, as one authority states, "the most prominent characteristic of the existing community health system is its complexity," and, further, "health care is characterized by a multiplicity of poorly coordinated subsystems, many of which operate almost entirely independent of each other." Because of its complexity and the existence of numerous subsystems, health care is, in fact, disorganized, creating problems in the distribution and quality of health care and the inefficient use of the services that exist (Hyman, 1976). Technology, if sensibly applied, may hold promise for correcting these problems (Brown, 1978). For example, the application of engineering systems theory to medical records, manpower needs, automation, and home-care needs could lead to a more homogenous system of health-care delivery.

Most medical care is performed today by the solo practitioner, the physician who is in practice by himself. The solo practitioner, although backed by consultation with other physicians, or laboratories, and by affiliation with a hospital, operates so small an enterprise that its very size may ultimately mitigate against efficiency. It has been found that until a certain scale of operation is reached, it is difficult to utilize scarce medical and technical skills to their best advantage. Thus, the solo practitioner ends up wasting much of his skill on tasks that do not require medical expertise. This decrease in efficiency can severely reduce the physician's ability to provide quality care to suit the varied needs of the patient population. Moreover, the workload of the average physician is often so great that it limits his time and narrows his perspective in keeping abreast of both medical advances and social change (Committee for Economic Development, 1973).

Entry into the system

Because the illness system is actually a "nonsystem" the following occurs: (1) The infant death rate is 75 percent higher for the poor; (2) Over one-half of adults need dental care, while less than one-fourth of the affluent adults need such care; (3) The incidence of heart conditions is four times as high for the lower as for the higher income group; (4) The percentage of persons with limitation of activity from chronic conditions is twice as high for those with low incomes as for those with moderate or high incomes; (5) Lower income groups have 60 percent more restrictive activity days from illness (Ward, 1976). These disparities result in part from difficulties experienced in entering the health-care delivery "system."

Every day three million Americans seek medical care. Some find it; others do not. Some are helped by it; others are not. Another twenty million Americans probably ought to enter the daily search for medical help, but are not healthy enough, rich enough, or enterprising enough to try. The obstacles are enormous. Health care is scarce and expensive and usually offered in an atmosphere of mystery and unaccountability. For many, it is obtained only at the price of humiliation, dependence, or insults (Ehrenreich, 1974).

Fractionation of services

An American may find himself suspended among a dozen or more individuals or organizations who deal with health care. There are the physicians, dentists, hospitals, laboratories, nursing homes, insurance companies, and federal social security and state and county offices which are supposed to help arrange or pay for family health care. Each of the following represents a component of the health-care system: drugstores, health insurance companies, hospitals, laboratories, physicians (a wide range of types), special nursing care, psychiatric consultation, ambulance service, special medical supplies or equipment.

Each of these organizations and individuals acts as an independent business. Each specialist and institution of the medical care system is geared to a particular kind of illness, while no one shows concern for the overall health care of an individual. The system does not organize itself around the individual to offer preventive care, nor does it adequately organize around him to meet his health-care needs if he falls ill. Each one of the specialists and institutions frequently assumes that his individual responsibility begins and ends when the person crosses his doorstep.

As noted previously, most physicians are in solo practice in America or in practice with other physicians with the same specialty. If you need the services of another specialist, one physician may refer you to another, but it is up to you to find that recommended physician, make and then wait for an appointment, and get to his office—frequently at some distance. You also frequently find that you have to start from the beginning in describing your health problems, since the first physician will not have called or discussed the case with the new physician or given him any of your background. Similarly, hospitals, laboratories, nursing homes, and other organizations frequently act as independent businesses. A physician may inform a patient he no longer needs hospital care but instead needs skilled nursing-home care. Often neither the physician nor the hospital will assure that the patient finds and gets to a nursing home. As with physician referrals, the patient may be given a list of names, but be left on his own to make all of the arrangements.

Financing arrangements are no less difficult. The health insurance statement is not obliged to describe exactly how the patient's policy fits in with Medicare or with the other policies that a person may own, or that he may buy if he chooses. In fact most families have no way of knowing the real value of their insurance until they fall ill (Kennedy, 1973).

1.5 Demand for Health Services

Elasticity

Studying influences of demand for any commodity is difficult because at any one time a number of influences are occurring and the effects on any one variable cannot be isolated. Changes in demand can be observed from two perspectives: changes in purchasing activity over time, and purchasing among different groups at the same time. In both cases, there are a number of differences which might account for the variations in price behavior. Over a period of time, there will have usually been changes in such important variables as income, relative prices, and age composition of the population. The concept of elasticity is useful as a reference point in judging demand. Elasticity refers to the sensitivity of demand to the variables in question, price and income in particular.

The income elasticity of demand measures the quantitative change in demand related to a given percentage change in income (percent change in quantity purchase divided by percent change in income). The change in quantity is measured in units, such as physician visits or days of hospitalization, which in health care is sometimes called the "use measurement."

Price elasticity measures the percent change in demand associated with the given percentage change in price. The change ideally should be measured in units, or use, but expenditures may be substituted. Since the change in expenditures necessarily includes the change in price, the price elasticity in terms of expenditures must be interpreted quite differently from use elasticity. Most discussions of price elasticity refer to use even though use may not be directly measured (Ward, 1976).

As with other goods, it is generally assumed that prices of medical care have a negative effect on demand, that is, the higher the price the lower the demand, other things being equal. It is also assumed that the extent of the response of demand is less than proportional to the change in price, so that demand for medical care is price-inelastic. This assumption is difficult to verify, not only because the problem of separating prices from other variables, as mentioned above, but also because of some special characteristics of medical care. An important one is the absence of measurement of the quantity of medical care.

However, crude evidence points to price inelasticity with respect to demand. This is consistent with the belief that medical care is a purchase over which the buyer has little discretion. At the same time that medical care expenditures have risen more than other prices, medical care expenditures have risen as a proportion of total expenditures.

Different components of hospital expenditures show different elasticities with respect to income. Hospitalization appears to be less sensitive to income than physician and dentist services, although the problem is difficult to analyze because of the fact that the patient bears less of the expense because of insurance coverage.

Factors affecting patient demand for health care

Medical care has been said to be solely a "need" rather than a "want." Economic considerations, such as income and price, therefore, have little or no influence in determining the amount used. For care that is expected to substantially improve well-being, and that can be obtained at a cost that does not strain a family's resources, this may well be the case. Often, however, the type of illness present and its prognosis are uncertain. For example, the distinction between a severe cold and pneumonia is not clear-cut, at least to the untrained person. Chest pains may indicate indigestion or a serious heart condition. In these and similar instances, income and price would be expected to affect decisions involving the purchase of medical care.

The incidence of illness. Need is generated by the incidence of illness while demand is generated by the interrelationship of illness with other factors. The unexpected nature of illness causes a fluctuating demand for health services.

Economic factors. Prices and incomes theoretically affect not only a person's decision whether or not to seek medical care, but also the extent of the care once treatment is undertaken. For example, the effect of prices or of income may or may not have much effect on whether or not a maternity patient goes to the hospital (although it may influence the choice of the hospital), but once she has been admitted, it may affect the length of her stay.

The effects of prices will be discussed in Sec. 3.1, but insurance has a complicating influence which merits considerable attention.

The widespread practice of providing health insurance through employment tends to increase medical demand because it makes insurance available to those who might not otherwise opt for it. Even when an employee contribution is required there is an element of subsidy in that the employer pays a portion, thus tipping the balance in favor of coverage.

One aspect of insurance tends to reduce medical demand: that is the administrative costs of the insurance. In 1962 the operating expenses for private health insurance were 51 percent of premium income. For the nonprofit Blue Cross plans, the ratio was a little less than 7 percent. The ratio for individual (as opposed to group policies) was 47 percent. These administrative costs require a diversion of funds that would otherwise be available for the purchase of medical services.

Demand and supply

To assess the level of demand, the actual amount of medical services purchased or received is normally observed. Such usage, however, does not solely reflect demand, but is instead a combination of both supply and demand. With most goods, demand is taken to be independent of supply. Forces determining demand and supply come together and establish a price that will clear the market. The actual amount supplied at any one time becomes equal to the actual amount demanded. With medical care, supply and demand are not independent because of the role of the physician (the supplier) in prescribing the amount of medical care the patient should receive.

Although the influence of supply and demand is of special importance in medical care, it is not unique to it. The automobile mechanic and television repair man can tell a customer what to buy and the plumber decides what needs fixing. In a still broader context, there are some economists who think that the industrial society decides what it will produce, and then, through advertising and marketing techniques, creates a demand for it. But in medical care, more so than other cases, the consumer is left without discretion in some instances, especially if his very life is threatened.

Excess demand

One disturbing characteristic of some health-care markets is the failure of price to reach an equilibrium level (a level where the quantity demanded and the quantity supplied are equal). For instance, the market for house calls appears to be characterized by excess demand (see Fig. 1.1).

The "going price," about $20 per visit, is not high enough to bring supply and demand into balance. The quantity (number of house calls) of patients that are willing and able to pay that price is much greater than the quantity physicians are willing to supply. Some observers believe that the market for physician services in general is characterized by excess demand.

Excess supply

The market for general surgery, on the other hand, can best be described as an example of excess supply (see Fig. 1.1).

At the going price for most general surgical procedures, the quantity which surgeons are willing and able to do is much greater than the quantity demanded. The condition of excess supply is also probably present for many types of specialty surgery such as ophthal-

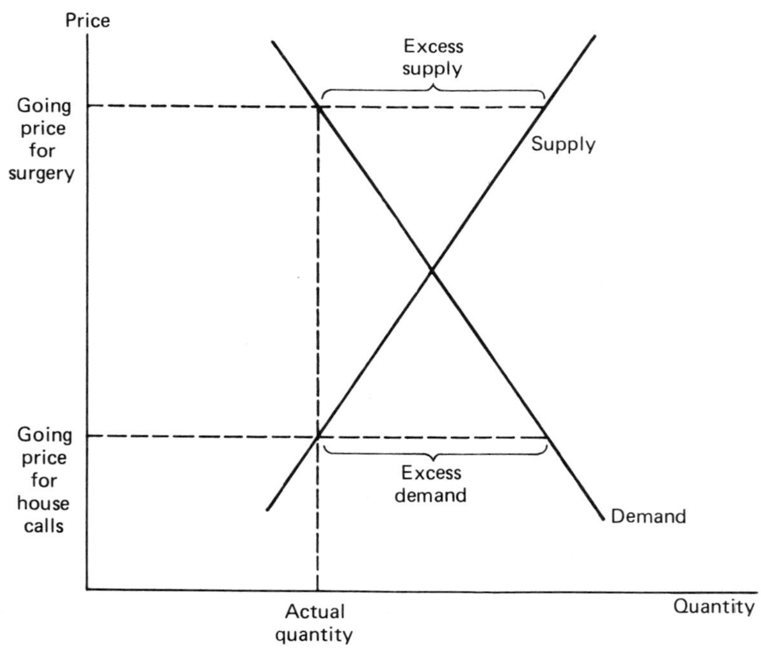

Figure 1.1 Excess demand and supply.

mology and gynecology. The persistence of a disequilibrium price is a clear indication that the market departs substantially from the competitive norm. This departure is further explored in Sec. 3.7 (McKinlay, 1973).

More precise estimates of the effect of supply and utilization were provided by Martin Feldstein (see Ward (1976)). He estimates that an increase in hospital beds per capita has a direct effect on utilization even in the absence of any effect on price. In any increase in beds, half of them would be used without a fall in price, but the remaining half would only be absorbed by a fall in price. This analysis is used to explain the observed phenomenon that hospitals tend to operate at the same occupancy rate after they expand capacity.

1.6 Outpatient and Emergency Care

Outpatient care

The shift in emphasis from inpatient, those admitted to the hospital, to outpatient or ambulatory care in the past fifteen years has been phenomenal. The figure commonly cited in the health-care field for outpatient visits is one in three. This ratio means that one person in three visits a hospital or clinic as an outpatient each year. This compares to one person in seven actually being admitted to a hospital each year.

Outpatient care is considered less costly compared with inpatient hospital care and that is one of the chief reasons for the emphasis on outpatient activities by third-party payers, government, and others.

The average hospital employs three to four people for each patient. Compare this to an outpatient department that in 1947 recorded 59 outpatient visits per clinic employee at a cost of $2.66 per visit. By 1966 this number had risen to 72 visits for every clinic employee at a cost of $6.93 per visit (Durbin and Springall, 1969). In 1975 the cost had increased to $10.45 per visit but the number of visits per employee remained relatively constant at 74 to one.

There are numerous other factors that are influencing the development of outpatient facilities and the shift to outpatient care. According to Durbin and Springall (1969), these include: (1) a young population which has been encouraged to have checkups; (2) the population explosion; (3) the rural-to-urban shift; (4) the manpower shortage (this makes it necessary to group physicians to greater advantage); and (5) the increased ability of equipment to provide quality and timely diagnostic reports. The latter point reflects the importance of technology in determining patterns of health care.

No one knows for certain where the trend toward outpatient

care will take the health-care system, but it is almost certain to accelerate in the future because of the support it is being given by third-party payers and planning agencies.

Emergency care

Emergency care has developed into a specialty of its own with greatly expanded hospital emergency rooms and statewide emergency-care systems. Unfortunately, because of the physician shortage in many areas, and the mobility of the population, emergency rooms are being used more for nonemergency care than for trauma victims. Patients who cannot find care anywhere other than the local emergency room are using it as they would a physician's office.

The emphasis in "emergency medicine" has created a whole new group of health-care professionals. The paramedic, the emergency-care nurse, and the trauma physician are now an integral part of our health-care system.

1.7 Health Maintenance Organizations (HMOs)

One attempt to stem the increase in health care costs has been the development of the Health Maintenance Organizations (HMOs). The HMO is a generic term that describes an organized system of health care which provides a full range of comprehensive services to an enrolled population, in return for a fixed sum of money, agreed upon in advance.

While HMOs may vary in structural design, they have three essential components that distinguish them from the more traditional health providers: (1) They are organized systems delivering comprehensive health care; (2) They service a voluntarily enrolled population; and (3) They are on a basis of fixed-contract cost.

A well-known expert in the health-care field defines the essential elements of an HMO in a somewhat different way. He states that they include the following:

1. A set of specific individuals for whom the delivery unit is responsible.
2. A prepaid comprehensive set of health services, preventive and therapeutic, ambulatory and institutional, available from the delivery unit.
3. A set of specific physicians and institutions to serve the delivery unit.
4. A combined single capitation payment for all covered services provided by the delivery unit (Brown, 1972).

Thus, the HMO concept represents one approach intended to correct the central criticism of the health-care delivery system which is that it is a "nonsystem." At the present time, the health-care delivery system is not bringing into a systematic relationship a set of users, a set of comprehensive services, a set of providers, and a given amount of dollars.

There are basically two models of HMOs as they pertain to physician practices. The first is a group practice model, in which physicians are organized in a common facility; and the second is the individual practice model, in which physicians continue to practice solo on a small-group basis. In the first model, the groups themselves bear the risk-sharing responsibilities for their subscribers while, in the second, participating physicians are reimbursed by a central HMO on a fee-for-service basis. In both models, the HMO promotes cooperation among physicians, hospitals, and patients (Moore, 1975).

Effect of HMOs

By some estimates, HMOs reduce hospitalization by as much as 50 percent. Widespread competition from HMOs has put a brake on the wasteful behavior of hospitals and of physicians who charge on a fee-for-service basis. But obstruction by the AMA—which called HMOs "medical soviets" back in the 1930s—has severely retarded the growth of such groups. The major barrier against them has been the "corporate practice of medicine" doctrine; organized medicine secured legislation or court decisions in many states to make it illegal for a corporation organized by laymen to directly employ physicians. Medical "ethics" have also formed a barrier by prohibiting among other things, contracts where the business is awarded to the lowest bidder, working for a layman on a salaried basis, and any kind of advertising (Ehrbar, 1977).

1.8 Health-Care Planning

The need for health-care planning stems from the comments that the "health-care system is a nonsystem," as discussed in Sec. 1.2. Today, the principle of state intervention and control in health matters is admitted; the only variation is in the greater or lesser efficiency of the intervention and in the greater or lesser frankness with which the role of the state is admitted.

Legislation

Some of the more important pieces of legislation and actions that have taken place since the 1900s as they relate to health-care planning are as follows:

- 1912: Creation of the Public Health Service provides legitimacy to public health as a national issue.
- 1920: The New York Academy of Medicine studies 180 hospitals in the first formal recognition of the need to plan hospital usage.
- 1930: Health-care delivery coordination and planning initiated by private nongovernment agencies at state and local levels.
- 1935: "The Need for More Hospitals in Rural Areas" by A. B. and P. Mills, concerned with service areas of hospitals, i.e., 50-mile radius for population base.
- 1946: Hospital Survey and Construction Act (Hill–Burton Act, PL 79-725) provided federal aid to states for hospital facilities. The act mandated the states:
  - To create a hospital planning council.
  - To survey existing facilities (bed count divided by population).
  - To coordinate new construction with demonstrated need for existing facilities.
- 1954: Hill–Harris amendments to Hill–Burton revised and expanded the program:
  - To include funding for modernization or replacement of public and nonprofit hospitals.
  - Direction changed from rural hospitals to altering urban facilities.
  - Emphasis from construction to planning of health services.
  - National health planning introduced because of vast expenditures of federal dollars.
- 1963: Commission on Heart Disease, Cancer and Stroke recommended steps to reduce the incidence of these illnesses through new knowledge and more complete utilization of medical knowledge.
- 1965: Heart Disease, Cancer and Stroke Amendments of 1965 (PL 89-239):
  - Establishment of the Regional Medical Programs (RMP).
  - Categorical in approach.
- 1966: Partnership for Health Act (PL 89-749) provided the opportunity for the consumer-dominated Comprehensive Health Planning (CHP) areawide agencies to be created across the major portion of the nation. Initially, the agencies were voluntary without authority to enforce the plans developed for the health system.
- 1966–1970: National Commission on Community Health Services study recommended:
  - Community health services need greater federal participation.
  - Comprehensive health planning must address all levels of health care.
  - Regional and areawide planning bodies must be established to address problem health areas.

- A single system must be established to provide personal health services.

- 1972: Social Security Amendments Act (PL 92-603) passed by Congress to insure efficient use of scarce health funds. Section #1122, Capital Expenditure Reimbursement, was to serve the triple purpose:
  - To give CHP agencies added authority.
  - To reduce the cost of health facilities through legislation.
  - To induce the health provider to become involved in planning.

  Planning was essentially reactive rather than proactive.

- 1975: National Health Planning and Resource Development Act (PL 93-641) passed.
  - Health System Agency mandated to develop a Health Systems Plan and an Annual Implementation Plan.
  - Improves the coordination between the areawide and state levels.
  - Hospital-bed expansion programs of 10 beds or 10 percent capacity must be reviewed.
  - State agency must issue written findings as to cost, efficiency, and appropriateness of a proposed inpatient service, compared with similar or existing facilities.
  - Organized ambulatory care, site acquisition, purchase of facilities, and home health care are exempt from review at this time.

Planning methodology and definitions

The usual method employed by health planners is first to define the term "planning" and discuss the dimensions of health it embraces. When applied to areawide planning, such planning encompasses the overall health needs of a geographic area to determine how these needs can be met in the most effective and economical manner with existing and future facilities and programs. This is accomplished by determining the population at risk, its level of health, the availability of health resources, and the utilization of existing health services, in order to identify unmet needs and develop programs to meet these needs. This definition and its clarifying statement emphasize the elements of a process by which health planning can be done, a need to coordinate the persons involved in determining the needs, the making of decisions by persons in a position to implement these actions, the future orientation of the planning, and its continuity with the present planning cycle. Planning may also be defined as a means of dividing tasks into manageable pieces, matching resources, and providing for their application in an orderly and timely basis for the purpose of accomplishing complex tasks (Hyman, 1976).

The planning process

In 1967 the U.S. Dept. of Health, Education and Welfare's (HEW) guidelines for comprehensive health-planning grants for the states outlined a number of functions a comprehensive health-planning agency should perform. Among these are undertaking surveys, determining problem areas, identifying needed services, selecting goals and priorities, developing short- and long-range policy and action recommendations, and designing evaluation criteria. These functions, in fact, embrace the traditional steps of the planning process. By stating them as functions, HEW necessarily does not perceive them as being linked to each other in a rational and logical cycle of planning. Different functions can be performed at different times and out of context with each other. If this should occur, the logical sequence of the planning process would be destroyed, because each step in a process depends upon the preceding step in a cause-effect relationship.

Goals of comprehensive health-planning agencies

Health goals are desired conditions of the health of people or directions the health system or related systems should move. Goals are future-oriented and they represent ideal states considered to be necessary for the positive health which society would like to attain. In addition, they are also value-judgments which society confirms it would like to have through its many devices and mechanisms such as government legislation, public forums, and mass media.

In Public Law 89-749, Congress stated its general goal in these terms: "fulfillment of our national purpose depends upon promoting and assuring the highest level of health attainable for every person in the environment which contributes positively to individual and family living." Many objectives are statea in general types of terms; however, it is useful for objectives to be stated in terms that permit measurement. With respect to a longevity goal, a negative objective might be stated in terms of reducing the number of deaths due to heart attack, stroke, and cancer from 630 per 100,000 population to 500 per 100,000 population by 2 years from now. If analysis concludes that the takeoff point for rapid increase of deaths due to these conditions is in the 45-to-64 age bracket, any emphasis would be more specifically defined by age.

1.9 The Government's Role in the Delivery of Health Care

Recently the medical profession has felt itself besieged by all sorts of pressures from the federal government. Why does the government intervene in the marketplace to change certain ground rules? Economists hold there are three justifiable reasons: One is to supply services—often called peer public bids—that simply cannot be provided in the normal market setting. We could not maintain a very good army if different persons decide to buy one day per month of a soldier's time or one shell for one tank or one life preserver for a ship.

The second reason why government intervenes in economic affairs is to affect the income distribution. Typically, in this country, such intervention is designed to benefit only those persons who suffer the grossest inequalities of income. Hence, we have social security programs and welfare programs which tax the more affluent and redistribute resources to the poor.

The third area of government intervention is in purchasing goods and services for the market, which operates at less than full productivity. These quasi-public goods are such things as highways and research. Certainly highways and bridges could be private ventures, and we could pay each time we drove over them. Research could also be purchased privately. There is a general feeling, however, that not enough research or highways or bridges would be produced to meet the public need. Therefore, the government intervenes and buys some of these goods and services from producers.

Defining medical care as a right, not a privilege, implies a number of long-run and serious changes in the organization of the delivery system for health services. Assuming that the government is serious in its purpose, there are several options open for action.

The first is for the government to set itself up as a major producer of medical services. As a second option, the government might shun the production route and become a new participant, acting in the consumer interest both for services directly and for production of medical services in general. The government appears to have selected the second option and is now moving rapidly and forcibly to enlarge this role as an advocate for the consumer of health services. In fact, the government seems to be reducing the producer role it has had, in part, in the past.

Traditionally, public agencies have provided health services through city health departments, county hospitals, and city clinics. The passage of various pieces of legislation in the past few years has literally worked to destroy many of these governmental health services, and the government now has undertaken to create incentives for private producers to replace these public organizations (Levy and Loomba, 1973).

1.10 Study Questions

1.1 List two results of the Flexner report.

1.2 List two recommendations of the Coggershall report.

1.3 Name two diseases that have decreased since 1900. Name two that have increased.

1.4 What are the two major factors that influence physician manpower and lead to an apparent undersupply?

1.5 List two problems encountered by the solo practitioner.

1.6 List eight individuals or organizations who deal with health care.

1.7 Define price elasticity of demand.

1.8 Is hospitalization or physician service less sensitive to income? Why?

1.9 List four factors that affect patient demand for services.

1.10 Give an example of excess demand.

1.11 Give an example of excess supply.

1.12 List five factors that have contributed to the shift from inpatient to outpatient visits in hospitals.

1.13 List three essential components that distinguish HMOs.

1.14 List six functions that a comprehensive health-care planning agency should perform.

1.15 Describe the health-care planning process.

1.16 What are the three reasons economists give for government involvement in the delivery of health care?

1.17 If medical care is viewed as a right, not a privilege, then what options does the government have?

1.18 What are the two major ways in which technology has affected health care?

1.19 Technology has affected health-care manpower distribution. How?

1.20 Give an example of how technology has changed the way in which diseases are treated and the mortality rates from various causes.

1.21 List three ways in which technology has affected health-care costs (both negatively and positively).

1.11 References

Brown, J. H. U. (1978). Technology in health services delivery. *IEEE Trans. Sys. Man Cybernetics*, SMC-8: 9–11.

Brown, R. E. (1972). An analysis of the health maintenance organization concept. *Medical Economics*, June, pp. 39–43.

Committee for Economic Development (1973). Building a national health-care system. New York.

Durbin, R. L., and W. H. Springall (1969). Organization and administration of health care. St. Louis: C. V. Mosby.

Ehrbar, A. F. (1977). A radical prescription for medical care. *Fortune*, February.

Ehrenreich, J., and B. Ehrenreich (1974). The American health empire: power, profits and politics. New York: Vintage Books.

Geddes, L. A. (1977). Historical background of interdisciplinary engineering. In C. A. Caceres (ed.), The practice of clinical engineering. New York: Academic Press, pp. 15–30.

Hyman, H. H. (1976). Health planning: a systematic approach. Germantown, MD: Aspen Systems Corp.

Kennedy, E. M. (1973). In critical condition. New York: Pocket Books.

Levey, S., and P. Loomba (1973). Health care administration: a managerial perspective. Philadelphia: J. B. Lippincott.

McKinlay, J. B. (1973). Economic aspects of health care. New York: Prodist.

Moore, T. F. (1974). An HMO financial model for the Dow Chemical Company and Dow Corning Corporation. Unpublished study.

Rushmer, R. F. (1972). Medical engineering. New York: Academic Press, pp. 33-47.

U.S. Public Health Service (1972). Physician visits, United States, 1969. Washington, D.C.

Ward, R. A. (1976). The economics of health resources. Reading, MA: Addison-Wesley.

2
Hospital Organizational Structure

GEORGE H. MACK

2.1 Learning Objectives

Upon completing this chapter you should be able to:

1. Describe the administrative organization of hospitals and the interrelationships between various functional groups.
2. Explain the major factors that guide hospital administrators in the direction of hospitals.
3. Explain the role of clinical engineering in hospital operation (from the administrator's point of view).
4. Explain what is meant by a cost center, a line department, and a staff department.
5. Explain the differences between and characteristics of primary, secondary, and tertiary care facilities.

2.2 Introduction

This chapter describes the unique features of the hospital which make it perhaps the most complex organization in our society. Understanding the influence created by the presence of multiple professional groups within the organization and the role of the chief executive officer (the hospital administrator) in coordinating these components is vital to the clinical engineer. Acquiring an appreciation of the interdependent roles of each manpower component and recognizing the need to develop interpersonal skills will prepare the clinical engineer to maneuver effectively within this environment.

2.3 Recent History of the Hospital

In 1899 the Association of Hospital Superintendents (later to become the American Hospital Association (AHA) in 1907) was founded. Its focus was on the improvement of hospital organizational management. The Flexner report mentioned in Sec. 1.3 upgraded medical education and provided the physicians' professional self-evaluation with the stimulus it needed. Standard setting became a necessity in evaluating the quality of patient care delivered. In 1918 the American College of Surgeons (ACS) published a "minimum standard" which only 89 of the hospitals in the United States and Canada could meet. By 1945, with the standard nearly unchanged, over 3100 hospitals of 25 beds and over were in compliance. This activity is now the function of the Joint Commission on Accreditation of Hospitals (JCAH) founded in 1950 and jointly governed by representatives of the American Medical Association (AMA), AHA, ACS, the Canadian Hospital Association, and the American College of Physicians.

Advancements in medical technology since the last half of the nineteenth century to the present day have had an enormous impact upon the management of the hospital as an operating enterprise. With each new development, a cellular division of sorts has taken place, giving birth to yet another discrete body of knowledge, each with its own cherished role, a diversified nomenclature, and more importantly, a self-image apart from those of its organizational ancestors. The core tasks of the first professional nurses are traceable today into dozens of separate professional groupings from dietetics, physical therapy, occupational therapy, respiratory therapy, and speech pathology to name just a few. The pathologist's role is now subdivided among the physician subspecialities of hematology, toxicology, microbiology, cytology, and surgical pathology; the nonphysician tasks are performed by a highly technical substrata. The same goes for virtually all of the more than 200 health occupations today. The division of labor directly paralleled the knowledge explosion.

What does this mean to the hospital organization generally? The emphasis on medical cure, the focus on the patient as the recipient of care, the institutionalization of diagnostic and treatment services in the hospital, and the multispecialty nature of the health manpower has made the coordinating role of the hospital administrator extremely complex. It is a truism that increased specialization breeds increased interdependence. It is also a fact, however, that professionals, by definition, cherish their "right" to self-regulation. Consequently, group conflict is unavoidable. Internal communications are impeded. Consensus is more apt to be absent. Goal achievement for the institution as a whole may not be congruent with the goal of a unit of the whole. Give and take must occur. Negotiation and compromise are mandatory. Increasingly, management decisions are being made in consideration not only of internal forces at work, but also an increasing number of outside pressures. The modern hospital, therefore has become a dynamic organization in which interpersonal skills, effective communications, and leadership styles play a significant role. These roles continue to be shaped and reshaped with each technological breakthrough. Each breakthrough, in turn, forces a re-evaluation of the organizational and the structural alignments of its departmental components.

2.4 The Hospital as a Delivery System

The American hospital delivery system can be generally viewed as having three underlying characteristics that account for many of the complexities society now faces in dealing with its concern for health care. (1) The system is primarily a voluntary system, that is, it represents a collectivity of independent, autonomous units; (2) The hospital component of the delivery system is primarily a nonprofit system, with the vast majority of all hospitals not required to maximize their social responsibilities in providing services to a defined population nor provide a return on the capital investment; (3) The hospital is becoming primarily a third-party financed system. This last feature is unique to hospitals in our society, for no other system of goods and services employs a similar economic mechanism. Third-party financing has its advantages and disadvantages. On the one hand, the system enjoys an assured source of financing; and the society at large benefits, since these costly services are financed to a growing extent by general tax dollars rather than through direct consumer levies. It hurts to pay, nonetheless, but it doesn't hurt as much as if you paid only at the cashier's window. On the other hand, however, third-party reimbursement has interfered to some extent with the factors of the marketplace which normally dictate what goods and services shall be produced, in what quantity, and at what quality level. Prepayment favors consumption. Demand for services can, and often does, bear no direct relationship to societal needs or the available supply of necessary delivery resources. (Brown, 1972). This subject will be discussed further in Chap. 3.

A pluralistic system

Pluralism not only describes the varied sources of hospital financing; it also can be used to describe the variety in the forms of ownership, type, and size of the nation's hospitals. Based upon 1975 national statistics, there were 7156 hospitals in the USA. Of that number, 6186 were short-term general hospitals. Short-term hospitals are those in which acute care, as opposed to chronic, long-term care, is provided. Of the nonfederal short-term hospitals, over half (3291) were nongovernmental, nonprofit institutions, 740 were investor-owned (proprietary) hospitals, and the balance, approximately 1800, were owned and operated by state and local governmental bodies. The Federal government operates 330 short-term general hospitals. Not included in this breakdown are short-term specialty hospitals, e.g., psychiatric, TB and other respiratory diseases, and all long-term facilities (American Hospital Association, 1976).

Diversity in ownership of US hospitals can be analyzed within a single ownership category. For instance, within the Federal government there are many separate agencies and bureaus involved directly in the operation of hospitals. The Veterans Administration, the Public Health Service, the Indian Health Service, and the Armed Forces together operated 382 hospitals in 1975. In the nonfederal category, private industry and organized labor, too, are running hospitals such as the mining company-owned hospitals and the coal workers' Appalachian Regional Hospital Network. There are osteopathic hospitals, church-owned hospitals, and those owned and operated by charitable organizations such as the Shriners burn centers across the country. With each ownership type there is often a different institutional objective to be accomplished, a vastly different reporting relationship between the chief executive officer and the board of governance (trustees), a different political environment in which to operate, a different constituency to serve and please, and a different set of problems to solve.

Primary, secondary, and tertiary care

Gradations of care ranging from a patient encounter in the private office of the physician to highly specialized care such as open-heart surgery have been categorized into three groupings: primary, secondary, and tertiary care (Fig. 2.1). Primary care generally describes the first-level care rendered by a single physician, commonly a general practitioner or family physician, a chiropractor, or nurse practitioner. The setting is an office, clinic or emergency room; the medical condition does not involve hospitalization. Secondary care connotes hospitalization requiring the routine services offered in the typical community hospital. It should be noted at this point that the average general hospital in the U.S. has approximately 165 beds, and that *half* of all general hospitals *have fewer than 100 beds.* The technological advances described earlier have resulted in the need to centralize to the extent possible highly specialized resources, be they specialty-physician manpower, or specialized equipment and facilities. The individual practitioner in a 100-bed rural hospital, or an urban community hospital may lack the expertise or access to the facility appropriate to the needs of the patient, and so he refers a patient to such a setting. This highly sophisticated specialty medical care is termed tertiary care and describes the function of the largest urban hospitals which are usually formally affiliated with medical schools. All of the university-owned or affiliated teaching hospitals fit into this category.

Excellence in training physicians and other health professionals requires excellence in the affiliated hospital's clinical facilities and services. The mission of the medical school, therefore, is intrinsic to the mission of the university-based teaching hospital. This is not to say, however, that a hospital, in order to render tertiary or third-level care, must be firmly entrenched in an academic environment. Indeed, there are many voluntary, nonprofit community hospitals which provide tertiary care services with no formal medical school ties. This does, however, mean that there is the opportunity for competitive influences to dictate what services are offered. These competitive motives, unfortunately, have served to create unnecessary and costly duplication of services in a specific community or region. This has contributed significantly to spiraling health-care costs and the increased regulatory intervention by local, state, and Federal governments (see Chap. 3).

Tertiary care implies a primary emphasis on high-technology specialty care, e.g., cardiac catherization, open-heart surgery, high-risk maternal and neonatal care, cancer therapy, and burns. The concern for quality in all hospitals creates a paradox: When does a community hospital (secondary care) become a tertiary hospital or stand to lose its tertiary services if the patient volume is insufficient to maintain the skills of the specialist physician as well as the team of specialized nurses and technologists? Conversely, what happens to the quality of medical education if a university teaching hospital (tertiary care) is competing with the local community hospital for the patient load necessary to maintain the clinical skills of its per-

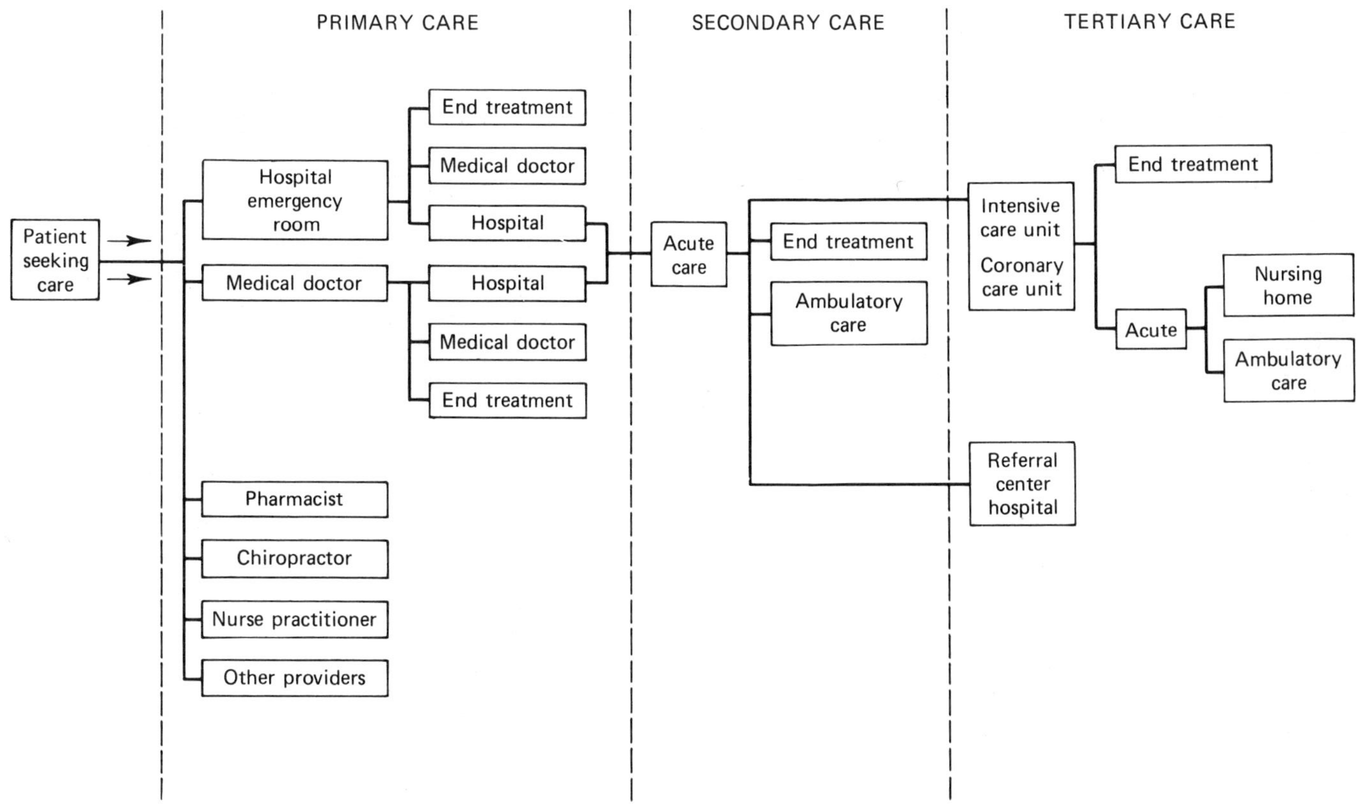

Figure 2.1 Interrelations of primary, secondary, and tertiary care.

sonnel and physicians while meeting the teaching and research mission requirements of the medical school? It is this area of institutional interface in which there needs to be far more cooperation between hospitals and less competitiveness (see Sec. 1.8 on planning).

Trends in hospital utilization and composition

Significant trends in hospital utilization can be seen as the industry responds to increased sophistication in biomedical technology. Increased expectations by the public, the beneficiaries of this improved scientific capability, and pressures to improve efficiency and lower or contain rising delivery costs have resulted. Recent figures for the period 1970–1975 show a significant reduction in the number of hospital beds as hospitals attempt to consolidate underutilized services (see Table 2.1). The rate of hospital admissions has increased, but the average length of stay has decreased. The volume of community hospital inpatient care based upon admissions and average daily census, was about 25 percent higher in 1975 than in 1965.

There is a general correlation between the size of the hospital and the array of medical services offered. Figure 2.2 graphically illustrates the fact that the larger the size of the hospital, the greater likelihood that the hospital offers services that involve complex equipment, a large capital investment, and more highly skilled per-

Table 2.1. Trends in Utilization in U.S. Hospitals 1970–75. [Reprinted, with Permission, from *Hospital Statistics,* 1976 Edition, Published by the American Hospital Association.]

	1970	*1971*	*1972*	*1973*	*1974*	*1975*
Hospitals	7,123	7,097	7,061	7,123	7,174	7,156
Beds (000s)	1,616	1,556	1,550	1,535	1,513	1,466
Admissions (000s) Note 1	31,759	32,664	33,265	34,352	35,506	36,157
Average percentage occupancy Note 1	80.3	79.5	78.0	77.5	77.2	76.7
Outpatient visits (000s)	181,370	199,725	219,182	233,555	250,481	254,844
Personnel per 100 inpatients	196	209	221	233	250	269
Payroll expense per inpatient day	$33.16	39.07	44.17	49.18	55.93	66.12
Total expense per inpatient day	$53.95	63.82	73.89	83.67	97.23	118.69

Note 1: Admissions and occupancy figures exclude newborns.

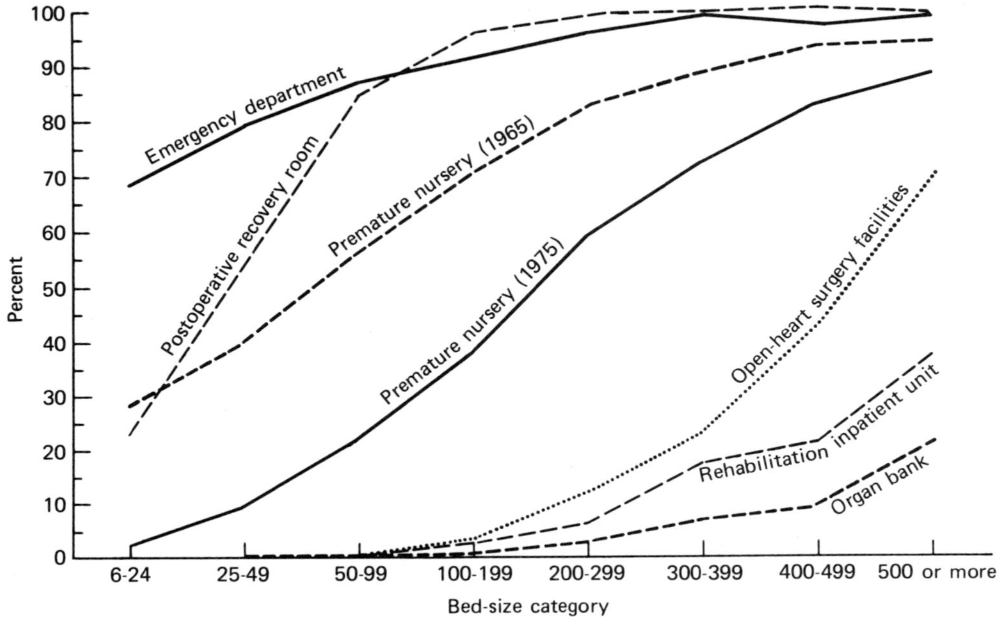

Figure 2.2 Percent of community hospitals reporting selected facilities and services, by bed-size category, 1975. Reprinted, with permission, from Hospital Statistics, 1976 edition, published by The American Hospital Association.

sonnel. Bear in mind, also, that the earlier figures shown in Table 2.1 represent *averaged* data and that one-half of all U.S. community hospitals are under 100 beds in size. Although the average of all U.S. hospitals in 1975 employed 2.69 full-time-equivalent employees (FTEs) for every inpatient, the average large community hospitals and nearly all tertiary care hospitals have nearly five (5) employees per inpatient as a result of their labor-intensive service offerings. In addition, a large number of employees in these settings are used to staff large outpatient clinics which are not generally included in the majority of U.S. hospitals except for emergency-room services.

Shared service. Mention should be made of a new organizational trend, namely, hospital mergers and the growth in the number of multi-institutional arrangements either through corporate reorganization under a single governing authority, or through formalized shared service agreements. This is a direct outgrowth of escalating costs and increased governmental regulation. Economies of scale, reduction of duplication of scarce resources, and concerns for cash flow problems in financing the enterprise have attracted a number of institutions to establish what is termed a consortia (shared) model of organization. Danielson (1977) distinguishes between a "health consortium" and a "hospital consortium" by stating, "The latter may respond more appropriately to the priorities of shared support services, financing, and administration, while the former deals primarily with clinical services, health education, health services research, and the management of patient care programmatically". With health-care costs receiving the greatest Congressional attention in our history, it is safe to say that the consortium approach to delivery of health care in the U.S. will be looked upon as an opportunity to deal more rationally with many of the health problems confronting our society, but it is by no means a panacea.

2.5 The Hospital as an Operating Enterprise

The governing authority of any organization carries with it the responsibility for defining the goals and objectives of the enterprise and directing it by means of broad policy decisions toward the successful accomplishment of the stated goals. In the hospital, the management group has evolved historically into a diffuse-authority structure, commonly referred to as the triad model. The three components of this model are the governing board (Trustees), the chief executive officer (hospital administrator), and the medical staff organization (Fig. 2.3).

The governing board

The early hospitals of the mid-nineteenth century had a very centralized governing structure. The trustees in those days were the people who put up the money and as the principal "owners," they handled the major problems. As Brown (1970) put it, "They sat alone in the governance saddle because they were responsible for the saddlebags." As scientific medicine brought the practice of medicine into the hospital at the end of the nineteenth century, the hospital organization quickly became the depository for the medical resources of the community. The trustees at this point sat not only on the corporate financial assets, but also now directly influenced both the pro-

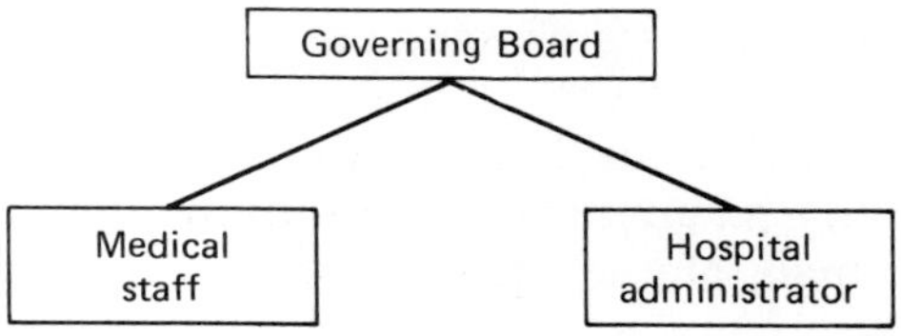

Figure 2.3 The management triad.

fessional and economic destinies of the physicians and collected funds from patients. This model was arrived at more by compromise by each of the three components rather than as the result of organizational logic. It served hospitals reasonably well for decades since hospitals at this time were growing rapidly in both utilization and societal acceptance. But within the past decade, this model has fallen short of the mark both from organizational and policy standpoints. New claimants to the governance action have been speaking their minds. Their criticisms and concerns have been fueled by rapid inflation, burgeoning technology, and increased utilization of facilities. Organized community groups, consumer advocates, and the third-party payers (the private insurance industry and the government) challenged the self-serving features of the governance model and demanded more open participation and responsiveness to their concerns. The courts, too, have had their influence in forcing hospital trustees to re-examine their contemporary role and commit themselves to new accountabilities.

What motivates a person to become a trustee? Surveys have repeatedly reported the foremost reason as being a desire to "serve the community" and "help people." Secondary motives are the challenging responsibility and the desire to improve the operating effectiveness of the enterprise (Cunningham, 1976). In order to cope with the forces of change, new trustees must shun the all-too-popular trustee posture of letting the administrators and the physicians bear the brunt of change and begin to exert the personal and collective leadership that comes with community accountability. The past "reacting" role must be reshaped into an affirmative, anticipatory role.

The functions of the governing board can be summarized as follows:

1. To determine the policies of the institution within the context of community needs.
2. To provide equipment and facilities to conduct patient-care programs consistent with the community needs.
3. To see that proper professional standards are defined and maintained (quality assurance).
4. To coordinate professional interests with administrative, financial, and community needs.
5. To provide adequate financing by securing sufficient income and by enforcing businesslike control of expenditures.
6. To provide for the safe administration of funds given in trust, e.g., gifts and contributions.
7. To maintain accurate records of its finances and activities.
8. To surround the patient with a safe environment, there-

by fulfilling a moral and legal responsibility to the community. This function entails more than the basic physical safety of the facilities. It entails three subelements:

(a) The selection of a qualified administrator, medical staff, and other professional and nonprofessional manpower.
(b) The appointment of medical staff members based upon the merits of their individual credentials.
(c) Assurances that no board member is in conflict of interest with any decision facing the board or, if so, that he not participate in the voting on any such matter.

Although the hospital governing board's functions are essentially universal, the reporting relationships between the hospital administrator and the board are more complicated in the case of the university-owned teaching hospital. In these organizations there are usually two intermediate officers in the administrative hierarchy through whom the hospital administrator must communicate with the board. These echelons are represented by the vice-president for health sciences and the president of the university. Although the titles may vary somewhat, the vice-president serves as the chief executive officer of the academic medical center comprised of the university hospital, the college of medicine, and in many institutions the colleges of nursing, pharmacy, and allied health professions. This organizational entity, commonly referred to as the Health Sciences Center, reports administratively to the university president and through him to the university board of trustees, or in the case of state universities, to the board of regents. This reporting relationship varies significantly from the community hospital in which the administrator reports directly to his board. In many respects, it is an unwieldy and inefficient governance model, given the contemporary issues threatening both medical education and hospital management. Today when both medical schools and hospitals are struggling with societal mandates to improve their effectiveness in a climate of increased cost-control regulation, federal cutbacks in biomedical research support, and increased public expectations, it is difficult and perhaps unrealistic for either the hospital administrator or his boss, the vice-president, to receive from the board their undivided attention on these vital issues so that the board is on the one hand kept informed of the impact these issues have on the medical center, and on the other hand, that the board communicate informed and timely patient-care policy directives to the administrator. The complexities of the hospital industry demand a more direct link to the hospital. This liaison is impeded since the monthly board meeting must address equally pressing, but nevertheless time-consuming agenda items of concern to the other elements of the university campus. The remoteness of the university board members serves to endanger the administrator's effectiveness in keeping the board informed on matters of vital significance to the successful operation of the hospital and advising them on policy matters. Often the university board hears only selective information (usually bad news) and deals with it in a crisis atmosphere, since time constraints preclude a more comprehensive presentation. Much of the information reaching the board is "filtered out" or presented by the president, which potentially could prevent the board from hearing the hospital administrator's personal perspective, thus serving to further isolate the board. Effective long-range planning by the board is hampered by this structure.

Traditionally, university boards have recognized this problem by establishing committees and subcommittees charged with dealing exclusively with medical center and hospital matters, but this structure is now being challenged as inadequate. In the past few years, several university-owned teaching hospitals have restructured a totally separate governing body independent of the university board for these reasons. A three-year study by the Kellogg Foundation sponsored by the Association for Academic Health Centers was announced in January, 1977, to examine these complex issues of organization and governance in these settings. It is likely that the internal and external forces affecting the hospital enterprise will result in an increased trend toward separating the hospital governance structure from that of the university main campus in order to permit the chief executive officer of the university hospital to deal more directly with his board on a parity with his community hospital counterpart. Figures 2.4 and 2.5 contrast the governing structure of the typical community hospital organization with the university teaching hospital. Because of the triad described earlier, the governance structure directly influences the structure of the medical staff organization and the hospital management configuration.

The hospital administrator

Whether he or she carries the title of director, superintendent, president, executive director or whatever, the administrator is the chief executive officer of the enterprise. His function is identical to that of the president of any corporation.

If the administrator didn't exist, he'd have to be invented. Why? Because the reason for administration is that human behavior can be caused to happen. Furthermore, it is the task of management at all levels to cause that behavior to occur in ways that assure the achievement of the purposes of the enterprise.

Management is both an art and a science. There are as many suc-

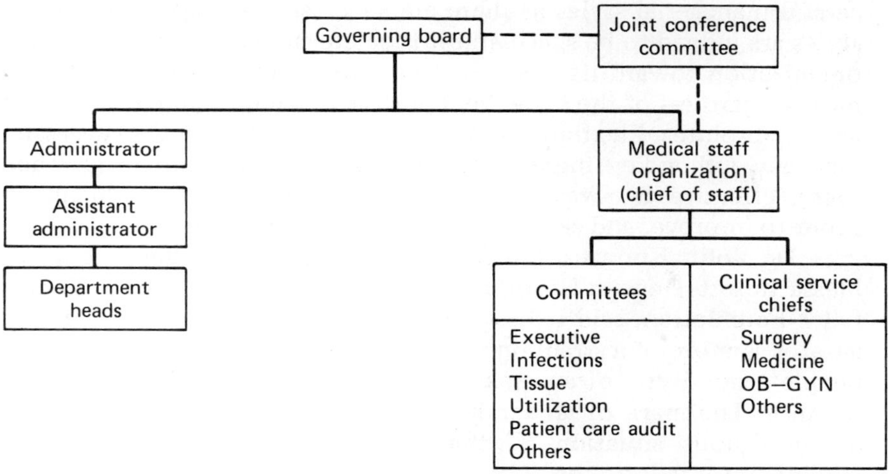

Figure 2.4 The community hospital.

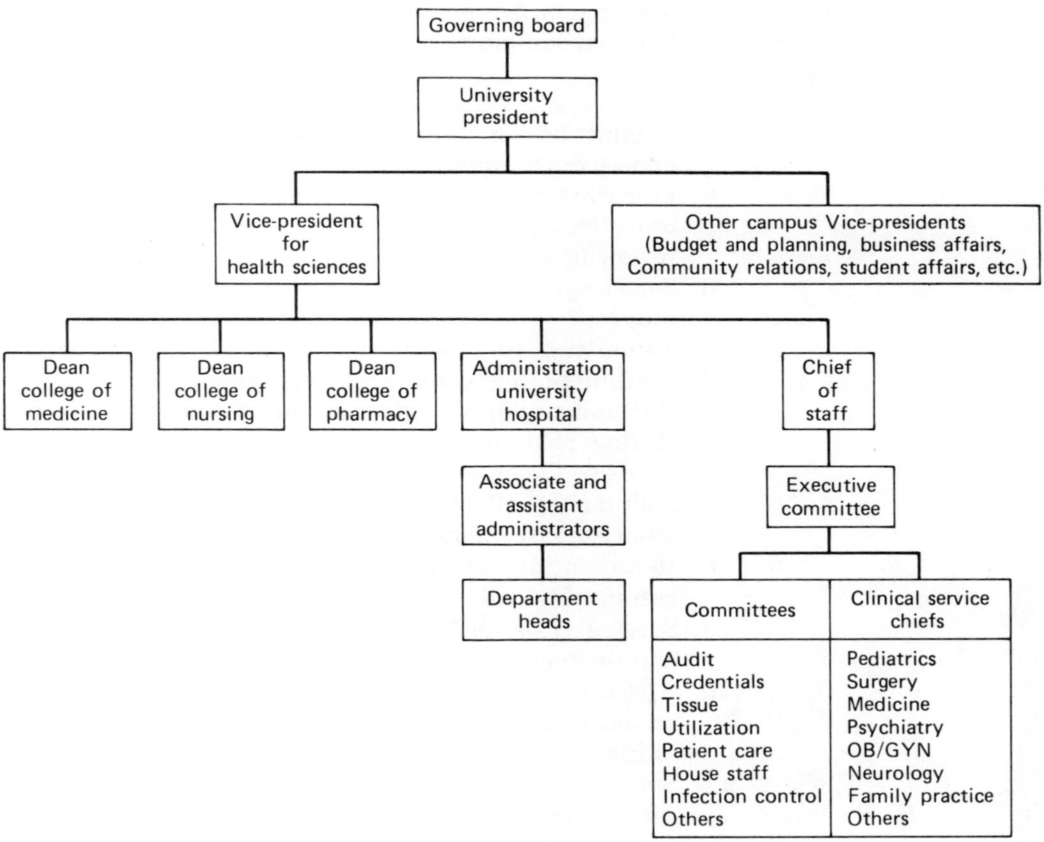

Figure 2.5 The university-owned teaching hospital.

cessful managerial styles as there are successful managers. Individual styles are judged to be successful if the determined results further the organization toward its goals. In a very real way, the ends justify the means regardless of the style, be it authoritarian or democratic. There are no absolutes. Flexibility is a necessary ingredient in the successful manager. Other key ingredients are negotiation (the art of give and take), effective two-way communication, a willingness to change in order to improve, and skillful delegation of responsibility.

Dr. John Knowles, the former administrator of the prestigious Massachusetts General Hospital and currently president of the Rockefeller Foundation, said it best when he stated that "being an administrator is more of a strain on the character than on the intellect. The people around you often make life uncomfortable and tax your endurance. The mark of an administrator is often his ability to tolerate an ambiguous situation. If the frustration tolerance is low, the administrator will not be effective" (Cunningham, 1976, p. 67).

As the chief executive officer, the administrator acts in the capacity of the board's agent in carrying out the day-to-day operations that reflect the broad policy formulations of the governing body. Cunningham has summarized the duties of the administrator as follows:

1. Submitting for board approval a plan of organization and recommending changes when necessary.
2. Preparing a plan for accomplishing the institutional objectives as approved by the board and periodically reviewing and evaluating it.
3. Selecting, employing, controlling, and discharging employees.
4. Submitting for board approval an annual budget expressing both expenditures and revenue projections.
5. Recommending the rate structure for hospital charges.
6. Having responsibility for safeguarding the operating funds of the enterprise.
7. Maintaining all physical properties (plant and equipment) in safe operating condition.
8. Representing the hospital in its relationships with the community and other health agencies.
9. Serving as liaison between the board or its committees and the medical staff.
10. Assisting the medical staff with its organizational and medical-legal administrative problems and responsibilities.
11. Submitting to the board annual reports which describe the nature and volume of the services delivered during the past year and a financial expression of the past

year experience and any special reports as may be requested by the board.

12. Advising the governing board on matters of policy formulation.

The medical staff

"Every industry has its problems, but nobody else has doctors," noted an anonymous hospital observer (Cunningham, 1976). Who are these people? How does their behavior create hospital organizational problems? What can minimize the problem?

Doctors represent the initiators of every action that results in the direct provision of the organization's "product," namely, patient-care services. They determine who will be admitted, when, where, what medical services are to be furnished, in what sequence, in what dosage, with what equipment and supplies, when, where, by whom, and they determine who is discharged and in what order. From their perspective the hospital is their workshop, and the employees their servants. They control the conditions of work for most of the 200-plus health occupations in the hospital while representing less than 10 percent of the total manpower mix. For all practical purposes hospitals are run by doctors. The patient, because he lacks the expertise or may be unable to express his desires at all to anyone, cannot select his purchase as he would in a supermarket, nor can he judge the appropriateness or quality of his care; as the payer, the patient subordinates his consumer rights to the physician's judgment. The physician, rather than the patient, therefore, is the true consumer of hospital services, and yet this key ingredient, the prime mover so to speak, is outside the organization. He is granted the privilege to treat patients in the hospital directly by the board, not the administrator. As a clinician, therefore, he is not an employee of the hospital. He is an independent contractor engaged in the private practice of medicine, and although the medical profession controls nearly 90 percent of the hospital dollar, he is not held responsible or accountable for costs.

The HEW Secretary's Advisory Committee on Hospital Effectiveness in 1968 included in their report the fable of the Martian who, having landed on Earth, interrogated a hospital administrator on the hospital enterprise, its functions, and administrative structure. Upon hearing that the doctor was such a powerful influence in directing the inputs and outputs but yet was placed outside the organization, the Martian replied, "But you must be joking! As anyone can plainly see, such an arrangement would be impossible to manage."

The administrator acknowledged that it was not easy. The intelligent visitor was heard to mutter as he was leaving: "Impossible—or very, very expensive." (Schulz and Johnson, 1976, p. 83).

Obviously, the effective administrator must make it his practice to involve the medical staff in hospital management matters to the maximum extent possible since medical participation in the decision-making process is in the best interest of both the hospital and the patients and community it serves. Physician involvement leads to the much-needed physician accountability. Neuhauser (1972) reported greater physician participation and involvement in hospital activities may have a beneficial impact on the actual quality of care rendered. Similar research studies found positive correlations between highly structured medical staffs and higher quality of hospital performance (Shortell, 1974).

But physicians, like nurses, engineers, architects or any other professional who is asked to become involved in a time-consuming activity for which he was not specifically trained, can be reluctant dragons when it comes to taking on new roles. Reluctance notwithstanding, Mechanic (1972) pointed out that differing perspectives are involved. The physician views the problems of medical practice from a personal rather than an organizational perspective. To him, the most essential need is for his patients to receive a high standard of care while maximizing his personal and professional satisfactions. The administrator or board member, on the other hand, is primarily concerned with the maximum utilization of available health resources and personnel.

"Indeed," Mechanic (p. 40) says, "the optimal organization of medical resources may require a degree of control and surveillance over the doctor's work which is threatening and unattractive to him. One needs no theory of professional conspiracy to explain conflicts between the health professions and other groups; Conflict is a natural product of the different perspectives from which they view the medical context."

The individual with whom both the governing board and administrator relate in dealing with medical staff matters is the chief of staff. He is the elected representative of the collective medical staff, those physicians who are granted privileges to admit and treat patients in the hospital.

The chief of staff appoints all of the committees other than the executive committee whose members are elected by the staff or appointed by the board or administrator or some combination of these. He speaks for the medical staff on all matters and is responsible directly to the board for the quality control and continuing medical education functions. The Joint Conference Committee is typically found in most community hospitals. This committee's members come from the governing board members and the medical staff. Its purpose is to formally link the two components of the "triad." However, studies have concluded that this committee structure, although organizationally sound, may not be as effective in practice as it was designed to be (Schulz and Johnson, 1976).

The functions of the medical staff organization are:

1. To advise the governing body on medical affairs.
2. To accept accountability for the quality of care rendered to patients in the hospital.
3. To request, review, and act upon reports of medical staff committees.
4. To scrutinize the professional ethics of its members and to initiate corrective action as indicated.
5. To develop, implement, and review medical staff policies.
6. To recommend action to the administrator on all medical-administrative matters.
7. To assure that the standards of the Joint Commission on Accreditation of Hospitals (JCAH) are followed as a basic guideline for standards of care.

Since this text is designed especially for clinical engineers, mention again must be made of the university-owned teaching hospital, for therein lives an altogether different creature, the physician academician. Although he is a product of identical basic training, his career interests extend into teaching and research activities to a far greater extent than the physician in the community.

The role of the clinical engineer in these settings is particularly challenging and gratifying since the biomedical research interests of the physician may lead to close working relationships. The university hospital's juxtaposition to the medical school brings the clinical engineer into an environment enriched by the two components of the medical school, the basic sciences department (e.g., anatomy, physiology, pathology, microbiology, pharmacology, biochemistry) and the clinical sciences (e.g., surgery, medicine, psychiatry, pediatrics, obstetrics-gynecology). Historically, medical breakthroughs have emerged from these settings. One can expect accelerated progress in these areas as clinical engineering and academic medicine get to know one another better.

Interrelationships among health workers

It is critical that the clinical engineer acquire an appreciation for the fine tuning necessary to effectively communicate among his coworkers in the hospital. Neuhauser (1972) in citing earlier studies, termed the hospital a differentiated organization in which complex tasks of highly skilled professional colleagues demand a participatory decision-making structure while the repetitive tasks performed by unskilled workers require a hierarchical formalized structure. The gamut runs from the loosely knit structure of the physicians to the highly centralized supervisory structure of the maintenance shops, housekeeping, central supply, and laundry services. Nurses, clinical engineers, pharmacists, and skilled technicians fall somewhere in between. The position in the spectrum is determined by the degree of discretion in independent decision-making as perceived by the employee. Table 2.2 shows the discretion-score rankings based on a 1965 study by Gerald Bell (Neuhauser, 1972). Clinical engineers may be ranked close to nurses and assistant department heads, while biomedical equipment technicians (BMETs) would be ranked about the same as lab technicians and shop tradesmen.

This social stratification explains many of the employee behavioral patterns and the managerial strategies employed in dealing with them. Highly structured and routinized (lower scored) tasks can lead to worker alienation and boredom. Loosely knit, associational activities of the highly skilled colleagues results in personal gratification to the individual employee but works against the centralized control and coordination needs of top management.

The matrix offered by Neuhauser shown in Fig. 2.6, illustrates the teamwork and interdependence necessary to maintain constructive interrelationships. We have adapted this figure to incorporate the new team member, the clinical engineer. The important point to stress in this presentation is that the health-care team approach to the delivery of patient care creates both a vertical and a lateral reporting relationship. One worker, therefore, has more than one boss. The nurse and clinical engineer may report hierarchically to their respec-

Table 2.2. Gerald Bell's Perceived Discretion Score for Hospital Occupations in One 180-Bed Hospital (The higher the score, the more discretion based on questionnaire responses by hospital employees.)

	Occupation	Score
1.	Housekeeping, laundry	1.3
2.	Dietary helpers	1.4
3.	Cooks	2.0
4.	Nurse's aides	2.2
5.	X-ray technicians	2.3
6.	Orderlies	3.0
7.	Lab technicians	3.0
8.	Pharmacists	3.0
9.	Secretaries	3.1
10.	Plumbers, carpenters (semi-skilled worker)	3.2
11.	Dietary supervisors	3.7
12.	Nurses (staff)	4.5
13.	Assistant department head (nursing)	4.9
14.	Assistant department head (other)	5.0
15.	Department head (nursing)	5.4
16.	Department head (other)	5.4
17.	Doctors	5.6
18.	Administrators	5.7

From D. Neuhauser, *Hospital Administration,* 17(4):8–25, 1972.

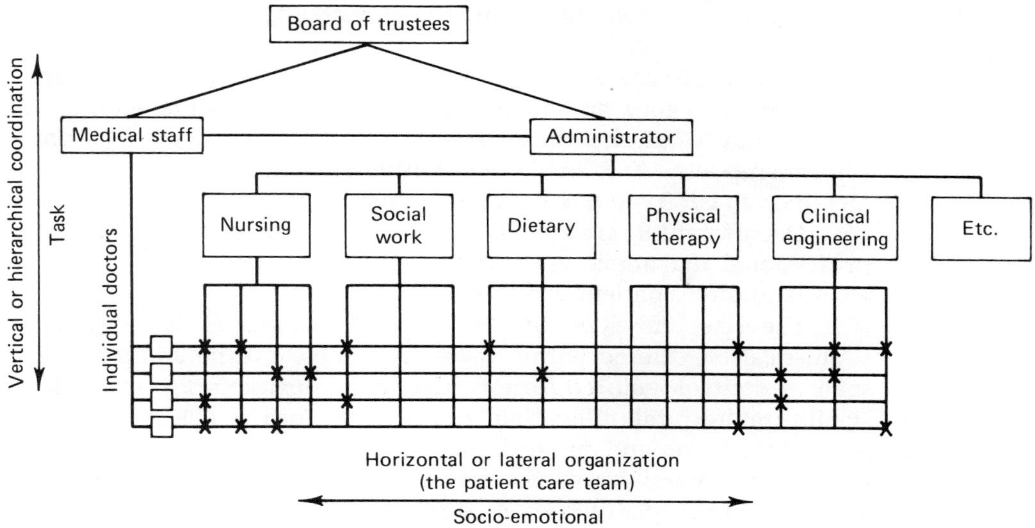

(x) indicates both a member of a department and a patient care team.

Figure 2.6 Necessary teamwork and interdependence. From D. Neuhauser, *Hospital Administration,* 17(4):8–25, 1972.

tive supervisor and laterally follow the directions given to them by the doctor(s) directing the patient's clinical management. The three of them, for example, will collaborate in planning laterally for the introduction of a neonatal monitoring system in the intensive care nursery, but each one of them will communicate information vertically for the necessary coordination and approvals among their administrative supervisors. The physician must lobby to the administrator for the funds to acquire the instrument, the nurse must plan for the staffing ramifications associated with acquiring a new array of equipment and must evaluate, select, and procure medical equipment as well as ensure that it is budgeted (assuming that the unit is a nursing department cost center); the clinical engineer must also evaluate, select, and procure medical equipment as well as coordinate with the plant engineer so that pertinent building and safety codes are met. Obviously this tightly woven interaction among roles creates a tense and stressful climate for role conflict and ambiguity. The atmosphere for stress is all the more charged due to the protectionist tendencies of any professional to challenge any perceived or bona fide threat to his domain. To him, change is an encroachment on his turf; therefore he will maneuver to maintain the status quo, often to the detriment of effective teamwork and goal congruence with the organization. See Kahn et al. (1964) for a more detailed description. See Jaco's analysis of how all this behavior intersects with the patient's role in coping with his illness and the hospital environment (Jaco, 1972).

2.6 Departmental Organization

Section 2.4 described the different types and sizes of hospitals by ownership category and function. Space limitations preclude an in-depth analysis of the departmental organization of each of these various models. For such an analysis, the reader is referred to McGibony (1969), and Schulz and Johnson (1976).

Departmental groupings can be generally differentiated into professional diagnostic and treatment services, general support services, and direct patient-care services (Fig. 2.7).

The size and type of the hospital determines the degree of sophistication offered within these groupings and the administrative span of control required for effective task delegation. For example, a small rural hospital of less than 100 beds will provide basic services in general medicine and surgery, pediatrics, and obstetrics-gynecology, but would generally not have a comprehensive emergency-room service with subspecialty backup services of neurosurgery, plastic sur-

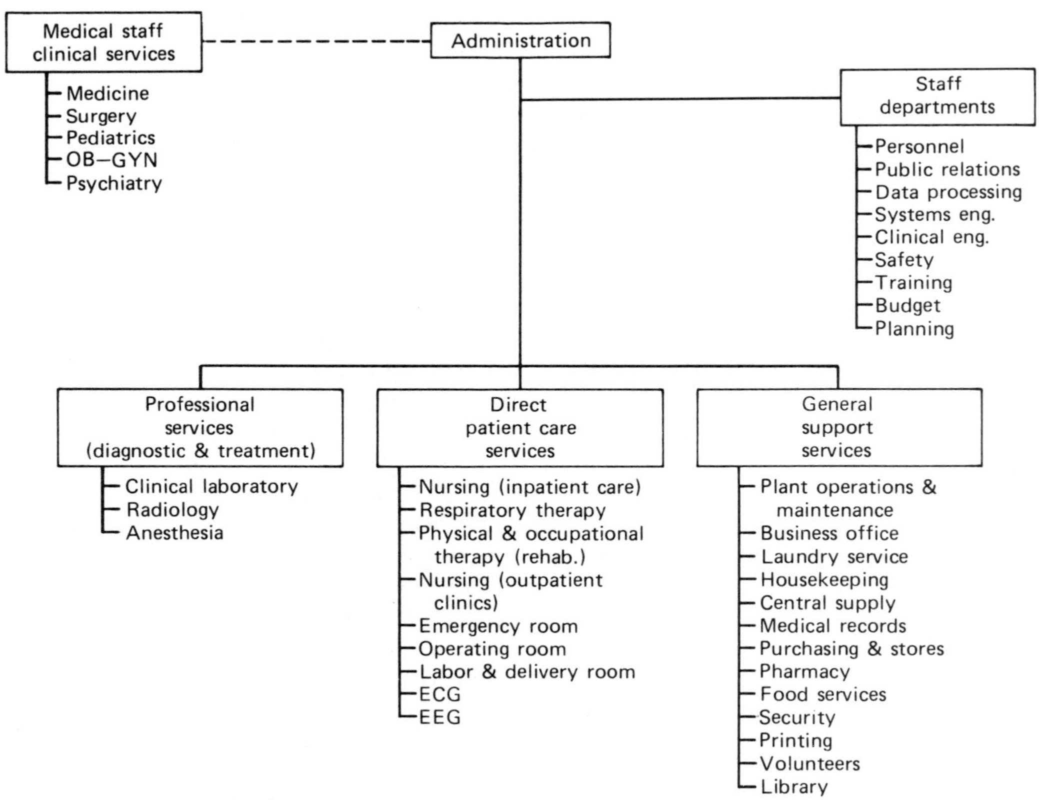

Figure 2.7 Hospital departmental groupings by general function.

gery, psychiatry or neurology, nor would it have elaborate coronary care and surgical intensive care inpatient facilities.

The scientific support services, laboratory, radiology, and anesthesiology would be smaller, offering fewer subspecialized components. Usually, due to manpower shortages, size, and geographic isolation, these hospitals utilize the part-time services of a pathologist, radiologist, and registered pharmacist and would more likely employ nurse anesthetists due to the scarcity of full-time physician anesthesiologists. The subspecialization within the medical staff and the nonphysician technical manpower categories are minimal. Patients requiring these facilities and services, are, therefore, referred to urban centers, to either larger community hospitals, or the academic health center.

Other diagnostic and treatment services, such as electrocardiography, respiratory therapy, and pulmonary function laboratory, would be offered as part of a single organizational unit or cost center termed the cardiopulmonary laboratory; newer and more elaborate technology such as clinical ultrasound and computerized tomography would not be found in the small hospital since it requires a physician expertise and a considerable patient through-put in order to assure quality and economic feasibility to the institution. Respiratory therapy and the blood-gas laboratory would be combined into one unit, whereas in the large teaching hospitals they are often two separate departments.

Ambulatory medicine in the small hospital under 200 beds is generally nonexistent since these patients still receive their ambulatory care in the physician's office or in free-standing, group-practice clinics.

The nursing department in the small hospital and still to a great extent in medium to large community hospitals is responsible for overall hospital operations during the evening and night shifts as an extension of the hospital administrator. This is accomplished through the use of the evening and night-shift nursing supervisor who must be responsible for coordinating not only nursing activities but also must be the common coordinating link between the admitting office, plant operations and maintenance, pharmacy, dietary, and central supply. In case of emergency, other necessary off-duty personnel are called in.

In the small hospital, the instrumentation preventive maintenance program, environmental safety, and other functions of interest to the clinical engineer would normally be assigned to the Director of Plant Operations who would also be responsible for general maintenance and repair, switchboard operations, security, and safety. He would usually report directly to the administrator or assistant administrator, depending again upon size. He would make use of the consultative services of a clinical engineer for assistance in developing an instrument-control program, equipment prepurchase evaluation and specifications, in-service training, safety-check criteria, preventive maintenance schedules, code interpretations, and advise the facility renovation and new construction through a shared services contract with a larger hospital in the area. Chapter 6 describes in more detail how such a program is structured in varying types of hospitals.

Line versus staff departments

The term "line management" refers to the decision-making responsibilities of administrators and their middle-management supervisors who report to the administrator and are responsible to him for decisions they make independently as part of their delegated authority and responsibility. Staff departments, on the other hand, serve an advisory function to management in the area of their expertise, e.g., public relations, personnel, and planning. These department heads make recommendations to top management which can be either accepted, rejected or modified. However, in most cases the administrator is well advised to follow the recommendations of these functional specialists because of their technical knowledge and background in areas of the business in which the administrator is not as fully trained. The responsibility for the final decision, however, rests with line management, with ultimate responsibility for all institutional management decisions resting with the hospital administrator.

2.7 Study Questions

2.1 List the five organizations that govern JCAH.
2.2 What are the three underlying characteristics that account for the complexities facing the health-care system?
2.3 In what ways is the hospital system a pluralistic system?
2.4 Define and describe the meaning of primary, secondary, and tertiary care.
2.5 What competitive factors affect the quality of tertiary care?
2.6 List six hospital services that involve complex equipment, a large capital investment, and highly skilled personnel.
2.7 Distinguish between "health consortium" and "hospital consortium."
2.8 Explain the triad model of authority.
2.9 List four functions of the governing board.
2.10 Why are reporting relationships between the hospital administrator and the board complicated for a university teaching hospital?
2.11 List three ingredients found in the successful manager.
2.12 List six duties of the hospital administrator.
2.13 Explain who initiates most actions in the hospital and the extent of his control over costs.
2.14 How is the chief of staff selected and what are his functions?
2.15 List four functions of the medical staff organization.
2.16 List three basic sciences departments. List three clinical sciences departments.
2.17 Give an example of both vertical and horizontal coordination.
2.18 List three department groupings.
2.19 Distinguish between line and staff departments.

2.8 References

American Hospital Association (1976). Hospital statistics. Chicago.

Brown, R. E. (1966). Judgment in administration. New York: McGraw-Hill.

Brown, R. E. (1970). Governance of the hospital—strictness and structures. National Forum on Hospital and Health Affairs. Durham, N.C.: Duke University.

Cunningham, R. M. (1976). Governing hospitals: trustees and the new accountabilities. Chicago: American Hospital Association.

Danielson, J. 1977. Health consortium responds to total health care needs. *Hospitals, J.A.H.A.*, 51:69-73.

Jaco, E. G. (1972). Patients, physicians and illness, 2nd ed. New York: Free Press.

Kahn, R. L., et al. (1964). Organizational stress: studies in role conflict and ambiguity. New York: Wiley.

Kralewski, J. E. (1974). Collective bargaining among professional employees. *Hospital Administration*, 19(3):30-41.

McGibony, J. R. (1969). Principles of hospital administration, 2nd ed. New York: G. P. Putnam and Sons.

Mechanic, D. (1972). Public expectations and health care. New York: Wiley-Interscience.

Neuhauser, D. (1972). The hospital as a matrix organization. *Hospital Administration*, 17(4):8-25.

Schulz, R., and A. C. Johnson (1976). Management of hospitals. New York: McGraw-Hill.

Shortell, S. M. (1974). Hospital medical staff organization: structure, process, and outcome. *Hospital Administration*, 19(2):96-107.

Somers, A. R. (1969). Hospital regulation: the dilemma of public policy. Industrial Relations Section. Princeton, N.J.: Princeton University.

3
Economics of Health Care

EARL A. SIMENDINGER and TERENCE F. MOORE

3.1 Learning Objectives

Upon completing this chapter you should be able to:

1. Describe the economic forces that control health-care delivery.
2. Explain the several ways of paying for health services.
3. Explain the government's role in the economics of health-care delivery.
4. Explain how clinical engineering affects or is affected by the economic factors that exist in the health-care system.

3.2 Introduction

The original definition of economics is traced to the early Greek word "oikonomia" meaning "prudent household management." Today, the central theme of economics is resource allocations; or the allocation of scarce resources (Samuelson, 1976).

There are two methods of allocating resources: one is through the price system and the other by the direct decision made by a group of planners. The goal of an economic system, in terms of maximum satisfaction, is to use the input of allocated resources so that they make contributions that are proportional to their cost. Also, the contribution of an industry to the economy must be measured in terms of what it actually provides as output for the economy.

Economists view the field of health-care services as an industry because it produces and sells goods and services as other business organizations do. It is the purpose of a free marketplace to work efficiently and to distribute and sell goods and services to the consumer at the best price, thus forcing producers to be competitive.

The consumer plays a major role in this activity as he chooses among goods and services of varying quality and at different prices. As discussed in Chap. 1, in a competitive marketplace, harmony exists between producer and consumer because strong competition forces a producer (provider) to supply these services at the lowest cost. Thus, when producers are highly competitive, the self-interest of the consumer or producer referred to earlier is served. However, it must be recognized that health services are unlike other goods and services. In the case of the health industry, the producer is actually in a very special position because the features of the system place the producer in a strong bargaining and controlling position compared to the "consumer" patient (Hepner and Hepner, 1973).

The cost of a day of hospital care is currently increasing at an annual rate of more than 15 percent. This cost doubled in the short span of 5½ years from the summer of 1970 to the end of 1975 in spite of the wage and price controls that were in operation for the hospital industry during half of this period. While the general level of consumer prices has risen 125 percent since 1950, the cost of a day of hospital care has climbed more than 1000 percent. An understanding of the cause of this abnormal rate of cost increase is a necessary first step in the development of appropriate government policies to deal with rising health-care costs and with the financing of health care (Executive Office of the President's Council on Wage and Price Stability, 1977).

3.3 The Financing of Health Services

The magnitude of health-care expenditures

In 1929, health care took only 3.6 percent of the gross national product; 41 years later, in 1970, that figure had doubled (HEW, 1970). In fiscal year 1976, the total outlay for health care took 8 percent of the gross national product: $135 billion.

Federal, state and local governments have paid at least one-fourth of the cost of health care since the 1920s, but government had little direct voice in the operational policies of the delivery system until recently. The provision of health services was largely in private hands—a coalition of physicians, hospitals, and private insurers—until the mid-1950s. By that time, however, spiraling costs, tight capital, unequal access, and inequality of care had badly shaken the coalition, and in 1965 the 89th Congress brought this private dominance to an end. In fiscal year 1965, only 19 percent of federal expenditures for health went for personal services. By fiscal year 1972, this proportion had risen to 82 percent and the federal money alone paid 25 percent of the nation's personal health bill. State and local governments paid another 12.5 percent. Private money still provided 63 percent of the nation's outlay for personal health services in fiscal year 1972 (still more than half of this coming directly out of the pockets of individuals and the balance from private health insurances), but this was down from 77 percent in fiscal year 1965 (Feldstein, 1975). Thus, the private sector still bore most of the costs.

Between fiscal years 1966 and 1971 the annual increase of national expenditures for health care averaged 12.4 percent. This peaked at 13.5 percent in fiscal year 1970 and in the two years following fell off to 11.1 percent and 10.3 percent, respectively. Nevertheless, the price of health services was still rising more rapidly than those of other industries at that time, with the possible exception of construction (HEW, 1970).

The role of insurance

National health insurance. The basic purpose of proposals for national health insurance is to make medical care available to those who cannot otherwise obtain it, and to transfer the costs of their care to the population as a whole. Society, rather than exclusively the sick, pays the cost of care. When sources of financing are considered separately from the problem of organization, the principal consideration is the distribution of the cost and benefits associated with a given plan.

The distribution of benefits cannot really be determined in advance, since it depends upon the incidence of illness and upon conditions for eligibility for medical care. For example, some plans would make care available only to lower income groups, so the distribution and benefits would necessarily be to them regardless of the incidence of illness (Ward, 1976).

The distribution of costs of national health insurance is analyzed in the same way as the distribution of the burden of any tax. A tax is considered "proportional" if high-income groups pay in tax the same proportion of their income as do low-income groups. Some types of national health insurance proposals require employers to make health insurance available to employees, at the employee's option, and then to pay three-fourths of the cost. The government would set minimum benefits. The employer portion of the insurance payment would be included as income to the employee for tax purposes, and the employee payments would be tax deductible. A government-sponsored plan offering similar benefits would be set up for low-income families and those not covered under employment plans. Employer plans would utilize deductibles (elimination of a basic annual amount of medical payments from coverage) and copayments by the insured as part of the cost above the minimum. Except in the government-sponsored program, insurance premium payments would be the same regardless of income. Insurance plans would be managed by private insurance companies, and employers would be required to offer an option, if available, by health maintenance organizations.

Some of the elements any national health insurance plan should include are as follows:

1. National health insurance should offer options, i.e., it should provide a choice between programs with different practice patterns. A person should be able to choose, for example, between care by solo practitioners or by a prepaid group practice.
2. National health insurance should contain economic incentives, i.e., it should provide incentives for a person to use those services that are the least expensive. This obviously applies to unnecessary "Cadillac" care, but it should apply particularly to those services that are least costly in the long run.
3. National health insurance should cause evolutionary, not revolutionary change.
4. National health insurance should encourage diffusion of responsibility, i.e., it should encompass those responsible for patient care, particularly physicians, and the construction of the operating and capital budgets (Freymann, 1974).

Politically, the primary thrust for national health insurance has come from organized labor and liberal democrats. The coalition wants to establish health care as a "right" that the government provides to everyone on a zero-cost basis. This approach is embodied in the Kennedy–Corman Bill (a National Health Security Act in 1977), the most radical of 18 national health insurance bills introduced in the last Congress. This bill would provide complete control of health care by the federal government (Ehrbar, 1977).

Most of the other bills are compromise proposals that have been put forward by health-care providers and consumers, who want to keep the financing and control of health care in the private sector. All of these plans call for mandatory private insurance, with less extensive benefits than the health insurance security act would provide. Kennedy maintains that the only way to contain health-care inflation is to put the entire system under government control.

Medicare and Medicaid. Medicare and Medicaid were launched in 1965 in the naive belief that the only barrier separating the aged and the poor from equal access to health care was the inability to pay. In retrospect, it is difficult to understand how we could have ignored the elementary economic principle that dumping millions into the demand side of the system without increasing supply would inevitably balloon costs. By 1970, the costs of these programs were double those projected by the Social Security Administration for that year in 1965. The results of these gross underestimations were that these two new programs preempted most of the new money the federal government was able to provide. In fiscal year 1972, for instance, the federal health budget increased by $1.5 billion, but $1.1 billion of this went to these two programs.

It's easy to understand why outside health insurance has become so important. Medicare was not meant to cover all hospital and doctor bills. Over the years, however, it has increasingly picked up less of the burden, a trend that continued to gather momentum with changes in the law that became effective January 1, 1977.

Both the hospital and doctor segments of the coverage work on a system of deductibles and copayments. A decade ago, for example, the deductible on hospitalization—the amount individuals had to pay before the federal coverage took over—was $40. Effective January 1, 1977, the deductible went to $124, up almost 20 percent from a year earlier (Phalon, 1977).

Copayments—the amount of money individuals have to pay out as their share of hospital bills—have been rising too. After the $124 deductible has been met, Medicare pays all the ordinary costs for the semiprivate room, covering drugs, laboratory tests, X rays and the like, for the first 60 days in the hospital. On the 61st to the 90th day, the patient has to make a copayment of $31 a day, up from $26 last year and from $10 a decade ago. Beyond the 90th day, a patient can draw upon a once-a-lifetime reserve of an additional 60 days. The copayment in those days rose January 1, 1977 from $52 to $62 each. A long, painful way from the charge of only $20 a day applied in 1968.

3.4 Mechanisms for Reimbursement

Fee for service

In fee for service, the patient pays a provider, usually a physician or hospital, in accordance with the specific treatment received. In this traditional style of medicine, the physician plays a kind of dual role. He personally treats the patient, but he also manages the care provided for the patient by others and recommends to the patient what he should buy. The physician orders hospitalization, prescribes drugs, administers laboratory work, and makes referrals to other professionals. Charges are made on an item basis except occasionally, such as in obstetrics, where a single fee is charged for management of a specified condition.

Some economists are rather avid in their criticism of fee for service, which they believe is the most serious flaw in the health-care system. Furthermore, they claim that it is a piecework system and a professionally undignified way to practice medicine. This criticism is not new, since many have contended that old patterns and models of practice, particularly the fee-for-service type, need to be replaced. One argument is that under the fee-for-service system the quality of care suffers, as doctors may work too quickly in order to see more patients. Seemingly the patient's welfare is in direct conflict with the physician's economic motive to see more patients. It is also believed that this method of payment encourages unnecessary procedures (Hepner and Hepner, 1973, p. 56).

Capitation

In the capitation method of reimbursement, the physician may be paid a straight sum for each patient in his continuing care, a method that is used in many foreign countries. The British National Health Insurance system pays general practitioners in this manner. The physician may take as many patients as he wishes, but he must provide complete care for all. Although the practitioner is responsible for his patients' care, he can refer them to the hospital-based specialist as necessary.

Salary

An organization may hire physicians on salary, paying them on a time basis. This is the method currently being used in some large teaching hospitals and most large industries with employee health programs. Other examples of its use are emergency-room physicians in hospitals, and physicians in federal government practice. Under some foreign health-care systems, the specialists who treat hospitalized patients are salaried employees and there is much competition for these positions. In sum, the salaried physician is similar to a salaried executive.

Group practice

The term "group practice" really denotes a type of organization rather than a method of reimbursement; however, it is often neither the fee-for-service, capitation method, nor salary basis as it relates to the physician's income. A group practice is defined as three or more physicians formally organized to provide services and distribute income among themselves. Studies indicate that the cost per patient visit is somewhat less for group practices than for solo practitioners.

3.5 The Medical Marketplace

Figure 3.1 shows how the medical-care dollar in the United States was spent in 1973. It shows that hospital and nursing care accounted for approximately 38 percent of the health-care expenditures at that time.

Based on the data through mid-1973, the nation's annual expenditures for medical care totaled $94 billion, which is about 7.7 percent of total expenditures. Annual expenditures per capita are $441. In the decade of the 1960s, there was a three-fold increase in medical expenditures and the Social Security Administration then projected that expenditures would approach $206 billion by 1980 ($906 per capita).

The size and shape of the U.S. health-care industry

In 1976, more than $135 billion for health care was spent in the United States. That is over $600 for every resident and accounts for 8 percent of our G.N.P.

Of this total, 47 percent was spent in hospitals and nursing homes, 25 percent went for physicians' and dentists' services, 10 percent for pharmaceuticals.

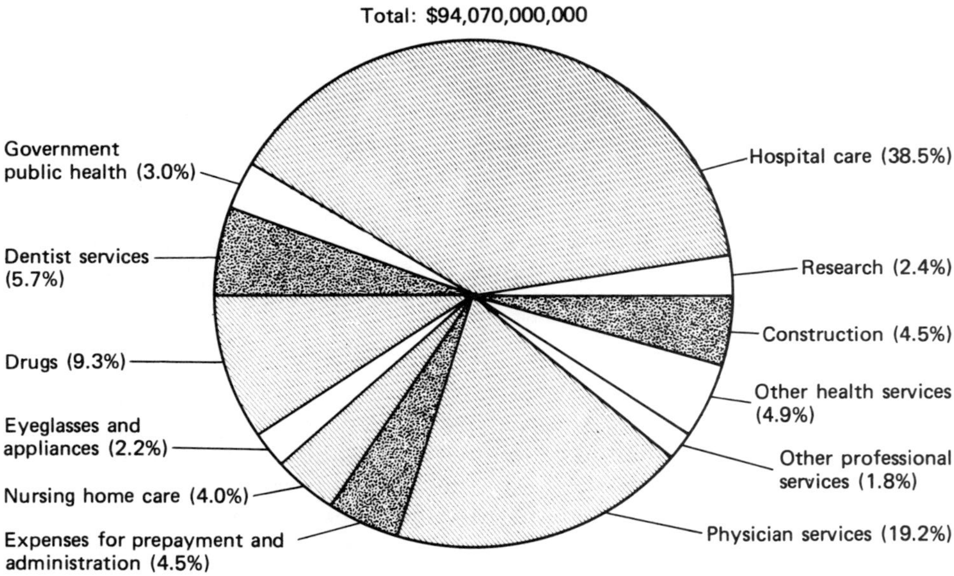

Figure 3.1 U.S. health expenditures, July 1972 to June 1973. From Social Security Bulletin 37, Feb. 1974:7.

The remaining 18 percent was spread over several items such as public health, insurance prepayment, research, eye care, and hospital construction. Though eye care gets only 2 percent of the budget, it amounts to almost $3 billion annually, but then over 100 million people use some form of visual correction. We spent $200 million on contact lenses, and that is at the factory level. Two percent of the national budget was spent on research.

This $135 billion represents an 11 percent gain over the previous year and a doubling in the past ten.

Hospital-room charges increased 16 percent in 1976, physician's fees went up 13 percent, and insurance rates also increased now that 90 percent of the population is covered.

Forty percent of our national health-care budget is spent in the hospital. In 1976, there were 8000 hospitals—85 percent of them nonprofit—with 1.5 million beds occupied by 1.2 million people any day of 1976 (which represents a decrease of 16 percent over 10 years due to a shorter stay in the hospital, a much expanded use of the outpatient clinic, and a 20 percent decrease in babies born in hospitals in the last 10 years).

Thirty-three million people were admitted to the hospital last year, with most of them admitted for surgery, extensive diagnoses and tests, which have increased 25 percent in 10 years.

Eight times more people visited the hospital on an outpatient basis than were "admitted"—250 million in 1976 (excluding military, government or VA figures). The outpatient load has doubled in 10 years, and one of the reasons for the doubling is that some surgical procedures that 10 years ago required hospitalization are now being done on an outpatient basis. This reduces the cost both to the patient and the hospital. These economies are often encouraged by insurance or prepaid health programs. The delivery of health care to the outpatient clinic has provided some excellent opportunities for nonmedical technology to be used, created hundreds of new (small) businesses, and thousands of new jobs.

The physician

There is little question that the physician remains the prime mover of medical care. However, there is serious doubt that the total solution to the problem of health care delivery rests solely on meeting the supposed shortage of 50,000 M.D.'s. While there may be a shortage of physicians (see Sec. 1.3), the real problem rests with poor distribution both geographically and by specialty. In the future, real progress for health improvement might very well come through social gains.

Allied health personnel

In taking an overview of the health manpower field, it is interesting to note that physician manpower comprises only 16 percent of the total health manpower, while those that are considered to be allied health categories comprise 84 percent (Hepner and Hepner, 1973).

No one would question the increasing demand for allied health manpower, particularly in the technical areas, as it relates to the delivery of health care. The advances of medicine have little meaning when society does not have the adequate manpower available to deliver this knowledge. At the present time approximately 200 careers make up the health industry, which employs 4 million people. About one out of every 20 of the nation's employed are involved in the health industry. From 1950 to 1966 the number of persons involved in health careers increased 90 percent while the nation's population increased 29 percent. In 1977 there were 20 or more health workers for every physician. In 1970 physician-directed health services increased 81 percent and dentist-directed services increased 47 percent. This is valid evidence that allied health workers are being accepted and better utilized. Without question, most health experts agree that fragmentation, specialization, maldistribution, and the increasing demand for health services have greatly added to the manpower supply problem.

The supply of college-trained health workers will fall nearly 25 percent short of the 1,776,000 needed by 1980 as predicted by HEW. It has also been estimated that by 1980 the demand for health workers requiring less than a college degree will be 580,000. However, a supply of only 475,000 workers is projected.

Severe shortages are also apparent in the dental workforce. It is estimated that by 1980 there will be a need for 250,000 dental hygienists, assistants, and technicians, but only 180,000 are expected to be available. An increased emphasis on environment and pollution has also created a scarcity of workers. It is expected that approximately 155,000 workers in environmental health care will be required by 1980, about 65,000 more than the supply.

Total health employment projections for 1980 are estimated at nearly 5.3 million workers. Figure 3.2 shows the various percentages of workers in the health-care field in 1969 as taken from a health resource statistics study. These figures were essentially the same in 1976.

The 1.8 million nurses, aides, and orderlies made up one-half of the nations health manpower in 1968—by far the highest component.

Insurance

The major difference between the financing of a hospital and a commercial enterprise is that the recipient of the services—

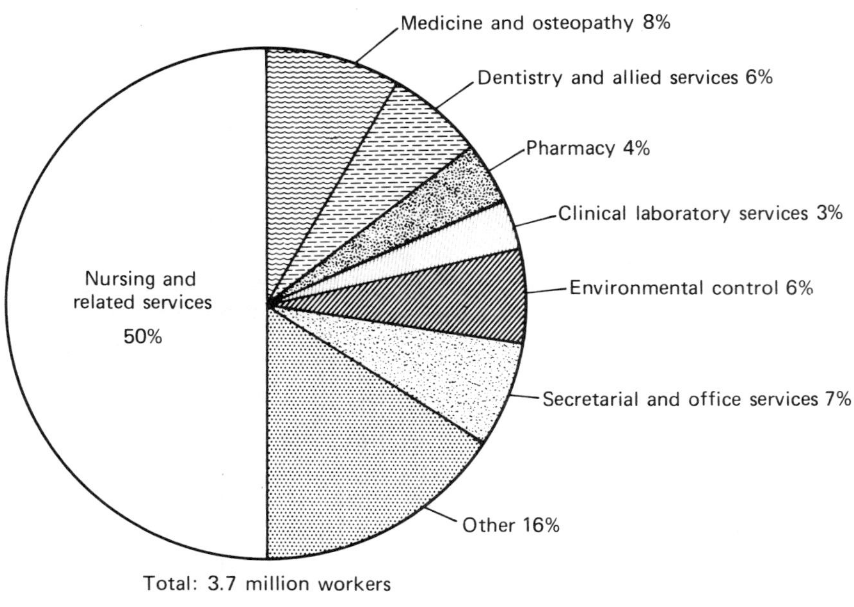

Figure 3.2 Percentage of workers in health care field. From Health Resource Statistics, 1969, Public Health Service Publication No. 1509.

the patient—in most cases does not pay directly for the services himself, but rather the hospital is paid by a third party. By mode of payment, patients are classified in the following categories:

1. Medicare.
2. Medicaid.
3. Blue Cross.
4. Commercial insurance.
5. Self-pay.

As noted in Sec. 3.3, the Medicare program underwrites the medical costs of patients over age 65 and of some needy persons under age 65. Medicare is divided into Part A covering hospital services, and Part B covering physician services. Medicaid pays the hospital costs of patients termed medically indigent and of many persons on government welfare programs. In addition to these government programs, some 1000 commercial insurance companies are third-party payers through group and individual health insurance policies covering about 122 million Americans. There are also some 500 independent nonprofit health insurance programs, with approximately 8 million members (Weltmann et al., 1974).

The growth of private health insurance is superficially one of the great success stories of American business. In 1940 only 12 million Americans had the most common form of medical insurance, coverage for hospitalization expenses. By 1967, no less than 175 million, or 83 percent of the civilian population were covered by private hospital insurance. One hundred million of these were covered by commercial, profitmaking insurance companies. The rest were Blue Cross, Blue Shield, and independent plans. Many of these subscribers were insured for medical expenses other than hospitalization as well. The private companies collected premiums totaling $5.86 billion and paid out $4.84 billion in benefits—administrative and selling expenses comprising much of the remainder (Ehrenreich and Ehrenreich, 1971).

Medical equipment industry and the clinical engineer

We can reasonably identify $8 billion of annual spending by the medical equipment industry for (essentially) hardware in the form of capital goods, expendable items (i.e., disposable syringes), and nonmedical research. This category employs 150,000 people who produce diagnostic, surgical, dental, optical appliances, and instruments. In 1977, the $8 billion spent in this sector was up 20 percent from 1976 and is projected by the U.S. Department of Commerce to go to $18 billion by 1985 (Bureau of Domestic Commerce, U.S. Department of Commerce).

Viewed in its absolute magnitude, $8 billion annual spending is a big number, but compared with the health-care industry as a whole, it only comes to 6 percent of the total. This exercise helps emphasize two aspects of the situation. It puts into perspective the size of the annual contribution from the nonmedical sector and re-emphasizes its subordinate economic role, without diminishing the technological contribution.

In 1976, 150,000 people were employed (U.S. Department of Commerce) to produce these $8 billion in 2000 companies that are categorized in Table 3.1.

These 2000 companies are spread out across the United States, averaging only 75 employees in each company. They make 8500 different items, but probably account for over 30,000 total items due to competitive duplicates. Sales for 1976 are up 20 percent over 1975 and are projected to go to $16 billion by 1985, which means there will be another 150,000 jobs including engineers and others in the health sciences. Many of those currently employed in nonmedical jobs will soon find their way into management. And by 1985, demand for well-educated personnel will double, particularly in the sciences and management.

Hospitals still depend upon preventive maintenance contracts and outside service agencies to handle medical equipment. A misconception is that the service representatives are highly skilled and only they can repair equipment safely. If administrators were to initiate a hospital investigation project and analyze the equipment history of, for example, clinical laboratory instruments, they would often find out that a number of these instruments required rebuilding due to servicing errors.

Table 3.1. Medical Equipment Companies

	Dollars (In Millions)	*Total Employment (In Thousands)*	*$ Produced Per Employee (In Thousands)*
Surgical and medical instruments	$1780	43	41
Electromedical equipment (incl. diagnostic inst.)	1110	17	65
Surgical supplies and appliances	2720	54	50
Dental equipment and supplies	1000	16	62
Optical instruments and lenses	1140	16	71

Time and travel also add to the cost of outside service agencies. Recent studies by the Case Western Reserve University Clinical Engineering Program show that the average cost per man-hour for outside service is $30.00. For an in-house clinical engineer it is $10.00, thus three times the number of man-hours are available for the same cost. The clinical engineer may spend more time on an unfamiliar piece of equipment, but his time is less expensive and he is a salaried employee who will be present if the equipment does not work properly after servicing.

The multiplicity of hospital equipment requires a variety of servicing engineers and agencies which once again leads to fragmentation of service. If both outside service and an in-house program are required, the clinical engineer can be the interface between hospital medical hardware service representatives and the hospital. Additionally, a medical facility operates 168 hours a week, while service agencies operate a regular 40-hour week. Night, holiday, or weekend calls for emergency repairs are usually available but at a grossly inflated price. An in-house engineer can be called upon at any time, a significant cost savings for the hospital.

While traditionally the main criterion used for justification was cost, it has become evident that more realistic criteria involve patient safety and more reliable medical care. Few hospital employees have been adequately trained to understand and maintain the increased number of automated and semiautomated electronic devices. The clinical engineer can:

1. Provide training programs on medical electronic equipment.
2. Provide equipment repair service.
3. Provide a comprehensive preventive maintenance equipment program—short- and long-range.
4. Inspect newly purchased equipment before official acceptance and payment.
5. Provide consulting information to physicians on patient equipment.
6. Provide input before purchasing all medical equipment.

Any director of a clinical engineering department must constantly be aware of the costs incurred by the department and the benefits that the hospital derives from the activities of that department. Hospital department heads often tend to focus upon the actual dollar cost of salaries and supplies for departments and ignore analyzing and documenting the services they are providing to other departments in the hospital. A shrewd manager will develop means of documenting the services provided by the department on a quantifiable basis and communicate these activities to his superiors and the other departments that receive these services.

A clinical engineer, in almost all instances, should be able to show that his/her department is cost-justified based upon the services it provides. A simple means of quantifying the value of the service being provided is to compare the charges with those that have been billed for similar services had they been provided by an outside organization. One 300-bed hospital that established a clinical engineering department was able to demonstrate that the savings in outside contracts alone accounted for more than three times the actual cost of operating the department. In addition, numerous services were provided that were not available before.

Third-party payers such as Medicare, Medicaid, and Blue Cross/ Blue Shield recognize any cost incurred for salaries, supplies, and equipment in a clinical department.

To date, various constraints being placed on hospitals by third-party payers do not directly affect the engineering department. The inability of some hospitals to purchase large pieces of equipment such as CAT scanners, reduces some of the electronic needs in these hospitals, but the effect is minimal, and is more than offset by the more stringent controls to monitor and calibrate all types of electrical equipment.

3.6 Microeconomics of Hospital Care

Historical perspectives

Hospital costs have increased more rapidly than costs for the economy as a whole for the past 35 years. There are numerous misconceptions regarding these increases and unfounded speculations about projected increases in the next few years. Section 3.6 focuses on hospital costs as a microeconomic unit.

Reasons for cost increases

There are three major reasons why health-care costs are going up and why they will continue to go up. First, the health-care industry is *labor-intensive.* Sixty percent of the costs go to pay wages and salaries. In most other industries, that figure is closer to 27 percent. Labor costs can be reduced, but to do so would require the elimination of services. Nurses could be eliminated; laboratory and radiology technicians could be eliminated; and night and weekend service could be eliminated; we could get rid of the emergency room and the outpatient department; we could do any of a number of things to cut our labor costs. But what responsible hospital is going to do it and what community is going to allow it to be done?

The second reason hospital costs are going up is that the cost of everything in the hospital market basket, the goods and services we purchase, is going up. The cost mix of the hospital market basket bears little resemblance to the cost-of-living market basket. To operate, hospitals must purchase those goods and services especially hard hit by inflation. Malpractice insurance is an example. Twelve hospitals in the Chicago area reported premium increases in 1975 that were up 400 percent higher than in 1974. The premiums will be even greater in 1976. In 1974, those twelve hospitals paid $1.60 per patient day for malpractice insurance; in 1975 they paid $6.17 for the same coverage. In some hospitals they are paying ten times more for coverage than they did just a few years ago. Increased costs for medical malpractice insurance, plus increased costs for food, energy, plastics, and the other items in the hospital market basket accounted for 45 percent of the hospital cost increases in 1975.

The third major reason for hospital cost increases is improvement of the product through the addition of services. Advances in medical technology and techniques are made every day, the advances are costly—intensive-care units, disposables for infection control, kidney dialysis, bypass surgery, etc. So long as advances are made, and so long as patients demand them, the costs of progress will have a decided impact on health-care economics. In the late 1960s, additions to services accounted for 50 percent of the increase in hospital costs. Today these additions account for 25 percent. If this factor is to be reduced, patients will be denied the advantages of new cures for illness and injury (McMahon, 1976).

Some authors state that while hospitals have no true incentive to restrain costs, they have plenty of reasons to raise them. For one thing, they have to keep their beds full in order to meet their budgets. Since doctors usually choose which hospital the patient goes to, and since price has effectively been negated as a relevant consideration, hospital administrators go after patients by catering to physicians. The catering often involves having all of the latest equipment, and leads to a lot of duplicated facilities that few patients need very badly and that most don't need at all. Competition to attract doctors also leads to the employment of interns, residents, and extra nurses, all of whom make a doctor's job easier but do little to make the patient's health better. The harried administrator has to please other constituencies as well. Nurses, technicians, dieticians, orderlies, and maintenance workers want to be paid more, and they all want to work shorter hours. But the strain of living within this year's budget prevents administrators from keeping all these pressure groups happy at once (Ehrbar, 1977).

The role of philanthropy

Philanthropy is important to hospitals, but can have deleterious consequences. Hospitals are like other nonprofit organizations in that many of them are going broke.

Philanthropy is usually applied to capital improvements. Capital investment by nonprofit institutions is self-destructive because it increases fixed costs. The managers of nonprofit institutions continue to throw up new buildings when they know perfectly well that there is no way that new facilities can do anything but lose money, even if they are filled to capacity. Of course, these managers will insist that the new facilities will permit them to improve or expand services, but that is true only in the rarest cases.

Much of the funding for institutions in the field of health, like education and the arts, continues to come from private philanthropy. In private philanthropy, at least a few foundations and individuals have had the courage to try an alternative—to favor the operating budget over capital construction. Fund-raising for capital projects always seems easier than fund-raising to increase endowment, which, in turn, is still substantially easier than the most direct and logical course—the earmarking of charity for operating funds (the strategy that plays a substantial role only in the arts). Because philanthropists are asked only for capital money, they give only capital money; because capital money is all they ever get, nonprofits ask only for capital money. What is done to provide funds for running new facilities? It is assumed by all involved that the recipient institution will pick up the cost of maintenance and operations. But it is the upkeep that dooms institutions to financial insecurity. More insidiously still, many donors, especially the large foundations, have adopted the practice of providing seed money on a matching basis, much of which goes for the initial provision of facilities, and then leaves still-struggling institutions to forage for themselves in the far more perilous world of operating funds (Vladeck, 1977).

3.7 Competition in Health Care—Monopolistic or Competitive?

The medical marketplace is more than deficient in free-enterprise practices, for there is a real lack of true competition among the producers (providers) of health care, thus placing the consumer (patient) in a highly compromising position. Also, there is a lack of productivity in the system, which results in inefficient use of available health resources (Hepner and Hepner, 1973).

In the field of health, many physicians have traditionally held erroneous beliefs about the allegedly high degree of competition existing in medicine. Economists, on the other hand, are prone to downgrade the free-competition idea of medical practice and to describe the industry as a distinct form of monopoly. Many of the reasons have already been cited in Chap. 1, but the reasons most commonly given by economists are:

1. In a highly competitive industry, prices are set by the forces of supply and demand as they interact in the marketplace. In the case of medical practice, prices are set by individual producers who have a high degree of latitude to discriminate in pricing.
2. In a competitive industry, price competition is used between producers as a means of attracting a greater number of consumers. In medical practice, the use of price as a competitive instrument is frowned upon and declared to be unethical.
3. In a competitive industry, the product or service that is offered for sale generally is homogeneous. In medical practice, the services offered are conventionally treated as heterogeneous (individual health care) with considerable differentiation occurring both in mix and number, and quality of services produced.
4. In a competitive industry, the consumer has the opportunity to be well-informed and is able to bargain among the producers to obtain the best terms of the sale. In the market for medical services, the consumer is not well-informed. He selects his physician on the basis of highly subjective criteria, and after the selection is made, he is in a relatively weak position to bargain on a price, or other services to be purchased, or to whom referral for additional attention will be made (since advertising is viewed as unethical).
5. In a competitive industry, there is ready ease of entry into the industry. In medical practice, there are numerous barriers to entry. Actually these barriers are erected through the educational process. At the state level, the barriers are maintained by the licensing of physicians. At the local level, medical societies and hospital staffs act as barriers to entry into practice.

Those who want to buy medical services and have the income to purchase them can do so (subject to being informed as to what is purchased). We have assumed that there is competition and, therefore, as a consumer enters into the marketplace to buy medical services, he does so in a setting where the resources supposedly are allocated effectively.

In summary, an overview of the medical practice industry reveals that the producers are in a very strong bargaining position relative to the consumers. For the last 20–25 years, as demand for medical care has grown for a variety of reasons, the power of the physician–producer in the marketplace has been strongly evident. Since consumers have not been in a good bargaining position, the physician–producers have not been subjected to pressures to be more efficient in their productive activities. This strong demand, moreover, has made it easier for physicians to raise their prices rather than to increase efficiency. If the industry were more competitive, such price increases would not be possible because the competitive incentives to increase efficiency and to reorganize production processes would work to the consumer's advantage, not to the advantage of the producers (Levey and Loomba, 1973).

3.8 Economic Fallacies of Health Care

General

One of the more prevalent beliefs of the American public is that hospitals are inefficient and mismanaged. They also think that because hospital costs to the patient have risen to unprecedented heights, profits are enormous. These prevailing notions have caused no end to the frustration in the hospital field and have periodically led the public relations program, aimed at telling the real hospital story, to address a disbelieving public. Some of the more widespread but faulty notions about hospitals include the following.

The lag in employee hospital wages is causing the rapid rise in hospital costs

Wages of hospital employees have increased significantly faster than wages in other sectors of the economy. In the period from 1966 to 1975, hospital wages rose at 8.7 percent a year while the earnings of all private nonagricultural workers increased at 5.8 percent. The higher rate of wage increase for hospital workers added approximately 1.8 percentage points to the rate of increase to the average cost per patient day.

It is important to keep in mind that this higher-than-average wage-rate increase does not contribute significantly to the increase in costs of hospital care. Even a 1.8 percent addition to the normal rate of cost increase raises the average cost by 20 percent in a decade. If this differential had not existed since 1966, the average cost per patient day in 1976 would have been $149 instead of $175 (Executive Office of the President's Council on Wage and Price Stability, 1977).

Unionization

Another recurrent hospital myth is that unionization of their employees in the 1960s fueled much of the cost increases. Wages of hospital employees did rise more than manufacturing wages, but they have been rising faster at least since 1949 and there was no jump in the differential during the 1960s (Ehrbar, 1977).

Profit centers all contribute to the institutions

Unlike industry, some profit centers in the hospital never produce a net surplus of funds, but continue to operate year after year because the services are needed in the community. In most hospitals, emergency-room services and obstetrical services lose money annually. If the hospital is deeply involved in education, the expenses for a medical education program, a nursing school, and paramedical training programs are also loss operations. If the hospital as a whole breaks even, the losing profit centers are continued because they are a necessary community resource.

Users of service determine the amount, kind, and quality of what they purchase

This point has been alluded to before, but it is worth elaborating. In a hospital the patient is usually not even asked if he wants the laboratory, radiology, or other diagnostic or treatment procedures ordered by the attending physician. He (or his insurance company) is expected to pay for these services without questioning their value to him. It is not clear whether the patient is a customer of the hospital, or whether it is the attending physician (who in many ways is a surrogate customer). In a very real sense, the user and purchaser of the services is a captive of his physician; he is not permitted to make decisions for himself as to which service he will select from among a whole array of available diagnostic and treatment procedures.

If in reality the attending physician is a customer, he too has many restrictions placed on his economic ability to select services for his patient. Once he admits the patient to the hospital he cannot order only selective services for the patient and have other hospitals perform these services on that patient. Even if the other hospital is located nearby and can perform services at a lower cost, they are not permitted to come into the patient's hospital and draw the necessary specimens. Once the physician admits his patient to a hospital, he loses his ability to purchase services at a cheaper cost on behalf of his patient. His only economic decisions are to withhold a procedure or service or to shorten the patient's stay in the hospital; he cannot pick and choose among those procedures he decides are needed.

Economies of scale

Unlike manufacturing, the more units produced, the higher the cost per patient day. Large hospitals have a higher average cost per patient day than do medium-sized hospitals, which, in turn, have a higher cost per patient day than smaller hospitals. This occurs because as the size of the hospital increases, so does the range and comprehensiveness of the services (Johnson, 1975).

3.9 The Impact of Health-care Economics on Business

Finally, escalating health-care costs, regardless of their causes, have a strong adverse impact on business. Health-care coverages were first negotiated through unions and were first granted in industry over 35 years ago. The typical large industry's coverages include:

1. One year of semiprivate hospital care, or two years in an approved nursing home.
2. Complete maternity care.
3. Extensive outpatient care including diagnostic and laboratory services, surgery, physical therapy, X ray, accidental and medical emergencies, and up to $800 annually for psychiatric services.
4. Payment for a wide range of prescription drugs, with the employee paying no more than $2.00 for each refill.
5. Dental benefits up to $750 per person per year, including examinations, cleanings, fillings, bridges, and dentures.

(These are relatively new benefits being negotiated by hospitals and even by hospital employees themselves.) It is estimated that General Motors paid $735 million in 1975 for hospital, medical, and dental expenses for its U.S. employees. That equals $1500 for every employee. The 1975 figure represents an increase in General Motors' annual health-care costs of $123 million (over 21 percent) compared to 1974. Of that figure, it is estimated that only about $15 million represented total improved benefits. A much larger share of the increase resulted from increased utilization and from higher prices (Morris, 1976).

By now it should be obvious that health care is not only big business, it is highly important to big business.

3.10 Study Questions

3.1 What is the goal of an economic system?

3.2 Describe the unique features of the producer/consumer relationship in health care.

3.3 How have hospital costs changed in the past five years compared to costs of other goods and services?

3.4 Eye care receives only 2 percent of the health-care budget. How much money is spent in this area?

3.5 What factors affect the distribution of benefits in national health insurance programs?

3.6 Describe several ways in which costs of national health insurance are distributed.

3.7 List the elements that a national health insurance program should include.

3.8 What are the two roles played by the physician in providing health care on a fee-for-service basis?

3.9 Describe the capitation method of reimbursement.

3.10 List the categories in which patient payment methods are placed.

3.11 List the three major reasons for increases in hospital costs.

3.12 What "incentives" do hospitals have to raise costs?

3.13 Is health care monopolistic or competitive? Explain.

3.14 Summarize, from an economics standpoint, the reasons for the view of medicine as noncompetitive.

3.15 Discuss the impact of health-care economics on business.

3.11 References

Ehrbar, A. F. (1977). A radical prescription for medical care. *Fortune*, February.

Ehrenreich, J., and B. Ehrenreich (1971). The American health empire: power profits and politics. New York: Vintage Books.

Executive Office of the President's Council on Wage and Price Stability (1977). The rapid rise of hospital costs. Staff Report, January. Washington, D.C.

Feldstein, M. (1975). The rising cost of hospital care. Washington, D.C.: Information Resources Press.

Freymann, J. (1974). The American health care system: its genesis and trajectory. New York: Medcom Inc.

Hepner, J. O., and D. M. Hepner (1973). The health strategy game. St. Louis: C. V. Mosby.

HEW (1970). Research and statistics note, October 30. Office of Research and Statistics, Social Security Administration.

Johnson, R. (1975). Hospital economics: 19 myths. *Hospital Financial Management*, January: 22–28.

Levey, S., and P. Loomba (1973). Health care administration: a managerial perspective. Philadelphia: J. B. Lippincott.

McMahon, J. A. (1976). Role of the health care industry in the health care economy. Chicago: American Medical Association.

Morris, G. (1976). The impact of health care economics on business. Chicago: American Medical Association.

Phalon, R. (1977). Medicare gap is widening. *Detroit Free Press*, March 12.

Samuelson, P. (1976). Economics: an introductory analysis. New York: McGraw-Hill.

Vladeck, B. (1977). Why our nonprofit institutions are going broke. *Medical Economics*, January 10: 281–282.

Ward, R. A. (1976). The economics of health resources. Reading, MA: Addison-Wesley.

Weltmann, A., B. O'Neil, and P. Taylor, (1974). A layman's guide to hospitals. New York: Coopers and Lybrand.

4

Codes, Standards, and Regulations

ROBERT J. CANGELOSI and JOHN G. WEBSTER

4.1 Learning Objectives

Upon completing this chapter you should be able to:

1. List the major codes, standards, and regulations that affect clinical engineering practice.
2. Describe the salient features of these codes, standards, and regulations.
3. Explain the application of the major codes, standards, and regulations to clinical engineering practice.
4. Differentiate between codes, standards, regulations, and specifications.
5. Explain the development and promulgation of standards.

4.2 Introduction

This chapter describes the codes, standards, and regulations that are becoming increasingly important in the practice of health care. It details the role of the government and other groups responsible for their development and implementation and presents details on radiological regulations and mechanical standards.

4.3 Purposes of Codes, Standards, and Regulations

Definitions

Codes, standards, regulations, and specifications must be defined and their functions carefully detailed in order to evaluate their proper use.

Codes. In the technological environment, a code usually is defined as a system of principles or regulations or a systematized body of law. Thus, codes represent the accumulation of a system of regulations which usually are enforced under an active law by a local, state, or federal government agency. One of the most familiar codes is the National Electric Code (NEC) which was developed by the National Fire Protection Association (NFPA), is promulgated as an American National Standard under the procedures of the American National Standards Institute (ANSI), and usually is enforced by localities as part of their regulations for the construction or building industries.

Standards. Standards are documents that are perceived to establish a minimum level of performance or to standardize such items as test methods or materials. Roy Towbridge, Director of Engineering Standards with the General Motors Corporation, defines standards as follows: "Standards are a distillation of engineering or technical experience, recorded in such a way that one schooled in the discipline for which the standard is intended may use the standard as a yardstick to govern designs, materials, functions, practices, or testing within the discipline. They are a solution to a problem, which may or may not be repetitive, which has been accepted by a duly constituted authority" (Towbridge, 1970). Standards are referenced in other documents such as specifications, codes, or regulations and therefore serve as a basis for these documents. Contrary to popular belief, standards are not utilized as purchasing documents but can be used as part of a purchase document.

Regulations. Regulations are the rules normally promulgated by the United States Government in the *Federal Register* and codified in the various United States codes. These rules or regulations are a method by which the government communicates with the public, industry, and other interested parties.

Specifications. Specifications are documents used to control the procurement of equipment and buildings. In most applications, specifications of a general nature usually reference other specifications, standards, handbooks, and drawings. Specifications usually cover design criteria, materials, processes, test methods, technical data, inspection, and other items necessary to procurement by establishing a technical agreement between the procuring activity and the contractor as to what is desired and what quality will be accepted.

Development and promulgation of standards

The main sources for the development and promulgation of standards, particularly for medical devices, fall into three general areas: voluntary, mandatory, or proprietary. *Voluntary standards* are those developed under the consensus process where manufacturers, users, consumers, and government come together voluntarily in open public sessions. In the consensus process, the document is made available to all interested parties. *Mandatory standards* (regulations) are those standards which are required by law, such as the 1976 Medical Device Amendments to the Food, Drug, and Cosmetic Act. This legislation and the development and promulgation of medical device standards result in mandatory requirements being imposed on device manufacturers. Probably the smallest percentage of technological standards are those of a proprietary nature. *Proprietary standards* are developed either by a company for its own internal use or by a trade association for use by its members. Some examples of proprietary standards can be found within the Compressed Gas Association. Proprietary standards can serve as a basis for voluntary or mandatory standards if the appropriate exposure is given to these standards and if there is consensus reached among all parties.

Sources — general

Standards organizations. Activity in medical device standards has accelerated rapidly since the late 1960s because of an increase in the activities of the Food and Drug Administration (FDA) in the area of medical device classification (Section 513 of PL 94-295). The classification project of the FDA, which is probably the primary mover for mandatory and voluntary standards, will be discussed in detail later in the chapter. In the voluntary area, the American Society for Testing and Materials (ASTM), the American Dental Association (ADA), and the Association for the Advancement of Medical Instrumentation (AAMI) have been the principal organizations concerned with voluntary medical device standards. AAMI is a recent entry into the standards area while the other two organizations have been in existence for years.

As interest was stimulated in the medical device area, the FDA responded by cataloging the various standards programs directed at medical devices and semiannually publishes the *Bureau of Medical Devices Standards Survey*. In January of each year, an international edition is published and in July, a national (USA) edition.

The NFPA and Underwriter's Laboratory (UL) standards activities have received widespread attention, primarily because of the listing for certification by the UL and by the adoption in many localities of the NEC which is developed under the auspices of the NFPA.

International standards promote the free flow of goods by creation of a common economic language among nations. The two major international standardization bodies, the International Organization for Standardization (ISO) and the International Electrotechnical Commission (IEC), have produced more than 2500 standards and the number is expected to double in the next five years.

The ISO was founded in London in 1946 when 64 delegates representing 25 nations met to consider forming an international standards organization. The object of the proposed organization was "to facilitate the international coordination and unification of industrial standards." At that time about 150,000 national standards were in existence, so the organization's early efforts were concentrated on the coordination of these standards. But, since national standards work had first priority following World War II, only limited interest was expressed in the organization. By the 1960s, however, international standardization was growing because of the proliferation of multinational corporations whose operations could be hindered by conflicting national standards. Today the national standards bodies of 73 countries (known as member bodies) are represented in ISO and are working in virtually every area of technology with the exception of electrotechnical matters, the province of IEC. In the medical field, ISO has six technical committees concerned with standards for medical devices: Technical Committee (TC) 121: Anesthesia Equipment; TC 76: Medical Transfusion Equipment; TC 150: Surgical Implants; TC 75: Stretchers; TC 106: Dentistry; and TC 84: Syringes for Medical Use and Needles for Injection. The USA member body in ISO is ANSI, which holds a voting membership on 107 of the total 150 ISO technical committees.

An ISO document goes through a number of stages before it is accepted as an international standard. After a draft proposal is written by a working group or a subcommittee, it is submitted to the responsible technical committee. Following agreement by the committee, it is sent to the secretariat for registration as a draft international standard (DIS); the DIS then is circulated to all member bodies for voting. If 75 percent of the votes are cast in favor of the DIS, it is sent to the ISO Council for final acceptance as an international standard.

ISO is financed through member dues and the sale of standards. The dues a country pays are determined by its gross national product. The USA, for example, pays approximately $150,000 per year through its representative, ANSI. (ANSI is financed largely through its member dues.) When ISO publishes a standard, a limited number of free copies are distributed to its members who, in turn, may reproduce and sell the standards. A predetermined amount of the selling price is returned to ISO headquarters.

IEC was formed in 1904 as the result of the adoption by the Chamber of Government Deputies at the St. Louis International Electrotechnical Congress of the resolution: "Steps should be taken to secure the cooperation of the technical societies of the world by the appointment of a representative commission to consider the question of the standardization of the Nomenclature and Ratings of Electrical Apparatus and Machinery." The members of the commission are from the national committees, i.e., one for each country represented. Currently, there are 45 such national committees.

IEC documents are called recommendations, not standards. They are processed, published, and sold in much the same manner as ISO documents. A limited number are distributed to the national committees who then have the right to duplicate them for sale, returning a small portion of the sale price to the IEC. However, the bulk of IEC's budget comes from the electrical and electronics industries which are the ultimate beneficiaries of the published documents.

The IEC Technical Committee 62 on Electrical Equipment Used in Medical Practice is the primary committee concerned with medical equipment. The IEC 62 has four major subcommittees of which the USA has the Secretariat for IEC 62D on medical equipment. European common market countries are the primary contributors to both ISO and IEC. Many of the countries have stated that they will forego national standards in favor of international standards developed by ISO and IEC, therefore providing great impetus to these two organizations to develop international standards.

U.S. Government organizations. The following are government agencies and programs for medical device standards:

1. The Bureau of Medical Devices within the FDA is responsible for the development of medical device and

in vitro diagnostic product standards and their promulgation as specified under PL 94-295, the Medical Device Amendments of 1976 to the Food, Drug and Cosmetic Act.

2. The Bureau of Radiological Health, also within the FDA, is responsible for administering the Radiation and Control for Health and Safety Act, PL 9602, and for developing safety standards for products that emit radiation, both in the medical and nonmedical fields. In the medical field, X-ray equipment, laser devices, and ultrasonic equipment are major activities in which the Bureau has engaged.
3. The Federal Communications Commission (FCC) has requirements that must be met by all devices that emit radio-frequency energy. Devices such as X-ray, diathermy, and ultrasonic devices fall under the jurisdiction of the FCC as devices that emit radio-frequency energy. The FCC also regulates the use of airways for medical communications.
4. The Occupational Safety and Health Administration (OSHA) of the Department of Labor focuses on safe working conditions for hospital employees; all hospitals must adhere to OSHA employee requirements.
5. The National Institute for Occupational Safety and Health (NIOSH), within the Department of Health, Education, and Welfare, is primarily a research agency with no regulatory authority. Its major function is to investigate the sources of safety hazards and to remedy them through OSHA programs.
6. The National Bureau of Standards (NBS) is a part of the Department of Commerce. One of the Bureau's principal activities has been to provide reference standards for calibrating medical and other instruments. NBS had one of the first medical device standards on mercury and glass thermometers. The Bureau has taken the position that if standards can be developed by voluntary organizations, then the Bureau need not engage in this type of activity. They do support technical activities within the various medical device standards organizations such as ASTM, ANSI, and AAMI to provide technical expertise for their committees.
7. The Veterans Administration (VA) primarily establishes standards for its purchasing of biomedical instrumentation. Because of the large purchasing power of the VA and its medical facility requirements, VA specifications have received considerable attention.

Trade organizations. Currently, there are several trade organizations involved in the review of medical device standards. The primary organizations are: the Pharmaceutical Manufacturers Association (PMA), the Health Industries Manufacturers Association (HIMA), the Scientific Apparatus Manufacturers Association (SAMA), and the Compressed Gas Association (CGA). Their primary role has been to represent manufacturers on the various voluntary standards committees and sometimes, as in the case of the CGA, they have developed their own standards within the trade organization. These organizations play a useful role in consolidating and providing a consensus of the manufacturers' viewpoints on various standards and activities.

Professional organizations. Various professional organizations play a role in the development of standards. Such organizations as the American College of Cardiology, the American College of Surgeons, and the American College of Neurosurgeons have committees that deal with medical device standards. These organizations primarily provide participants to voluntary organizations such as ASTM and AAMI.

4.4 Use of Standards for Procurement

Current medical device standards activities

Efforts to develop standards and specifications for medical devices in the USA are divided between government and voluntary agencies. Standards efforts by voluntary organizations have been fragmented and did not become a major force in the medical device standards area until the early 1970s. In a few areas cooperative efforts between the government and voluntary organizations evolved. The Voluntary Product Standards Program of the NBS is one such example. A full-time director and technical staff at the NBS examines dental materials and equipment. The American Dental Association and the ANSI Dental Standards Committee have been active since the mid-1960s developing ANSI dental standards.

Federal programs

The medical device standards and specifications activities sponsored by the federal government are extensive. They represent the largest standards effort in the nation under a single organizational structure. However, federal efforts are not a single program but consist of a variety of separate programs within different departments, agencies, and military commands. The principal government programs in the field of engineering and product standardization are those conducted by the Department of Defense (DOD) and the General Services Administration (GSA).

Other agencies that are active in the medical device standards area include the Veterans Administration (VA) and the FDA. The Defense Personnel Supply Center (DPSC) of the DOD is the largest purchaser of medical devices within the government and prepares a majority of the specifications and standards used in its procurement activities.

The General Services Administration, Federal Supply Service, Office of Standards and Quality Control. The GSA produces few of its own specifications and standards in the medical device area, but acts as the clearinghouse and coordinator for federal specifications and standards. GSA also acts as a cataloging and pricing agency for the other government agencies. The GSA is charged by statute with the responsibility for developing specifications and standards used for equipment, supplies, and related engineering services procured and used in common by more than one agency of the federal government. This responsibility relates to the role of GSA in providing administrative and housekeeping services to the entire federal establishment. Specifications and standards developed under the GSA normally relate to commercial items used in common by the government and the private economy. The level of quality established by federal specifications and standards is intended to assure that the government realizes the maximum value of each dollar expended in the procurement of its supplies. Under this program, more than 4000 specifications and 200 standards have been developed.

The GSA standardization program for materials, supplies, and equipment procured by the federal government is centralized in the Office of Standards and Quality Control. This office provides nationwide leadership in the development and execution of the programs of the Federal Supply Service (FSS) concerned with commodity control. To provide greater speed and flexibility in the preparation of new and revised standards, the concept of "assigned agency" was adopted. Under this method, federal agencies are given responsibility for specifications and standards development projects for which they have specialized knowledge.

As agencies accept assignments, they coordinate specifications and standards projects to meet the most pressing needs of the agencies. Federal specifications and standards are promulgated by the GSA. Before approval and promulgation, they are reviewed to assure that suggestions by other federal agencies and suppliers have been incorporated or reconciled. In some instances, it is necessary for the assigned agency, or GSA, to hold conferences with the agencies and industry for further development of the document before the specification or standard is ready for promulgation.

Although the GSA charter appears to be all-encompassing, the GSA presently acts only as a cataloging and pricing organization in the area of medical device standards. Most of the standards and specifications for medical devices listed in the *Index of Federal Specifications and Standards* were generated under the assigned agency concept by the Defense Personnel Supply Center of the DOD and the Division of Medicine and Surgery Supply Service of the VA. The GSA, not having adequate expertise in its own organization, has developed only a limited number of standards and specifications for some of the less complex medical devices. These also are listed in the *Index of Federal Specifications and Standards.* Since it is the coordinating buyer for the government, the GSA's primary function is to negotiate the most favorable terms for purchases.

Veterans Administration. The Department of Medicine and Surgery's Supply Service Division, an arm of the VA, is one of the largest government buyers of medical devices. Medical device expenditures exceeded $50 million in fiscal year 1971 with the largest part of the money being spent on X-ray equipment valued at approximately $25 million. This X-ray equipment was purchased under comprehensive VA-generated specifications and standards, and on-site inspections by VA quality-control personnel were conducted.

Section III of the VA supply catalog is an index of specifications which includes applicable specifications for items currently acquired by stations from sources other than VA supply depots. Most of these items initially are furnished by contractors as part of construction projects. This index includes VA Specification X-1414, Biomedical Monitoring Systems (Veterans Administration, 1970), which was the first document written in detail to specify the characteristics of biomedical monitoring systems. The VA's Prosthetic and Sensory Aids Services Division is active in developing specifications and standards for prosthetic, sensory, and rehabilitative aids.

The VA has construction standards which provide criteria for electrical safety, e.g., "X-ray Protection and Control Room Requirements," "Isolating Transformers for Anesthetizing Location," and "Electric Receptacles-Special Requirements." These documents, generated by the VA's Department of Architecture and Engineering, deal primarily with building construction but also include pertinent data that manufacturers of electrically powered medical devices should consider in the design and application phases of their product development programs.

Department of Defense, Defense Supply Agency, and Defense Personnel Support Center. The Defense Standardization Program is directed by the Office of the Secretary of Defense with the operational responsibility assigned to the Departments of the Army, the Navy, and the Air Force; the Defense Supply Agency (DSA); and other agencies of the DOD. The principal objective of the Defense Standardization Program is to provide uniform definitions of technical requirements for parts, equipment, and systems in which the various elements of the DOD have a common interest. Nearly 40,000 specifications and standards defining the technical requirements of the DOD have been developed through this program. By far this is the most extensive standardization program in the nation—government or private.

Under the Defense Standardization Program, the Defense Personnel Support Center of the Defense Supply Agency (DPSC, the assignee agency) is responsible for standardization of the following federal supply classification categories: Group 65—Medical, Dental, and Veterinary Equipment and Supplies; Group 66—Instruments and Laboratory Equipment; Subgroup 6630—Chemical Analysis Instruments; and Subgroup 6640—Laboratory Equipment and supplies. Within DPSC, the Directorate of Medical Material is the organization responsible for the preparation of all medical device specifications and standards. It is responsible for the technical aspects of the approximately $250 to $280 million annual procurement of drugs and devices for the DOD. Approximately 45 percent of this is spent for drugs and 25-30 percent on medical materials at a local level (hospitals and bases). DPSC tries to handle items that are common, can be processed in bulk with significant savings, and can be controlled rigidly for quality. Most electronic instrumentation is purchased on a local level.

The development of a specification or standard normally starts with an initial request for a device, channeled from the Surgeon General's office of the respective service (Army, Navy, or Air Force) to the Defense Medical Material Board (DMMB). The essential characteristics (EC) for the device are generated by DMMB and describe the performance characteristics and, in some cases, the types of materials to be used. Then, the ECs are sent to the Technical Services Branch, Division of Technical Operations, Directorate of Medical Material, DPSC, which prepares a technical data package for the specifications, drawings, packaging, and instructions. After the specification is prepared, it is sent to the DMMB for review. During the procurement cycle, any technical questions or requests for deviations are referred to the Technical Services Branch and if it concerns an EC, to the DMMB. When an item or test procedure is common to a large group of items, a general specification or standard is written.

Approximately 14,000 items are handled by the Directorate of Medical Material. Over 2000 specifications and a small number of standards have been generated. DPSC can write either federal or military specifications and/or standards. When the decision is reached to write a specification or standard, DPSC determines if other federal agencies are interested in it. If so, a request is made to the GSA for permission to write the specification or standards. (DPSC has written most of the federal specifications and standards in the medical device area.) Federal specifications are preferred because they have wider distribution and use. A civilian agency can reference a military specification or standard.

Department of Health, Education, and Welfare: Food and Drug Administration. Standards-setting activities for medical devices within the FDA are delegated to the Bureau of Medical Devices and the Bureau of Radiological Health (BRH). The Bureau of Medical Devices' statutory authority is provided by the Food, Drug, and Cosmetic Act as amended; the 1976 Medical Device Amendments will be discussed later. The BRH, through the Radiation Control for Health and Safety Act passed in 1968, is authorized to set performance standards for radiation-producing electronic products. Some of these products are used for medical purposes.

BRH has developed performance standards for television receivers, cold-cathode gas discharge tubes, and microwave ovens. It recently promulgated a performance standard for medical X-ray equipment. BRH will continue to examine products that fall within the purview of the Radiation Control for Health and Safety Act, i.e., to set standards for those safety aspects related to radiation only.

Specifications developed by the federal government represent the definition of technical requirements for products and services to be purchased by the largest consumer in the world. Aside from the statutory obligation of the government to assure procurement of optimum value, the sheer volume of government consumption and its mandate to buy competitively from broad supply sources required organized definitions of its technical requirements in the form of standards and specifications.

State and local government

State and local governments utilize various codes and standards as part of their regulations for controlling products and systems, particularly in the building area. Some of the most prominent examples used in this area are the UL listing; the NEC, which is developed by the NFPA; and other independent laboratory evaluations of proprietary standards or existing American National Standards.

Underwriters Laboratories, Inc. UL is an independent laboratory which was founded in 1894 by a group of insurance underwriters. It is no longer sponsored by insurance interests and draws its membership from many areas. The basic mission of UL is to investigate devices, products, equipment, and materials as to their hazards to life and property. Their investigations focus on the safety of items, not on their efficacy. UL conducts testing and evaluation programs which result in a listing of the device in accordance with standards that it develops. For biomedical equipment UL Standard 544 (for safety of medical and dental equipment) is utilized (Underwriters Laboratories, 1976); it initially was issued in May of 1972. Several existing UL standards being used to evaluate medical devices were combined with some additional provisions. UL 544 generated some controversy in the professional and manufacturing communities when it first was introduced. Since that time it has gone through several revisions and was made effective for listing products by UL in September 1974. UL has attempted to qualify the standards as an American National Standard under the procedures of ANSI. So far, this has been unsuccessful. In addition, in 1974 UL issued an initial proposed draft standard for laboratory equipment—UL 1262, "Proposed Safety Standard for Laboratory Equipment." This standard relates to safety, not accuracy of performance, and appears to cover most items used in the clinical laboratory.

Manufacturers contact UL for testing and evaluation with the intention of listing their devices if they meet the requirements of the standards. The cost of the evaluation is paid by the manufacturer submitting the item. Once an item passes the UL test, it is listed and the UL listing mark can be attached to production item models as long as they continue to meet UL requirements. As previously stated, listing does not indicate a level of clinical performance or efficacy. UL has a factory follow-up service to insure continued compliance of listed items with UL requirements. This service follows periodic inspections of production items at the factory and may involve the selection of items for testing at UL. This follow-up service is required as long as the UL listing mark is used; the cost for this service is paid by the manufacturer.

National Fire Protection Association. NFPA is a nonprofit association founded in 1896 to promote fire protection and safety. Membership in NFPA is open to individuals and organizations. NFPA standards are prepared by technical committees and adopted by their memberships after extensive periods of public review, sometimes requiring years. The most prominent NFPA document that is utilized by state and local governments is the NEC (National Fire Protection Association, 1978). It is referenced by many state and local governments and other regulatory bodies, making it a mandatory requirement in most locales throughout the U.S. The NEC covers hospital electric power, wiring, emergency power, receptacles, electrical appliances (equipment), and a large number of other items.

NFPA standards in the medical field usually are facility-oriented, but in some cases can involve specific medical devices. The following NFPA standards are applicable in the health care field.

1. Inhalation anesthetics standard NFPA 56A (National Fire Protection Association, 1973a) specifies safe practices. If flammable anesthetics such as cyclopropane are used, it requires conductive floors, isolation transformers, etc., to prevent sparks that cause explosions. Nonflammable anesthetics such as halothane require correct storage, piping, and ventilation. All compressed gases possess a large amount of pneumatic energy and present danger. Nitrous oxide and oxygen fittings should not be greased or oiled; otherwise an explosion may result.

2. Respiratory therapy standard NFPA 56B (National Fire Protection Association, 1976) discusses treatment with oxygen, artificial ventilation, etc. There are fire hazards created because of the increased combustibility of materials in an oxygen-enriched atmosphere.

3. Essential electrical systems standard NFPA 76A (National Fire Protection Association, 1973b) details safe practices for hospital electric power, wiring, and emergency power. An alternative source of power such as a motor–generator set requires periodic inspection.

4. Electricity in patient-care facilities standard NFPA 76B-T (National Fire Protection Association, 1973c) recommends safe practices for hospital electric wiring, installation, electrical appliances (equipment), and inspection. For example, Table 4.1 shows patients divided into different classes of electrical risk. Chapter 7 presents appliance tests that are taken from this standard as well as other means of preventing electric shock.

5. High-frequency electricity in health-care facilities standard NFPA 76C (National Fire Protection Association, 1975) presents safe practices for use of electrosurgery and diathermy equipment. Two safety suggestions of the standard are a large electrosurgical dispersive electrode to prevent burns to the patient and a monitoring circuit to warn of the disconnection of the dispersive electrode.

Other NFPA hospital standards available from the NFPA are:

• 3M	Hospital emergency preparedness	1970
• 50	Installation of bulk oxygen systems at consumer sites	1973
• 56C	Laboratories in health-related institutions	1973
• 56D	Hyperbaric facilities	1976
• 56E	Hypobaric facilities	1972
• 56F	Nonflammable medical gas systems	1974
• 56G–T	Inhalation anesthetics in ambulatory care facilities	
• 56HM	Manual for the home use of respiratory therapy	1973
• 56JT	Standard for combustible solid supplies in health-care facilities	
• 56K	Recommended standard for medical/surgical vacuum systems	

• 101	Life safety code	1973
• 801	Facilities handling radioactive materials	1970

Canadian Standards Association. The Canadian Standards Association (CSA) publishes a series of codes that cover subjects similar to those of the NFPA (Canadian Standards Association, 1973).

Table 4-1. Classes of Patient Electrical Risks. Reproduced with permission from NFPA 76B-T (Tentative Standard for the Safe Use of Electricity in Patient Care Facilities), Copyright 1973, National Fire Protection Association, Boston, MA. NOTE: Tentative means that the Standard has not finally been adopted by the Association.

Risk	*Hazard Level*	*Current Path*	*Typical Situations*	*Hazardous Currents (60 Hz)*	*Design Goal for Safety (Note 3)*
I	General	Accidental contact with skin or mucous membrane.	Most ambulatory patients (except during certain medical procedures), many bed patients (minimal contacts with electric appliances) and outpatients. Hospital personnel are subject to similar exposure.	*Skin Contacts:* 80 mA — Ventricular fibrillation 4.5 mA — "let-go" current limit 0.5 mA — sensation level (Note 1)	500 μA
II	Increased	Intentional prolonged contact with skin mucous membrane or subcutaneous tissue.	Those bed patients (some ambulatory) subject to diagnostic or therapeutic procedures involving electric appliances. Some very ill patients in special locations.	*Skin Contacts* 4.5 mA — "let-go" current limit (Note 1)	100 μA
III	High	Accidental contact with a conductive path to the heart.	Patients with externalized conductors, diagnostic catheters, or other electric contact with the conduction system of the heart.	*Epicardial Contact* 100 μA — ventricular fibrillation *Endocardial Contact* 20 μA — ventricular fibrillation (Note 2)	10 μA

Note 1 — Currents are ½ percentile values developed by Dalziel, i.e., ½% of population is affected by current equal or less than that shown, 99½% are not affected. The "let-go" current is an estimate for children.
Note 2 — Currents for Risk Class III are values extrapolated from animal studies and very limited reports of human measurements.
Note 3 — These currents are lower than known dangerous levels in order to provide a margin of safety.

Laboratory evaluation. There are various laboratories throughout the U.S. that test and evaluate medical devices. To meet the requirements of regulatory bodies for examination of medical devices prior to use, most states have rules that apply to purchases by their state hospitals. In addition, many states have adopted regulations applicable to medical equipment. In the case of California, an explicit state law governs medical equipment. Many cities also have general regulations that cover the purchases of medical equipment; New York, Los Angeles, San Diego, Chicago, and other large cities have adopted special requirements. For example, Chicago requires that a device be listed through UL, the Canadian Standards Association, or must have been tested independently by the Illinois Institute of Technology under a contract arrangement with the city of Chicago. A new entry into the laboratory evaluation area is the Maryland Electrical Testing Laboratory (MET) which has produced a document entitled "Standard for Performance and Safety Certification of Medical Equipment" (MET Electrical Testing Co., 1976) which deals with medical devices. MET provides a listing of devices similar to UL and is applying for accreditation in such cities as Los Angeles and Chicago.

4.5 Legal Aspects of Standardization

The following quotation, taken from the *ASTM Standardization News* (ASTM, 1977), describes the legal aspects and the viewpoints of the ASTM towards standards (Reprinted by permission of the American Society for Testing and Materials, Copyright, 1977).

Legal aspects of standardization

"The heart of a standard, of course, is its technical credibility, but a standard must also rest on sound legal foundations if it is to stand for very long.

"In the United States, there are four legal principles that standards must consider: due process, restraint of trade, authority and responsibility, and liability. Standards that are technically sound and consider these principles will most likely be upheld by the courts if they are challenged in litigation. On the other hand, a standard may be technically excellent, but if it fails to consider any one of these principles—especially due process or restraint of trade—it is liable to be struck down and may even jeopardize the organization responsible for its development.

"These legal principles are critically important to the consensus standards community, so let's inspect each one of them in depth.

"**Due process.** The term *due process* stems from an article in the U.S. Bill of Rights which stipulates that no one can be deprived of the benefits of his property except by due process of law. That right includes the ability to speak in one's own behalf and to seek redress. In ASTM this means that everyone on a committee has a right to his voice in the development of a standard, and if a person is dissatisfied with a standard he may appeal at any step in its development.

"**Restraint of trade.** Adherence to the principles of due process not only promotes the interests of all involved but also precludes discriminatory, anticompetitive effects, which leads us to the second element of standards legality—restraint of trade.

"Safeguards have to be maintained to prevent standards from possibly restricting competition by stifling innovation or excluding potential competitors from established markets. Also, care must be taken to assure that the interests of those who consume materials and products are not sacrificed to the interests of those who produce them.

"**Authority and responsibility.** The third legal principle observed by ASTM is authority and responsibility, which stipulates that:

1. Standards must be promulgated by organizations duly authorized through expression in their charters to engage in standardization activities. Such authority is often granted to professional societies, trade associations, etc.
2. Procedures for development and promulgation of standards must be spelled out in the organization's bylaws.
3. These organizations are responsible for maintaining their standards as well as upholding them against technical or legal attack.
4. The organizations must have competent sources of technical expertise.

"**Liability.** The fourth legal principle is liability, an important one that is especially of interest to people who are considering joining in the development of standards.

"We are frequently asked: 'What happens to me if a standard I helped develop is misused and someone sues for damages?' In Pennsylvania, where ASTM is chartered, the Nonprofit Corporation Law flatly states that the members of a nonprofit corporation shall not be personally liable for the debts, liabilities, or obligations of the corporation.

"Liability could be imposed upon the corporation, however, because it is the corporation that holds itself out as an expert on standards and publishes these standards for use by the public. Thus, if a suit were instituted for negligent misrepresentation (and that is the essence of legal liability), the corporation and not the individual would be the primary target." *Cangelosi and Webster*

4.6 Government Development of Standards

Medical device regulation by standards — United States Government

The key organization for the development and promulgation of medical device standards is the Department of Health, Education and Welfare, FDA, Bureau of Medical Devices. Under PL 94-295, the Medical Device Amendments of 1976 to the Food, Drug and Cosmetic Act provide in section 514 for the development and promulgation of medical device standards. The Bureau of Medical Devices has the responsibility for administering PL 94-295 which was enacted and signed by the President on May 28, 1976.

Under the medical device amendments, three device classes are established for the control of devices intended for human use (Food and Drug Administration, 1977; Willard et al., 1977). Class I is considered general controls (controls at the lowest level) and applies mainly to provisions reflecting the labeling of a device, device registration of device manufacturers, the meeting of Good Manufacturing Practices, and the keeping of records and reports. The next higher class of regulatory control is Class II performance standards. If the provisions of Class I are not sufficient to provide for the safety and efficacy of the device, it is placed under Class II performance standards. The highest regulatory control class is Class III, premarket approval. Devices are placed in Class III because of insufficient information; a standard would not provide reasonable assurance of safety and effectiveness; or the device is a life-sustaining or life-supporting device. Class III devices must undergo premarket approval *prior* to being placed on the market.

Expert extramural panels, known as classification panels, are established in specialty areas to determine the appropriate placement of devices into one of the three regulatory control classes. The panels are composed of members from outside the government with an executive secretary from the Bureau of Medical Devices assigned to assist the panel. Once devices are classified, the Bureau will issue a proposal in the *Federal Register* which will be open for comment. After all comments are evaluated, a final order of the regulatory control class of the device is published.

Devices placed in Class II may be subject to the development of performance standards under section 514 of the Food, Drug and Cosmetic Act. A performance standard may be developed to provide reasonable assurance of the safe and effective performance of a device. It may include provisions respecting construction, components, ingredients, properties of a device, its compatibility with power systems, and its connections to such systems. Provisions for testing on a sample or individual basis and for the management of performance characteristics of the device may be included. Where appropriate, the standard may specify the form and content of labeling for proper installation, maintenance, operation, and use of the device.

The Food, Drug and Cosmetic Act provides procedures for initiating performance standards and has provisions whereby an existing standard may be submitted for consideration. There are provisions for financial assistance to develop standards and for the periodic review of existing standards. Once a proposed standard has been developed, it is offered in the *Federal Register* as a proposal with 90 days for comment. Once the comments are resolved, a standard will become effective one year from the publication date of the final order. A standards advisory committee to resolve questions of a scientific nature which may result from the standards proposal are specified in the Amendments.

Various medical device standards have been under development, both intramurally and under contract, prior to the enactment of legislation. The following draft standards and research studies have been undertaken by the Bureau of Medical Devices:

Performance and safety standards

• dc Defibrillator	FDA-MDS-021-0001
• Insulin syringe	FDA-MDS-122-0003
• EMI (Electromagnetic Interference)	FDA-MDS-201-0004
• Environmental performance	FDA-MDS-202-0005
• Diagnostic ECG	FDA-MDS-021-0006
• Incubators/infant warmers	FDA-MDS-021-0028
• Implantable pacer	FDA-MDS-021-0032
• Electrosurgical units	FDA-MDS-081-0033

Studies underway

- Blood-Gas exchange — oxygenators
- Cardiac monitoring
- Electroconvulsive therapy
- Infusion pumps

The FDA publishes the *Bureau of Medical Devices Standards Survey* to provide a comprehensive listing of current national and international standards promulgation activities in the fields of medical devices and diagnostic products. These standards include voluntary and regulatory standards, tentative and recommended standards, recommended practices, purchasing specifications, policy statements, glossaries of technical terms, and proposed standards. The standards identify criteria such as the safety and performance of medical devices and diagnostic products and therefore will serve as a basis for the continuing review of the adequacy of existing standards and as guidelines for the development of new standards. The *Survey* is published in two parts: the International Edition (Food and Drug Administration, 1978a), in January and the National Edition (Food and Drug Administration, 1978b), in July.

4.7 Standard for the Operation of Hospitals

The Joint Commission on Accreditation of Hospitals establishes standards for the operation of hospitals. Hospitals that volunteer to be accredited request an accreditation survey. They are expected to meet the standards set forth in the Accreditation Manual for Hospitals (Joint Commission on Accreditation of Hospitals, 1976). Compliance with standards is assessed by statements given by hospital personnel, documentary evidence, answers to detailed questions, and on-site observations by surveyors.

Points covered by the survey include patients' rights, quality of professional services, building and grounds safety, functional safety, and sanitation and infection control. A wide variety of services such as medical records, emergency services, special-care units, etc. are also covered. Of most importance to the clinical engineer are the two standards below, which are taken from the manual. These standards require a documented program for equipment inspection, for periodic evaluation of the electrical power distribution system, and for other areas of safety.

Standard I

"There shall be a multidisciplinary hospital safety committee to adopt, implement, and monitor a comprehensive hospital-wide safety program. The program shall contain requirements related to the staffing, equipping, operation, and maintenance of the hospital, which shall be designed to produce safe characteristics and practices and to eliminate, or reduce to the extent possible, hazards to patients, hospital staff, and visitors.

"**Interpretation.** Regardless of the hospital's size or mission, there shall be a hospital safety committee. The functions and responsibilities of the safety committee may be assumed by an already-established committee, or subcommittee thereof, that has related interests and responsibilities. While its size and composition may vary with the individual facility, the committee shall be multidisciplinary, having representation from the administration, the medical staff, nursing services, engineering and maintenance department, housekeeping, and dietetic services. In-house specialists such as the radiation physicist, clinical engineer, or biomedical engineering technician should also participate at least in a consultative role. The chairman of the committee shall be appointed by the chief executive officer of the hospital; the safety director or officer shall be a member of the committee. When a major safety service is performed by an outside contractor, the contractor should provide consultation to the committee as needed.

"The chief executive officer shall designate an individual as safety director or officer who should be qualified by training and/or experience in safety. This individual should have the time to direct and coordinate all phases of the program. In a small facility, the director may be expected to perform his safety-related functions in addition to other duties.

"The committee shall meet as required, but not less than monthly, and shall maintain written minutes of each meeting. The pertinent committee findings and recommended actions shall be reported in writing to the governing body, administration, medical and nursing staffs, and all departments and services involved.

"The responsibilities of the safety committee and the safety director or officer shall include, but are not limited to, the following:

1. The development of written policies and procedures designed to enhance safety within the hospital and on its grounds to the maximum degree possible.
2. Coordinating and cooperating in the development of department/service safety rules and practices.
3. The establishment of an incident reporting system that includes a mechanism for investigating and evaluating all incidents reported and for documenting the review of all such reports and actions taken.
4. Liaison with the infection control committee, with a mutual exchange of information.
5. The provision of safety-related information to be used in the orientation of all new employees and in the continuing education of all hospital employees; this may include the use of suitably located safety information bulletin boards.

6. Conducting of a hazard surveillance program at specifically defined intervals.
7. The establishment of methods of measuring results of the safety program and periodic analysis to determine its effectiveness, including a review of all pertinent records and reports.
8. Familiarization with local, state, and federal safety regulations applicable to the facility.
9. Familiarization with major safety-oriented agencies, both governmental and nongovernmental.
10. Development of a reference library of pertinent documents and publications dealing with all facets of hospital safety. It is strongly recommended that copies of all applicable building and safety codes and standards be included.

"The authority of the safety committee, through the chairman or the safety director or officer, to take action when a hazardous condition exists that could result in personal injury to individuals or damage to equipment or buildings shall be defined in writing and approved by the governing body or chief executive officer, as appropriate, and by the medical staff.

Standard II

"Comprehensive safety systems shall be installed, and practices, policies, and procedures instituted, to minimize hazards to patients, hospital staff, and visitors.

"**Interpretation—electrical safety.** A continued effort shall be made at all levels to provide an electrically safe environment within the hospital. All personnel must be instructed that the use of electricity introduces the hazards of burn, shock, explosion, fire and power failure, with the possibility of injury, disability, or death, and damage to or malfunction of equipment or facilities. The increased variety, quantity, and complexity of electrical and electronic equipment in use in diagnostic and therapeutic patient support, particularly when multiple units are used on the same patient, have magnified the safety problem and have increased the need for expanded knowledge and additional caution.

"Written policies and procedures shall be established for, but not be limited to, the following:

1. Special precautions that must be taken when the care of a patient requires the use of any type of electrically operated device.

2. Methods and frequency of testing, and verification of performance and use specifications, for all electrical and electronic patient care equipment, based upon established safety requirements, performance criteria, and manufacturers' claims. Such testing and verification shall apply to both fixed and mobile equipment, with particular emphasis on life-support equipment such as respirators, defibrillators, physiologic monitoring systems, infant incubators and warmers, as well as devices with a high hazard potential such as electrosurgical equipment. All new equipment shall be so evaluated prior to its use. The testing interval for each device or category of equipment shall be consistent with the manufacturer's recommendations, standards promulgated by recognized technical organizations, and frequency of use. In no case shall the testing interval exceed six months.
3. Systematic, periodic evaluation of the electrical power distribution systems and all electrical and electronic nonpatient-care equipment. Evaluation shall be made of all elements of electrical power distribution systems, including receptacles, following new construction, renovation, or replacement, and at least annually thereafter. Receptacle inspection shall include testing for polarity, quality of grounding, and mechanical security. Inspection and testing of nonclinical equipment shall be performed at regular intervals to be determined by the chief of the engineering and maintenance department. Such nonclinical equipment includes, but is not limited to, electrical beds, lamps, television sets, radios, and all appliances, including microwave ovens. There shall be a system for visually identifying defective equipment that must remain in place and thus be accessible to potential users.
4. Monthly testing of conductivity in all areas with conductive flooring, and recording of results. All furniture in such areas shall be included in the conductivity testing. Where flammable agents are no longer permitted for anesthesia or skin preparation of the operative field, conductive flooring is not required. In such cases, the conductivity characteristics of the floor may be eliminated by a means approved by the authority having jurisdiction.
5. Provision of an isolated power (ungrounded electric) system in each anesthetizing location. Each isolated power system shall employ a continually operating

line isolation monitor which shall have both audible and visible signals. Line isolation monitors shall be tested on at least a weekly basis and only when the circuit is not otherwise in use. A permanent record of the test results shall be maintained.

6. The requirement for written records of all inspections performed on electrical and electronic systems and equipment, including any action taken or recommended.
7. Development and implementation of an effective system for determining current leakage, ensuring proper grounding and other device-safety requirements, and providing appropriate corrective measures.
8. Specific restrictions on the use of extension cords and adapters, and precautions to prevent overloading of any electrical systems. Extension cords should be used only in emergency situations, and shall be 16 gauge or heavier and meet the requirements of pertinent technical standards.
9. Distribution of the findings from all electrical and electronic surveillance activities and recommendations for corrective action to responsible department heads.
10. The education of personnel in the resuscitation of victims of electrical shock.
11. Prohibition of the use of personal electrical equipment by patients and staff, unless such equipment has been approved as safe in writing by the appropriate hospital personnel prior to its use. Exceptions as to specific items or categories of items, or as to restricted areas of use, may be designated by the safety committee.

"Qualified advice and service concerning electrical and electronic systems and equipment shall be available at all technical levels. Testing apparatus and inspection procedure forms shall be available and adequate for the types and quantity of electrical and electronic equipment present in the facility. Both personnel and testing apparatus may be provided through consultative or contractual sources when not available in the hospital.

"All personnel involved in direct patient care shall be educated as to the existence of patients who are particularly sensitive to electrical hazards because of debilitating medical conditions, a major loss of skin resistance due to the use of wet dressings, the presence of indwelling conductive catheters and probes with externally exposed intracardiac leads, or because of severe electrolyte imbalances."

4.8 Radiological Regulations

Radiological regulations can be classified into three general groups: federal, state, and professional bodies.

Federal regulations

For those states that have not executed an agreement to assume from the federal government the responsibilities of the regulations of radiation sources, the federal regulations are the applicable rules. These rules apply to X-ray sources, nuclear reactors, accelerator sources, and all radioactive isotope sources with the exception of radium. In this special case, radium is not licensed unless the state has regulations specifically for radium. The applicable sections of the Code of Federal Regulations is Title 10 for the Nuclear Regulatory Commission rules on isotopes sources and Title 42 for the Food and Drug regulations concerning X-ray machines and electron accelerators (Code of Federal Regulations, 1977). An additional source of guidance are the Regulatory Guides published by the Nuclear Regulatory Commission (United States Nuclear Regulatory Commission, 1977). At present, the number of guides of medical interest is quite small but there is an indication that the number will grow. In addition, the technical reports published by the Bureau of Radiological Health are frequently referenced for accepted practice and equipment functions (National Council on Radiation Protection, 1977).

State regulations

Where the state has assumed responsibility from the federal government, the rules will vary somewhat depending on the state. Generally the state regulations closely follow the federal regulations. The only source we might find in a hospital, which would still be federally controlled because it is a transuranic element, is californium-252, a neutron-emitting isotope used in implant therapy.

Under state regulation, all isotopes including radium are controlled and the users are licensed by the state to possess and use the sources for specific purposes. The best advice is to contact the state health department for a copy of its regulations and specific guidance in any unusual situation.

Professional bodies

Various accrediting and professional bodies have rules and recommendations for good practice which are usually quite general in nature. Examples are the Joint Commission on Accreditation of

Hospitals (1976), National Council on Radiation Protection (1977), International Council on Radiation Protection (1977), and the International Commission on Radiation Units and Measurements (1977).

4.9 Mechanical Standards

The Maryland Electrical Testing Laboratory (MET) has produced a MET standard 501 which deals with mechanical aspects of equipment (MET Electrical Testing Co., 1976). The following excerpts illustrate its contents.

Mechanical strength

"Enclosures housing live parts, including any access covers forming part of the enclosure, shall be capable of withstanding a 10-lb direct force applied over an area of 1 in^2 without showing appreciable damage or reducing protection clearance.

"**Openings.** Enclosures shall not restrict ventilation thereby causing excessive heat. Outer enclosures shall have only those openings which are indispensable to the functioning of the equipment. No opening shall provide accidental access to live parts, enamel insulated wiring or to moving parts.

"**Water-resistant properties.** Equipment likely to be used near or subjected to contact with various liquids or containers are classified in one of three groups:

1. *Drip-proof equipment* provides an enclosure that will prevent the entry of large amounts of falling liquid while maintaining safe and satisfactory operation. Equipment that possesses a level surface that is likely to be utilized as a support for a liquid container that, if upset, may cause a hazardous condition, must be classified as 'drip-proof.'
2. *Splashproof equipment* must prevent the entry of large amounts of liquid splashed from any direction to the extent that safe operation is not hampered. The enclosure must be 'splashproof' if the equipment is normally subjected to a splash or spray of fluid which could result in a hazardous condition if the fluid enters the enclosure.
3. *Watertight equipment* must provide enclosures that will prevent the entry of any amount of liquid even when submerged either during routine use or during cleaning.

"The relative humidity of the health care facility varies widely and the humidity range of the area of intended use must be considered. Therefore, equipment must be capable of functioning safely in an area of high humidity.

"**Structure and frame.** Normal operation of an apparatus shall not be affected by vibrations or mechanical shock, which may be experienced during normal use. 'Normal use' includes being dropped from normal operational heights. No hazards to either the patient or operator shall result from these drops.

"**Stability and transportability.** Mechanical stability of an apparatus shall be reviewed in all conditions of clinical use, and appropriate measures shall be taken to insure the maximum protection against physical abuse during operation. Live cords on mobile cart mounted equipment shall be anchored from a low point on the cart to minimize the chance that it will be tipped if someone trips on the cord while it is plugged in. Mobile equipment shall be stable while being moved on various types of floor surfaces and will not tip on deep carpet or gaps in the floor as in elevator doors.

"**Support of permanently installed equipment.** Equipment designed for installation in a permanent manner shall contain mounting facilities to withstand additional stresses as may be encountered during normal use. All equipment parts serving for support and/or fixture of patients shall be manufactured so that physical injuries are avoided and that fixations cannot become loose as a result of patient movement.

"**Temperature rise.** Enclosures shall not produce temperatures capable of harm to either the operator or the patient. The acceptable temperature rise measured at various locations would depend on the composition of the item in that area.

"**Indicators, controls and alarms.** Indicator lights will be used to indicate power to nonluminous heaters, power supplies above 15 V, and output circuits that could be hazardous after prolonged operation. The color code of various indicator lamps will comply with the International Standards which utilizes a red color as an indication of danger, blue or white as an indication of safe conditions, amber as caution, and green as an indication of functioning. Control knobs will truly indicate the level of the parameter that is being controlled. Knobs will be held in place by means of a pin through the shaft or some equivalent means to prevent misalignment of the control. Operation of controls will be consistent with the normal practice in which a clockwise rotation indicates an increasing value. Mechanically operated equipment capable of physical injury to either the operator or patient shall have at least two independent controls, for quick stop, located in a readily accessible area on the unit. If overheating can be a danger to either patients or operating personnel, two separate temperature-limiting devices, one of which must be preset, shall be provided. The preset limiting device shall be nonadjustable by the user and set no more than 10% above the maximum normally attainable temperature of the device. Audible alarms shall operate in the 100-Hz to 500-Hz frequency band and have at least a sound level of 65 dBA at a distance of 10 feet. The alarm level shall not be adjustable without the use of a tool unless a remote alarm is active or the monitored function is not essential for life safety. Alarm lights shall stay on when the audible alarm is silenced."

4.10 Study Questions

4.1 Define the terms "code," "standard," "regulation," and "specification."

4.2 Describe the three major methods by which standards are developed and promulgated.

4.3 What are the primary organizations that influence and establish standards regarding medical devices?

4.4 What is the major international standards organization, and what is the USA member body in this organization?

4.5 Describe the procedure by which an ISO document becomes an international standard.

4.6 List the six ISO technical committees that are concerned with standards for medical devices.

4.7 What is the IEC, and how does it differ from the ISO?

4.8 List the government agencies involved in medical device regulation.

4.9 What is the role of trade and professional organizations in medical device regulation?

4.10 What are the major federal agencies that are involved in medical device standards?

4.11 Summarize the activities of the GSA in medical device standards.

4.12 How does the VA utilize medical device standards?

4.13 What is the procedure by which a standard or specification is generated by DOD?

4.14 Which bureaus of the FDA play a role in medical device regulation?

4.15 Describe the primary activities of Underwriters Laboratories as they relate to medical devices.

4.16 List the NFPA standards that apply to the medical field.

4.17 What are the four legal principles that must be considered in establishing standards?

4.18 List the three classes of medical devices intended for human use, and briefly summarize each class.

4.19 Describe the make-up and function of the FDA classification panels.

4.20 Describe the procedure by which medical device standards are implemented.

4.21 What are the two JCAH standards that relate most directly to clinical engineering practice?

4.22 What are the responsibilities of the safety committee as specified by JCAH?

4.23 Summarize the JCAH standard relating to electrical safety.

4.24 List the medical devices that come under federal radiological regulations.

4.25 Under what conditions do state laws supersede federal radiological standards?

4.26 List the professional bodies that are involved in radiological safety.

4.27 What are the major categories of mechanical safety addressed by the MET standard?

4.11 References

(Section 15.7 provides addresses for these regulatory and standards agencies.)

Association for the Advancement of Medical Instrumentation (1975). Safe current limits standard. AAMI SCL-P. Arlington, VA.

Association for the Advancement of Medical Instrumentation (1976). ECG prefilled disposable electrodes. Arlington, VA.

ASTM (1977). The people, process, and principles of standards development. *ASTM Standardization News*, 5(3): 21–25, 60.

Canadian Standards Association (1973). Electromedical equipment. CSA standard C22.2 No. 125-1973. Canadian electric code.

Code of Federal Regulations (1977). U.S. Title 10 and Title 42, Washington, DC: U.S. Government Printing Office.

Food and Drug Administration (1977). Everything you always wanted to know about the medical device amendments . . . and weren't afraid to ask. HEW Publication No. (FDA) 77-5006. Silver Spring, MD.

Food and Drug Administration (1978a). Bureau of Medical Devices standards survey, international edition. Silver Spring, MD.

Food and Drug Administration (1978b). Bureau of Medical Devices standards survey, national edition. Silver Spring, MD.

International Council on Radiation Protection (1977). ICRP Reports. New York: Pergamon Press.

International Commission on Radiation Units and Measurements (1977). ICRU Reports. 7910 Woodmont, Washington, DC 20014

Joint Commission on Accreditation of Hospitals (1976). Accreditation manual for hospitals. Chicago.

MET Electrical Testing Co. (1976). MET standard 501: standard for performance and safety certification of medical equipment. Baltimore, MD.

National Council on Radiation Protection (1977). NCRP Reports. 7910 Woodmont, Washington, DC 20014

National Fire Protection Association (1973a). Standard for the use of inhalation anesthetics. NFPA 56A. Boston.

National Fire Protection Association (1973b). Essential electrical systems for health care facilities.

National Fire Protection Association (1973c). Electricity in patient care facilities. NFPA 76B-T (tentative). Boston.

National Fire Protection Association (1975). Safe use of high-frequency electricity in health care facilities. NFPA 76C. Boston.

National Fire Protection Association (1976). Respirator therapy. NFPA 56B. Boston.

National Fire Protection Association (1978). National electric code. NFPA 70. Boston.

Towbridge, R. (1970). *The Magazine of Standards*, April, 1970.

Underwriters Laboratories (1976). Standard for medical and dental equipment. UL 544. Chicago, IL.

United States Nuclear Regulatory Commission (1977). United States Nuclear Regulatory Commission Regulatory Guides. Washington, DC.

Veterans Administration (1970). Biomedical monitoring systems. Specification X-1414. Washington, D.C.

Willard, P. W., J. E. Kuphal, D. T. Santore, C. H. Swanson, and D. F. Juncker (1977). Impact of medical device legislation. *CRC Crit. Rev. Bioeng.*, 4(2): 363–374.

5

Hospital Information Flow and Handling

LEE E. OSTRANDER and STEPHEN E. THOMPSON

5.1 Learning Objectives

Upon completing this chapter you should be able to:

1. List the types of information that are processed in a hospital.
2. Describe the flow of and quality of information in hospitals.
3. Explain how the medical record is established and maintained.
4. Describe procedures for hospital financial record keeping and billing.
5. Describe the major roles that computers play in hospital information systems.

5.2 Introduction

According to available estimates, as much as 35 percent of total hospital costs are associated with the handling of information (Collen, 1970). Since engineering technology is used in the communication and processing of information, the importance of general knowledge concerning information management should be readily apparent to the clinical engineer. In the sections that follow, an overview of the information flow through a "typical" hospital is presented.

5.3 Hospital Information Systems

A method for obtaining, transmitting or storing information is an information system. The system may use computers and data lines or a set of paper forms hand carried from one department to another, or some other physical form. Hospital information systems are generally characterized by the requirement for rapid and highly accurate data transfer, as well as by the diversity of the information to be transmitted.

Types of information

Many types of information are present in hospital systems. The majority of this information is printed material, including graphs, narratives, numerical data, forms, and legal documentation. Telecommunication systems (e.g., telephones, the public address system, "beepers") supplement printed communications.

Perhaps less obvious than this type of information flow is the transfer of physical material. This includes pharmaceutical preparations, blood samples, and other laboratory specimens that are normally hand carried and that also convey information.

Information users

The users of hospital information are as varied as are the types of information. The primary users of this information are physicians, nurses, and technicians for patient diagnosis and treatment; and the business office, for patient billing. Other secondary users of hospital information are the hospital administration and a number of regulatory agencies outside the hospital. These secondary users evaluate accumulated data for long-term planning, auditing, quality control, preventive maintenance, and legal protection.

As insurance companies and government agencies have become more heavily involved in the financing of health care, their requirements for hospital information have increased. Finally, some types of processed hospital information are used by researchers in epidemiological studies, such as cancer research.

5.4 Clinical Records

Information concerning the individual patient can be divided into two functional categories: the patient's clinical record and the patient's financial record. The clinical record includes all printed information pertaining directly to the medical status of the patient.

As long as the patient remains in the hospital, the clinical record (more commonly referred to as the "patient chart") stays with the patient. When the patient leaves the hospital, the chart is moved to a central location, edited, and stored in a permanent form known as the "medical record." The exact form of this medical record and method of storage vary considerably from hospital to hospital.

Contents of the patient chart

The patient's chart is essentially a running record of the patient's condition and treatment. It may include numerical data, graphs, notes, narratives, forms, and legal documentation. A few general statements concerning the content and procedures for entering information follow.

Physicians' notes. The core of the patient chart is the physicians' notes. This section usually contains a set of progress notes (which are written in a narrative style) and a set of orders to the nursing staff. The nurse on duty reads these orders and initiates the appropriate diagnostic and treatment services. Virtually all clinical activity is a result of this basic procedure. If there is more than one physician seeing the patient, the consulting physician also writes progress notes and orders. In many instances, these notes constitute one of the primary means of communication among attending physicians.

Nurses' notes. Equally important to the welfare of the patient are the nurses' notes. Throughout the day, the nursing staff carefully records treatment services performed (e.g., inhalation therapy, drug administration, dressing replacement); the patient's vital signs (e.g., temperature, blood pressure); and the patient's general condition (e.g., skin color, activity). Since the physician may see the patient for only a few minutes each day, the physician must rely heavily on the nursing observations in the patient chart.

Clinical test reports. In addition to the professional notes in the patient chart, there are the results of the diagnostic procedures ordered by the physician(s). These may include laboratory test results, radiology (X-ray) reports, electrocardiograms, electroencephalograms, tissue biopsy reports, etc. As laboratory test results, electrocardiograms, and the like accumulate in the chart, the physician can monitor the patient's progress.

Legal documentation. The patient's chart also contains all of the necessary legal documentation. Upon entering the hospital, the patient or a member of the patient's immediate family is required to sign a "treatment consent" form. At various times throughout the patient's stay at the hospital, any number of additional legal forms may be required, ranging from the well-known surgery consent form to the disposal of dismembered tissues form to the autopsy consent form. All are stored permanently after the patient's discharge from the hospital or death.

Hospital charting policy

Note that the entry of data into the patient chart is not necessarily restricted to physicians and nurses. Depending on the charting policy of the hospital, ward clerks, laboratory technicians, and others may enter data into the chart. Furthermore, since the chart accompanies the patient through the hospital, it is also likely that some information generated outside the nursing unit would be included (e.g., emergency room notes, recovery room notes).

Patient chart illustration. Figure 5.1 shows sample physician's progress notes, a physician's order sheet, and the corresponding nurses' notes. The hypothetical patient described here is in acute abdominal distress probably caused by a small bowel obstruction. (This has been verified by a previous radiologic report not shown.) On this particular morning the physician has left a set of orders instructing the nursing staff to give demerol as needed for restlessness, to irrigate the Levine tube (naso-gastric tube), and to order the listed laboratory tests. A second physician has agreed to perform surgery on the next day. The nursing staff then carried out both sets of orders, as indicated on the nurses' notes.

5.5 Financial Records

The handling of financial records includes the generation, posting, and payment of charges accumulated by the patient, and the control of overall hospital income and expenditures by the hospital management. At present, this area shows the greatest variability among hos-

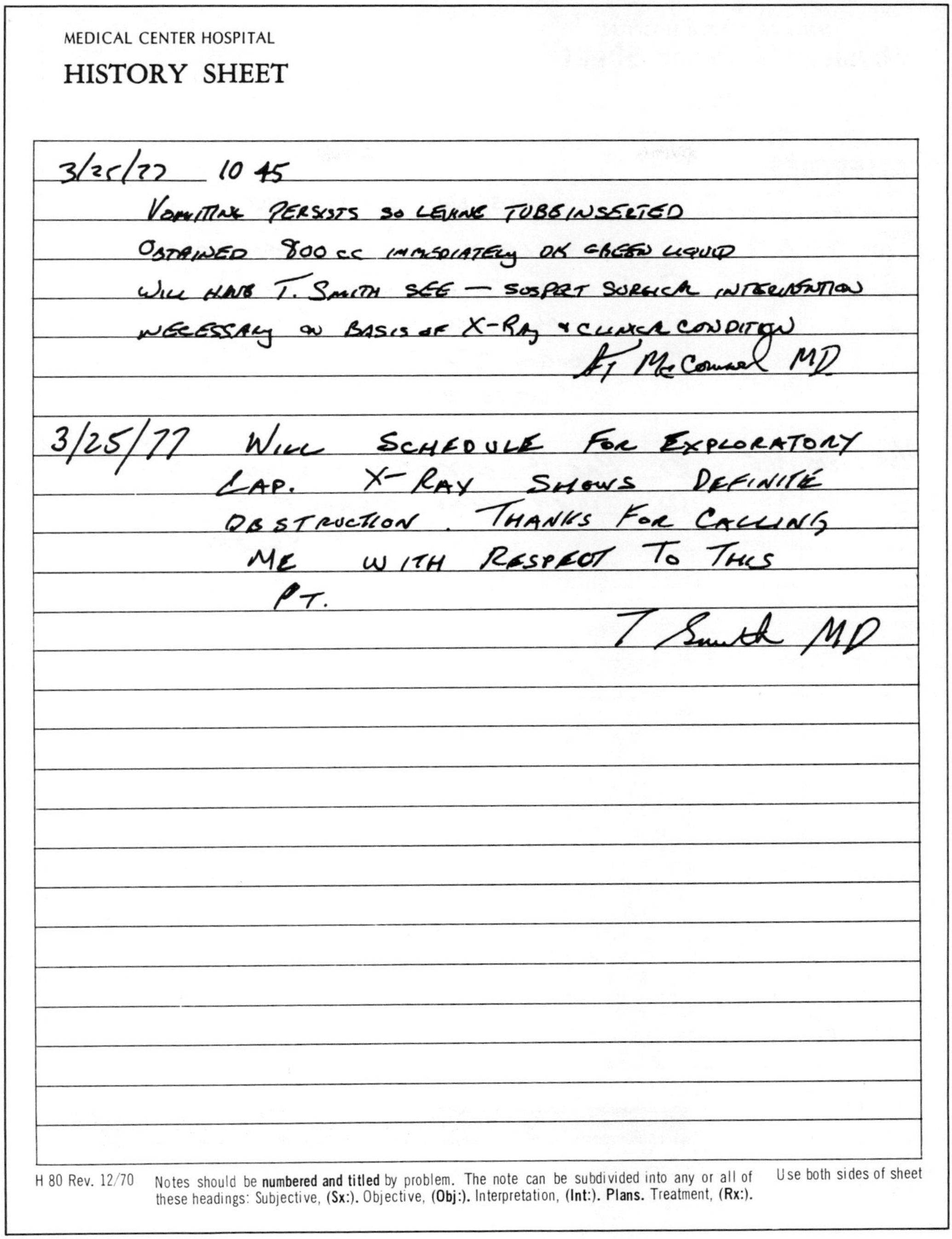

MEDICAL CENTER HOSPITAL

HISTORY SHEET

3/25/77 10 45

VOMITING PERSISTS SO LEVINE TUBE INSERTED
OBTAINED 800 cc IMMEDIATELY OF GREEN LIQUID
WILL HAVE T. SMITH SEE — SUSPECT SURGICAL INTERVENTION
NECESSARY ON BASIS OF X-RAY & CLINICAL CONDITION

AJ McConnell MD

3/25/77 WILL SCHEDULE FOR EXPLORATORY
LAP. X-RAY SHOWS DEFINITE
OBSTRUCTION. THANKS FOR CALLING
ME WITH RESPECT TO THIS
PT.

T Smith MD

H 80 Rev. 12/70 Notes should be **numbered and titled** by problem. The note can be subdivided into any or all of these headings: Subjective, **(Sx:)**. Objective, **(Obj:)**. Interpretation, **(Int:)**. **Plans.** Treatment, **(Rx:)**. Use both sides of sheet

Figure 5.1 Patient chart illustration: (a) sample physician's progress notes; (b) physician's order sheet; (c) nurses' notes.

MEDICAL CENTER HOSPITAL

Physician's Order Sheet

INSTRUCTIONS:
1. Imprint patient's plate before placing in chart.
2. After each set of orders are written, remove first yellow copy and send to PHARMACY.
3. "X" out remaining unused lines after last copy is used.
4. Imprint new set and place in chart.

ALLERGIES:

Date Ordered	Time Ordered	Time Executed	Time Posted	USE BALL POINT PEN ONLY
3/24/77	1045	13 00		① IRRIGATE LEVINE q 4° & PRN TO KEEP PATENT.
				② PUT LEVINE TO LOW GOMCO SUCTION
				③ LAB - Na, K, CO_2, CL IN A.M
				Rx DEMEROL 25 mg IM q 3-4 PRN
				RESTLESSNESS Al McConnel MD
3/25/77	1205	1445		① PRE SURGERY LAB WORK - AM
				② GET LAP PERMIT SIGNED
				T Smith MD

"ALL ORDERS SHOULD BE PRECEDED BY THE NUMBER AND TITLE OF THE PROBLEM TO WHICH THEY REFER. NUMBER ONE (# 1) IS RESERVED FOR ROUTINE ADMISSION AND MAINTENANCE ORDERS."

Figure 5.1 continued.

MEDICAL CENTER HOSPITAL

NURSES NOTES

DATE	HOUR	
3-25	1300	Levine irrigated c̄ 30cc N/S and left to drain. Levine draining dk. brown liquid.
	1335	Turned to side. C/O about tube hurting nose. Med given for restlessness.
	1405	Sleeping occasionally. Turned q 2h. Levine draining well.
	1445	Dr. Smith's orders noted.
	1510	OP permit signed & lab called.
		Fair Day ...
		– A. Thompson, RN 1545 hrs.

H1 REV. 5/66

PAGE ______

Figure 5.1 continued.

pitals, especially in the type and design of the information system. Financial information processing can range from relatively simple "adding machine" systems to highly complex computer systems. Consequently, the supporting information flow systems must also vary considerably, making it difficult to talk about "typical" systems or to give "general" examples. Nevertheless, it is important to note that although larger hospitals usually have a higher volume of information flow and require more sophisticated information processing technology than do smaller ones, the type of information to be processed and the function performed must be similar.

The patient's financial record is handled independently of his or her clinical record; and generally is inaccessible to all but hospital business office employees. Prior to admission, the business office begins a running account of all charges incurred by the patient. These include room, board, laboratory tests, X rays, medications, and all other patient services.

Posting of charges

The logistics of charge-posting vary greatly from hospital to hospital, ranging from simple "hand" posting to fully automated charge-posting systems. Many hospitals use a charge-card system, (e.g., the Addressograph-trademark system), in which the patient is issued an embossed plastic card containing his name, identification number, attending physician's name, and other pertinent information. This card accompanies the patient throughout the hospital and is usually stored in a convenient location at the nurses' station. If treatment or diagnostic services are needed, a member of the nursing staff uses the card (and a special machine) to imprint patient information onto the appropriate multicopy charge slip or requisition form. These copies may be hand-carried by messenger or sent by pneumatic tube to the service area indicated. After the service has been performed, one copy is sent to the nurses' station for inclusion in the chart, one copy is retained by the service for its records, and one copy is sent to the business office (or more specifically, to the data processing department). Some hospitals have, to varying degrees, automated this basic procedure.

Billing the patient

At the end of each day, the business office adds the sum of the nonspecific charges (room, board, etc.) to the sum of the day's charge slips to determine the daily bill. Additionally, there are a number of independent support services, e.g., anesthesiology and radiology, that may bill the patient separately.

Upon discharge, the patient is presented with a bill reflecting the total amount owed the hospital. Of course, if the patient is eligible for Medicare, Medicaid, Blue Cross, or other third-party payments, he will have to pay only the balance. Depending on the degree of automation in the business office, third-party payments may or may not be reflected in this total.

Administrative uses

The overall control of hospital income and expenditures is a complex topic involving accounting, management economics, and law. This subject is covered thoroughly in hospital management textbooks and will not be discussed further here (McGibony, 1969).

5.6 Interdepartmental Communication

A tremendous volume of information must flow between the individual departments of the hospital. In addition to forms and other written records, verbal communications must also be considered in this category. While most interdepartmental communication is patient-related, a substantial percentage is concerned with hospital support services, such as maintenance and purchasing.

Standard forms

Standard forms constitute the vast majority of the interdepartmental paperwork. Those that pertain directly to the patient are initiated by the nursing service in response to a physician's verbal or written order and are routed to the appropriate treatment or diagnostic service by messenger. In some hospitals, pneumatic tubes or other mechanical devices may be used in place of messengers. Those forms not dealing directly with the patient usually pertain to one of the many hospital support services. These include purchasing, housekeeping, laundry, security, clinical engineering, central supply, and maintenance. In the event that no standard form is applicable, an interdepartmental memo may occasionally be necessary. These are handled in a like manner.

Telecommunications

The transmission of printed or written material often takes hours. When speed is the primary concern, hospital telecommunication systems become important. The telecommunication system includes the telephone, pocket page, and public-address system. Each system has its own set of advantages and disadvantages, and the choice of one particular system over another depends largely on the specific application. Relatively elaborate telephone systems have evolved in most hospitals. Since hospital telephones are often needed in life-or-death situations, hospital telephone systems must be capable of handling emergency calls to and from almost any location without significant delays. The telephone supplements but seldom replaces written messages, however, since a printed copy of any transaction is usually required for permanent records. For example, emergency laboratory tests are frequently requested by phone although the appropriate forms must eventually be filled out.

Paging

A prerequisite for effective use of the telephone system is that the party at the receiving end must be easily located. In the hospital environment, this is often not the case; hence the need for the public-address system and/or pocket-page system.

The familiar hospital public-address system is still one of the most effective means for reaching an individual in a hospital. However, in large hospitals where paging is frequent, the heavy use of the public-address system can lead to considerable noise pollution.

Recently the use of the pocket-page or "beeper" system has become more widespread. The pocket page is a "candy-bar sized" electronic device that emits an audible tone when activated at a central console, usually at the telephone operator's desk. If a person is "beeped," he or she calls the telephone operator for a message. Beepers can considerably reduce the need for extensive public-address system use, thus greatly reducing hospital noise pollution.

5.7 Structure of the Hospital Information System

The hospital information system is essential to the coordination of patient care under the direction of the attending physician(s). One way to appreciate the many aspects of the system is to examine Fig. 5.2. Six primary units of the hospital are shown: (1) nursing, (2) hospital support, (3) diagnostics, (4) treatment, (5) ancillary professional services, and (6) the business office. Additionally, the hospital will have a number of other functional units, e.g., medical records, personnel, and data processing, that support one or more of the primary units. Each unit is under the direct control of the hospital administration. The interconnecting pathways may represent the flow of printed data, of information-bearing physical materials, or even of patients as information carriers.

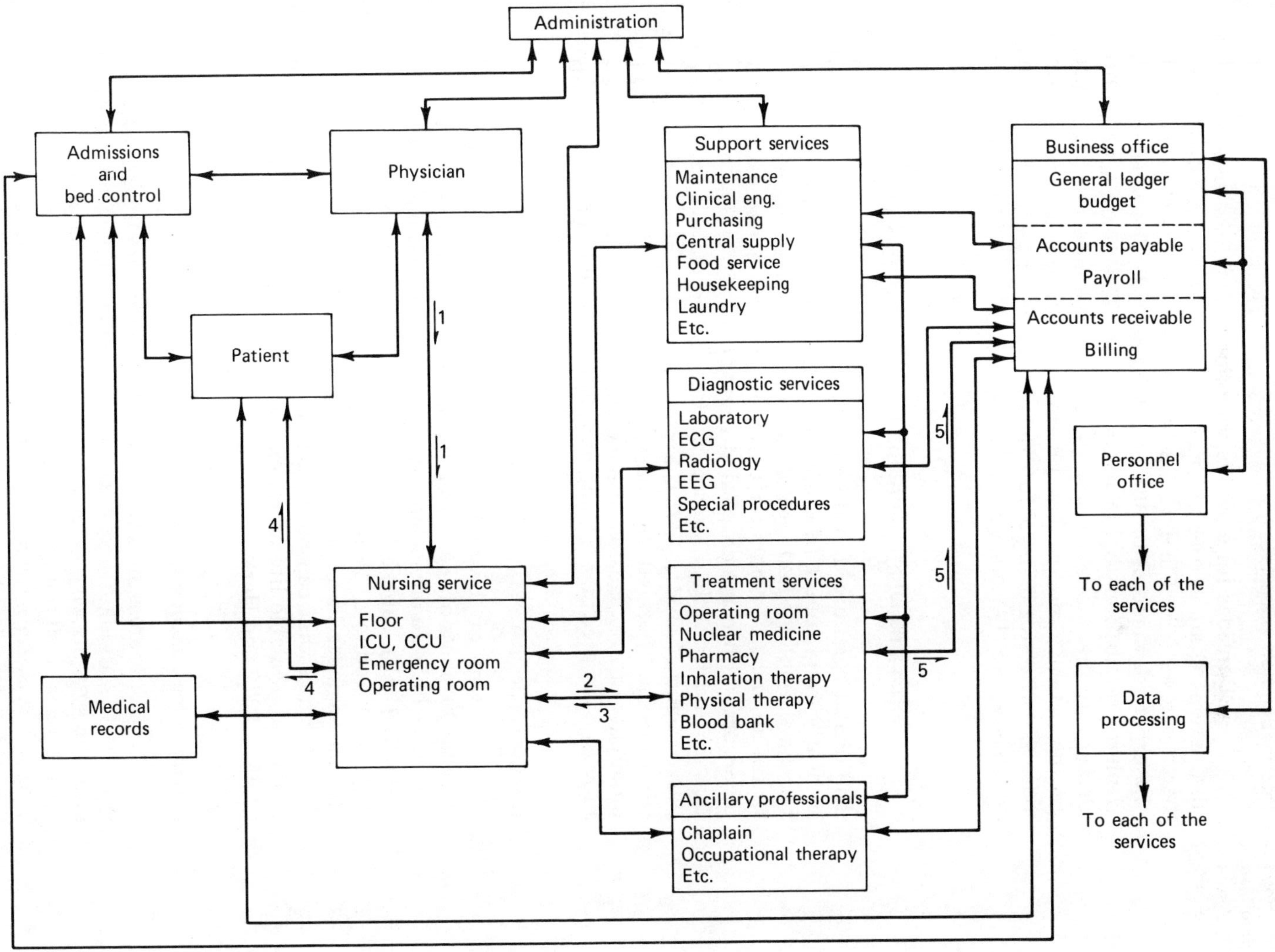

Figure 5.2 Hospital information flow diagram.

Information flow example

Consider for example the information flow associated with the patient chart entries in Fig. 5.1. The flow is indicated by numbered arrows ($\xrightarrow{n}$) in Fig. 5.2 as follows: (1) The physician enters the order for demerol on the patient chart; (2) The nurse sends the demerol requisition to the pharmacy; (3) The pharmacy sends the demerol to the nurses' station, (4) The nurse administers the demerol to the patient; and (5) The pharmacy sends the charge slip to the business office for billing. More detailed flowchart examples are available elsewhere (Petters, 1972).

5.8 The Digital Computer in Information Systems

The digital computer has the potential for being an integral element in nearly every aspect of hospital information systems. Some hospitals have implemented "total" automated information systems in which an array of computer terminals and automated instruments is incorporated into the hospital at strategic locations. The information flow to and from these devices is coordinated via a centralized computer facility (AAMI, 1972). More commonly, however, a hospital will have incorporated the computer only into limited areas with the possibility of an expansion of the computer role with time. Figure 5.3 shows a computer that is being used in a hospital information system.

Areas of application

One of the first targets for computer automation is typically the business office. Other areas, such as admissions and bed control must be automated later in order to be economically justifiable (Petters, 1972). Table 5.1 lists many of the areas in which a computer may be introduced. A few of these areas of special interest to the clinical engineer have been detailed in Sec. 5.9 to illustrate further the functions that can be expected of the computer. Others are described in available literature, e.g., issues of *Medical Instrumentation* (AAMI, 1972 and Harken, 1974).

Describing the computer system

The areas in Table 5.1 can be automated using computer systems of varying degrees of size and sophistication. Commercially available systems vary greatly as to their capabilities and methods of operation. For all computer systems there are certain considerations involved in sorting through the various alternatives, such as the following (Petters, 1972).

Figure 5.3 A central hospital computer facility at Albany Medical Hospital, Albany, N.Y.

Turnkey versus modular systems

A potential application area can be computerized by using one of the two basic approaches. The first is the *turnkey* approach in which the purchaser buys a functioning, ready-to-use system from the vendor. The second is the *modular* approach, in which the pieces of the system are bought individually and assembled to meet the specific needs of the purchaser. Each approach has its merits and disadvantages, and the choice depends upon the requirements of the hospital. The turnkey system generally has less flexibility and is more expensive, while the modular approach places more responsibility upon the hospital and may take longer to implement.

Software considerations

The importance of good software is often underestimated, perhaps because it is less tangible than a piece of equipment. The software for a physiological monitoring system requires typically five

Table 5.1. Computer Application Areas

Administrative Functions

Program planning
Budgeting

Business Functions

Patient billing
Accounts receivable
Payroll
Accounts payable
General ledger

Support Services

Engineering scheduling
Preventive maintenance and repair
Equipment control program
Inservice training
Inventory control
Menu planning

Diagnostic Services

Laboratory test scheduling
Laboratory test analysis
Laboratory reporting
ECG analysis
EEG analysis
X-ray scheduling
X-ray analysis

Treatment Services

Operating room scheduling
Patient monitoring
Drug indexing for pharmacy
Inventory control
Inhalation therapy scheduling
Blood donors
Index and catalog

Nursing Service

Personnel scheduling
Patient chart: Physicians' orders
Nurses' notes
Continuing education
Physiological monitoring
Patient management (diagnosis, therapy, planning)

Other

Preadmission and admission
Automated patient history-taking
Patient transfer (room)
Census
Bed availability
Medical records
Epidemiology and statistical services
Multiphasic health screening
Outpatient reporting
Personnel records
Poison control
Organ registry

man-years of effort to develop, while a total hospital information system may require 200 man-years of effort (Petters, 1972). Because of the effort required to produce software, there is an advantage in using the same software at multiple installations. This may reduce total cost and allows exchange of information and sharing of experiences. A file-oriented programming language called MUMPS is in use in many locations for handling medical records. The system has been standardized among many of its users (MUMPS Users' Group, Biomedical Computer Laboratory, 700 S. Euclid Ave., St. Louis, MO 63110). It takes just as long to develop software in MUMPS as in more widely used languages such as FORTRAN or COBOL.

Software packages are sometimes available from vendors. In other cases they may be developed through collaborative arrangements among several medical institutions.

5.9 Clinical Data Processing and Computers

An integral part of the hospital information system is the acquisition and processing of clinical data. Four areas of particular interest to the clinical engineer are described below: physiological monitoring, patient records, multiphasic health testing, and the clinical chemistry laboratory. Further considerations necessary for implementing an automated computer system for clinical data processing may be found in Sec. 3 of the Comprehensive Hospital Computer Applications Program Report (Petters, 1972).

Physiological monitoring by computer

Physiological monitoring by computer involves the three steps of data acquisition, computer processing, and data display. Data acquisition consists of collecting data from transducers and monitoring instruments, transforming the data by analog-to-digital converters where necessary, and feeding the digital data into the computer at specified sampling intervals. The computer processing may be done "on the fly," as samples are collected, or the data may be stored and processed as separate procedures. The processing generally consists of (1) the conversion of data according to calibrations entered at an earlier time into the computer, (2) the processing of waveforms, and (3) the extraction of parameters from waveforms. The data display generally consists of both graphic and digital output shown at the bedside and/or at a central nursing station. A display of trends with time for selected patient variables is often available.

The simplest of computerized monitors are built around the microprocessor and handle just a few variables such as the ECG, blood pressures, and/or airway flow.

Depending upon the system, provision may be made for long- or short-term monitoring and for various alarms to alert personnel to life-threatening conditions in the patient or to failures in the monitoring system. The physiological monitoring system can be part of a larger system which, for example, assists in nursing clerical functions. The system may also be tied in with a larger hospital information system.

There have additionally been some attempts to use monitoring systems to initiate therapy automatically. Some aspects of this may be seen in automatic ventilators which respond when the patient attempts to breathe. In general, however, computer-controlled therapy remains a research subject rather than a regular part of clinical practice.

Physiological monitoring occurs in the hospital in many areas, including the intensive-care unit, coronary care unit, operating room, delivery room, surgical recovery room, pulmonary care unit, and catheterization laboratory. A special application of physiological monitoring now in use is the analysis of the electrocardiogram by computer (NCHSR&D, 1973).

Medical records by computer

Computers have been used to assist in the acquisition of patient histories, in the organization of the medical record, and to a limited extent in the diagnosis of patient problems.

In the acquisition of patient histories, the computer function is a natural extension from the situation in which the patient fills out a questionnaire or in which the physician interviews the patient and takes notes. The computer can be programmed to select questions on the basis of the patient's previous answers or known condition, thereby providing some flexibility in the questions asked. Since data stored in the computer can be retrieved in any order, there exists the possibility of reorganizing the medical record and printing it out in an easily readable format. This is particularly important since the patient chart can become voluminous as data accumulates. Computer algorithms can identify patient problems. For example, a program may identify pulmonary function test data that are out of range according to classification of the patient's age, sex, and height (Sullivan et al., 1969).

Despite these advantages in information handling, the computer introduces the new problem of confidentiality of patient information. The very factors that make it easy to enter information make it easy for unauthorized retrieval of information. In an attempt to overcome this problem, several approaches such as security codes and user identification have been employed.

Multiphasic health testing by computer

Multiphasic health testing is a mechanism for providing preventive medical care. A multiphasic health screening program utilizes a facility designed for rapid clinical testing of large numbers of people (Hsieh, 1972). The person who enters the facility is asked for his or her patient history and then moves from area to area while undergoing various tests. The resulting data may be entered into a computer which performs calculations of test results and then prints out test summaries. The information generated enables the physician to determine the appropriate follow-up procedures.

Clinical chemical laboratory

Most of the analytical testing procedures in the clinical chemistry laboratory are suitable for automation because the analyses are routine, similar, and large in number (McLendon, 1974; Pollycove, 1972).

Both control and information-gathering functions may be performed by the computer. The computer can control the starting of pumps, switching of reagents, adjusting baseline sampling, etc.

Associated with the clinical laboratory are some inherent problems that must be considered by persons involved in laboratory automation. First of all, procedures for identifying the donor of a blood sample or other specimen must be reliable. Secondly, since equipment failure may inject errors into test results, maintenance standards and quality control are essential to the effective utilization of results from the laboratory. Assistance in clinical chemistry lab inspection, surveys, and evaluation is provided by the College of American Pathologists (CAP). In some locales, CAP inspections are required. The Joint Commission for Accreditation of Hospitals also sets minimal standards for safety, training, testing, and maintenance. Excerpts from the Accreditation Manual for Hospitals (Joint Commission for Accreditation of Hospitals, 1976) are given in Table 5.2.

5.10 Economic and Health-care Impact of the Information System

As noted earlier, information costs are a significant portion of total hospital costs. There are cost limitations, however, and costs must be balanced against the needs for health care. The result, then, is a matching of the quality and capability of the information system with the health-care needs, but modified by cost.

Distribution of information services

Certain areas of the hospital have special information needs. For example, the information relating to the critical-care units is usually of higher priority than information concerning other general ward areas of the hospital.

Additionally, the distribution of information services is different for large and small hospitals. Cost factors dictate these differences, even though the functions to be performed are the same.

The actual needs for a *complex* information system are less in the small hospital. For one thing, the various hospital facilities tend to be located physically closer to each other in the small hospital. The quantity of tasks of any type amenable to automation or computer handling in the small hospital may be too small to allow economic advantage from the use of a computer. On the other hand, sharing of computer services may be possible.

Information systems are expensive items in a hospital budget, even in the best of circumstances. A careful matching of the selected

Table 5.2.

The following excerpts are from the *Accreditation Manual for Hospitals,* (JCAH, 1976) and pertain to the Pathology Services Section.

"All tests performed by the pathology services must be carried out by competent personnel, using approved procedures of acceptable high quality. A complete description must be available for each test procedure, including sources of reagents, standards, and calibration procedures. These descriptions should be immediately available to the laboratory analyst in a card file, in a notebook, or in a set of flow charts, that includes pertinent references. Each procedure should be reviewed by the director or supervisor at least annually. A list of analytical methods employed by the laboratory, and information concerning the basis for the listed 'normal' range, shall be available to members of the medical staff.

"A record should be maintained of the daily accession of specimens, each of which should be numbered, or otherwise appropriately identified. This record should contain at least the following information:

- Laboratory procedure number or other identification
- Identification of the patient
- Name of the practitioner
- Date and time the specimen was collected
- Date and time the specimen was received
- Date, time, and by whom the specimen was examined
- Condition of any unsatisfactory specimen
- Type of test or procedure performed
- Result and date of reporting

"Orders or requisitions for inpatient and outpatient services must clearly identify the patient, his doctor, the tests required, and the date and time the request reached the laboratory, and must indicate any special handling required. Requests for examinations of surgical specimens should contain a concise statement of the reason for the examination.

"Written procedures for those who are collecting specimens should be developed to ensure satisfactory specimens for the tests to be performed. These procedures must be approved by the director, and should include at least the following:

- Ordering of tests
- Standard and special methods used for the collection of specimens, as well as precautions to be followed for special procedures
- Proper identification, storage, and preservation of specimens

"Communication systems, within the laboratory and between the laboratory and other departments/services, should be such as to accomplish efficiently both the urgent and regular transfer of information. It is desirable that an intercommunication system exist between the pathologic anatomy laboratory and the operating rooms of the surgical suite.

"The director must review reports of tests to assist him in evaluating the quality of work performed in the laboratory. Performance of instruments and equipment must be evaluated frequently enough to ensure that they function properly at all times. An effective system of quality control should include a preventive and corrective maintenance program. Appropriate records should be maintained for each piece of equipment, showing the date of inspection, validation, or performance evaluation, along with significant actions taken in response to revealed defects.

"There shall be suitable programs of validation of new testing methods and recalibration for all methods, a dependable reagent and glassware evaluation system, and a continual surveillance of the results obtained, using suitable reference samples. Automated analytical devices should be recalibrated and checked according to the manufacturer's directions.

Table 5.2. (Continued)

"The laboratory shall participate regularly in one or more proficiency surveys and/or reference sample services, covering all types of analysis performed to verify competence. These proficiency test services must equal, or exceed, the requirements of pertinent federal, state, or local licensing agencies, with respect to variety and frequency of testing as well as to criteria for satisfactory participation. Further proficiency testing may be conducted by the Joint Commission on Accreditation of Hospitals through check samples introduced at the time of the survey.

"All solid and liquid reagents, reagent solutions, and standards should be prepared, stored, dispensed, and periodically checked, so as to ensure accuracy and precision of results. All reagents and solutions must be properly labeled, and such labels must include appropriate accessory or precautionary statements when such information is indicated. Monitoring systems for reagents and materials should be such as to demonstrate satisfactory performance, by parallel testing with preparations of known acceptable quality.

"Each procedure in clinical chemistry must be rechecked at least once on each day of use. Data must be available to document the routine precision of test results and the schedule of recalibration. At least one standard, or one reference sample, should be included with each run of unknown specimens. Acceptable limits for standards and controls shall be established, as shall the course of action to be instituted when results are outside the satisfactory control limits. Control limits on all tests should be set so as to produce results that are commensurate with meaningful clinical applications."

system with individual needs for information is important in order to minimize costs.

Benefits/costs for automation and computers

In any information system, whether computerized or not, there is a balancing of benefits against costs. For the computerized system, the usual expectations are that it will provide accurate, precise, and reproducible results in high-volume situations and will allow better monitoring and control functions for administration, business, and clinical purposes. Precautions must be taken, however, that such a system will not sacrifice flexibility in handling emergency situations and that there will be backup in case of computer failure. In case of malfunction or changing needs, the system must be easy to readjust. It must also be operable by the available personnel with appropriate training.

The question that must always be asked for any system is—"Is it worth it, given the time and expense involved in setting up the system?" In making a specific choice of system, we should consider whether it is an already field-tested system or a new system with potential "bugs". In other words, what is the track record of the system? Also, what are the hazards of the system?

Costs of computer-based information systems are rapidly changing with new technology. The introduction of new hardware such as the microcomputer has decreased acquisition costs. However, many systems become nearly obsolete within as few as five years of purchase. Annual maintenance costs can run at 5 to 10 percent of the initial system cost. Software costs are usually the major expense associated with the development of a computerized medical information system. For this reason, overall costs of computerization remain high.

Quality of information

Information is conveyed in many forms; as numbers, as physical samples, as phone messages, etc. Implicit in the definition of information is the concept of value to the receiver of the information. In the strict interpretation of value, the information must be reliable, accurate, and reproducible. In the broader sense, one must consider the desirability of actually communicating the given pieces of information. Too much information (e.g., too many numbers) can be dangerously overwhelming, just as too little can be dangerously misleading. For user acceptance of information and of an information system, the value of the information must be favorably perceived.

In the clinical area, we can talk about the sensitivity and specificity of test results as a statistical measure of their informational value. Sensitivity is the probability of detecting by test a pathologic condition in someone who has that condition. Specificity is the probability of identifying a person as normal by test when that person does not have the condition that the test is intended to detect. Tests that are high in specificity and sensitivity are most desirable.

The importance of information value is seen in multiphasic screening, where strict criteria for abnormal test results (overly sensitive tests) can overwhelm the physician with patients suspected of abnormalities but who are in fact normal. On the other hand, loose criteria for test results (overly specific tests) may allow too many persons with pathologic conditions to remain undetected, thereby frustrating the beneficial effects of the testing program.

A similar problem can occur with ECG alarms to detect arrhythmias. A too-sensitive detection and alarm system is as bad as an insensitive one. While an alarm may provide information, it is the value of that information to the user which is important.

5.11 Conclusion

The medical devices for which the clinical engineer has substantial responsibility generate data. These data and the clinical engineer, along with the patient, are all elements of the hospital information system. It is important to the clinical engineer that this complex system be understood. He must likewise understand the uses of the data.

As a technological expert, he may be asked for evaluation of various proposals for automation or for introducing a computer into the information system. The trend is clearly toward increased automation and computer usage. However, the decision to introduce technology introduces problems as well as benefits, and both must be weighed in the final evaluation.

5.12 Study Questions

5.1 List three forms of printed material that contain hospital or medical information.

5.2 List three forms of physical material that contain information.

5.3 Describe three users of hospital information.

5.4 Distinguish between the patient chart and the medical record.

5.5 What are three items that appear in the patient chart?

5.6 List three charges that appear on the patient's financial record.

5.7 Explain the need for each copy of a multicopy form that requests a diagnostic service.

5.8 List three uses for forms that do not deal directly with the patient.

5.9 List three forms of rapid communication in a hospital.

5.10 When a physician orders a prescription for a patient, list the additional items of information flow required.

5.11 List four major functional or service areas where the computer can be applied.

5.12 Compare the advantages of the turnkey versus the modular computer system.

5.13 What is the typical effort required to develop software for (1) a physiological monitoring system and for (2) a total hospital information system?

5.14 List three advantages of physiological monitoring by computer.

5.15 Give an advantage and a disadvantage of generating medical records by computer.

5.16 Give two problems associated with automating the clinical chemistry laboratory.

5.17 Give two advantages and two disadvantages of computer-based information systems.

5.18 Distinguish between sensitivity and specificity.

5.13 References

AAMI (1972). Proceedings of the AAMI conference on the effectiveness of on-line biomedical computing, 1971. *Med. Instrum.*, 6(1), Jan.-Feb. 1972.

Collen, M. F. (1970). General requirements for a medical information system (MIS). *Comp. Biomed. Res.*, 3:393-406.

Harken, D. W. (ed.) (1974). Computer Issue. *Med. Instrum.*, 8(5), Sept.-Oct. 1974.

Hsieh, R. K. C. (1972). Toward a measure of effectiveness of the Public Health Service Health Evaluation Center. *Med. Instrum.*, 6:4-15.

Joint Commission for Accreditation of Hospitals (1976). Accreditation manual for hospitals. Chicago.

McGibony, J. R. (1969). Principles of hospital administration, 2nd ed. New York: Putnam.

McLendon, W. W. (1974). Clinical laboratory automation and data processing. *Med. Instrum.*, 8:275-280.

NCHSR&D (1973). Demonstration and evaluation of a computer-assisted analysis and interpretation of the electrocardiogram. Available from National Technical Information Service, Springfield, VA as PB218-976.

Petters, J. R. (Prin. Invest.) (1972). Comprehensive hospital computer application program (Vol. I: A guide to automation for hospital administration; Vol. II: Annotated bibliography 1965-1971; Vol. III: Flow diagrams of a typical community hospital; Vol. IV: Study methodology and recommendations; prepared for the National Center for Health Services Research and Development). NTIS No. PB 211 690-692.

Pollycove, M. (1972). Benefits and disadvantages of clinical laboratory computerization. *Med. Instrum.*, 6:42-46.

Sullivan, S. F., R. W. Patterson, and E. M. Papper (1969). Evaluation of pulmonary function data using computer time-sharing. *Proc. IEEE.*, 57:1996-9.

6
The Clinical Engineering Program

BRUCE H. BARKALOW

6.1 Learning Objectives

Upon completing this chapter you should be able to:

1. Describe the major functions of a clinical engineering program.
2. Describe three formats for the organization of these functions.
3. List clinical engineering functions needed in health care.
4. Define clinical engineering.

6.2 Introduction

In previous chapters dealing with the history and development of health-care delivery and hospital systems operation, combined with economic factors, codes, and information flow constraints, we set the scene for hospital-based clinical engineering programs. The role of technological support and the quantitative approach of engineering find increasing applications in the clinical environment. One such application of engineering, *clinical engineering*, is a relatively new and growing field that is being shaped and developed by the changing needs of practicing health-care professions (Oakes, 1975). Therefore, clinical engineering is dynamic in nature and attractive to those who enjoy the challenges of an emerging discipline.

This chapter is meant to provide a transition from the basic material concerning hospital function to the specifics of clinical engineering practice. Overall guidelines for clinical engineering practice and in-hospital program development are presented. The necessary details for the various phases of actual practice are more fully developed in subsequent chapters.

6.3 Clinical Engineering Program Functions

Overall, the function of clinical engineering is to effectively apply technology to the solution of clinical problems. Since health-care delivery at the clinical level involves a large and varied number of the sciences and the arts, clinical engineering is complex and interdisciplinary. The complexity and diversity of delivering health care in a hospital demands a realization sooner or later that a *clinical engineering program*—not merely an individual or collection of individuals practicing clinical engineering—is indicated, with strict priorities and protocols established for maximum productivity.

This chapter deals with a successful approach to the practice of clinical engineering through a managed program. Adaptations of this basic program make it applicable to large or small hospitals whether they be community-based hospitals, medical research centers, or government hospitals. Central concepts include organization, recognition of skills and limits, and most important, teamwork.

In order to approach the development and staffing of a clinical engineering program, we begin with an analysis of the functions to be performed by such a program. Each of these functions is qualitatively described here and examined in detail throughout subsequent chapters.

Clinical engineering has a support role in the planning and development of facilities, technology, and technical methods as they relate directly to health-care delivery. The boundaries or guidelines for these support roles are not clear and well-defined. What is applicable to one institution does not necessarily apply to another. A principal reason is that clinical engineers have varying skills. In order not to over-extend clinical engineering support, clinical engineering programs must vary in different institutions. Therefore, the functions as described herein are roles that can and are fulfilled by clinical engineers to varying degrees depending upon their skills, availability, and accessible support staff. Each clinical engineer must use professional judgment to gauge the level of support to be provided in the areas discussed. In general, it is wise for a clinical engineer to restrict his activities to the application of engineering to clinical applications. He should be less concerned with administrative, nursing, physical plant, or other areas represented by professional colleagues.

Adapting to technological change in patient care. If there has been any one factor in health-care delivery that has remained constant, it is the growth and change directed to the improvement of patient care through technology. The clinical engineering program must, as one of its major roles, participate in adaptations of technology to accommodate growth and change in patient-care regimes that improve health-care delivery. To this end, the clinical engineer must continue professional–technical growth in order to be able to effectively utilize new technologies. This requires continuing education for all concerned as discussed in Chap. 12.

Areas of changing technology in patient care are many. These include but are not limited to: computer-aided diagnosis, medical records, and therapy; vital organ augmentation (i.e., pacemakers, stimulators); organ/functional replacement (i.e., artificial limbs, communicators, heart valves, etc.); monitoring of vital signs; physiological transducers; behavior modification devices; resuscitation devices and methods; life-support devices and methods; and predictive mathematical models of physiological functions. As new methods and devices become available within the hospital environment, the clinical engineer has a multifaceted role. His duties and responsibilities include: assessing the needs for improved patient care in the institution served and matching improvements in technology to these needs; reviewing and screening requests for new technology as to safety and efficacy; applying, developing, or modifying existing technology to meet current demands for the institution served; and continually striving to remain current on all new developments having clinical application.

It is important to point out that as technology increases in importance and relevance in health care, not only is the clinical engineer faced with finding and applying new technology as requested or demanded by his clinical colleagues, but also to prevent growth and technological change when not fully warranted or in conflict with current codes, regulations, or accreditation requirements. This may take the form of prepurchase screening of new devices that are not yet proven, for example by studying the methods and ramifications of involving computers, or evaluation of new techniques producing new types of clinical data. These aspects are covered in detail in Chaps. 7, 8, and 9.

New facilities. In many hospitals, there are plans or developments underway to create new facilities. These new facilities may be designed to accommodate normal growth expansion of existing services or to provide a new service. Clinical engineering plays a role in any expansion that involves health-care delivery per se. For example, clinical engineering need not be involved with remodeling the payroll department, but definitely participates in clinical areas such as pediatrics or dialysis units. Clinical engineering should be represented in all phases of planning of such units and facilities. A stereotyped notion that clinical engineering merely assists in specifying the medical equipment once the facility is designed is not usually productive and leads to underutilization of available skills. In the development of such a new facility, the cost of the equipment may be a sizeable percentage of the overall project cost. If not properly accounted for in the beginning, the entire project may turn out to be fiscally impossible after considerable time and effort may already have been spent in planning. More important, selection and utilization of appropriate equipment at an early and iterative planning stage, can alter the staffing patterns for the nurses, the efficiency of medical interaction with the patients, or even environmental factors such as ambient temperature and lighting.

If clinical engineering representation in new facility planning sessions is the only engineering support available, it is important that to the extent possible, engineering planning techniques such as PERT or CPM methods be used (Thomas, 1974). The clinical engineer, participating in such a facility planning session, should see to it that good meeting protocols are maintained including minutes, subcommittee structures, medical and nursing representation, and time schedules for future commitments.

One other important factor to remember is that many of the medical and clinical colleagues may not work or relate well with engineering prints, drawings, and technical specifications. If, for instance, an intensive-care unit is being planned, it is a prudent idea to build a mock-up or physical model of such a unit or a repeating unit (such as a one-patient station) for the nurses and physicians to observe in order to make suggestions for improvements or modifications. Chapter 11 contains details on facilities and facilities planning for which these general guidelines apply.

Renovation. Perhaps more challenging than new facility design, are the projects that involve renovation of existing facilities. Here, the questions of codes become more complex with respect to the type of alteration to be made, grandfathering clauses of current codes and regulations, the methodology whereby the work is to be accomplished, i.e., in-house or outside contracts, and what is to be done to continue the patient-care routine in the areas to be renovated. To accommodate this latter concern, often the process of new facilities renovations requires complex work schedules so that patients in the existing facilities are minimally disturbed or are transported to other areas. Here again, contributions from clinical engineering for assistance in this planning and scheduling may be required to establish interim patient-care units with the necessary instrumentation.

In the planning of both new and renovated facilities, when there is a demand for clinical equipment, it is quite possible that equipment from other areas of the hospital may have to be sought, modified, or upgraded to meet different requirements. Rental equipment that has to be interfaced with existing equipment may have to be ordered. Technical support such as this may become an integral part of the overall plan with time provided in the schedule for this clinical engineering activity. Usually, this type of task must be considered as a project and as such may require temporary staffing by, for example, biomedical engineering interns, consultants, or possibly subcontractors.

Codes, laws, and regulations. The pertinent codes, laws, and regulations as known to date are presented in Chap. 4. They are mentioned again here because the clinical engineer provides assistance to the administration and medical staff members of the hospital through a detailed analysis of codes, laws, and regulations as they apply to clinical technology. For example, in California under state regulations licensing health-care facilities, certain pieces of equipment are indicated for special-care units. In some areas these specifications may not be clearly understood. For example, in a respiratory intensive-care unit, respiratory monitoring is required. Clinical engineering combined with medical input must determine what type of respiratory monitoring is really meant and what would be applicable for the type of patient in question. In order for this question to be answered, the clinical engineer must develop a general profile of patients expected into the unit and assess technological modalities to meet respiratory monitoring needs that will also be legally acceptable. In this example, respiratory monitoring could run the gamut from apnea monitors to on-line computer measurements of lung compliance.

The clinical engineering program must include activity to remain current on codes and regulations as they affect electrical safety, instrumentation, in-service education, and several other aspects of applied technology. In planning sessions, it is imperative that these codes and regulations be pointed out so that newly created or renovated facilities have minimal problems with subsequent inspections. As a matter of fact, if the administration agrees, it is wise to develop a working relationship with local representatives of various agencies enforcing current codes and regulations. Often, as part of a clinical engineering program, a mutual and beneficial exchange of information with local authorities can be accomplished. If this is accomplished, local authorities on codes and regulations can become a technical resource and an asset for developing or renovating patient-care facilities.

Acquisition and application of technology

The first phase of acquiring and applying technology is prepurchase planning. In most hospitals, an annual budget is developed. If, as part of the overall hospital budgeting plan, clinical engineering is included, then certain economies can be accomplished through planning. Prepurchase planning is a complex subject which is treated in Chap. 8. As an overview, acquisition and application of technology involves determination of need, translation of medical or clinical requirements into technical specifications, integration with financial planning for the hospital, technical evaluation both from an engineering and clinical view, and consideration of in-service education, and subsequent preventive maintenance and repair. It is important that the clinical engineering programs have overall control of the technical factors associated with purchase of new equipment or the application of new technological methods. In fact, in some hospitals the clinical engineering department owns all the medical instrumentation which is then consigned to other units. In a program such as this, utilization figures, maintenance figures, depreciation schedules, and other economic and functional data are used to determine cost effectiveness, medical effectiveness, and utilization of each and every device. Important progress in hospital cost, containment requires the clinical engineering program to provide this information (see Sec. 13.4).

In addition to the purchase of medical devices, there is another way in which technology can be acquired and applied in the hospital. This is through the acquisition or retainment of personnel with specific technological skills and expertise. Often what is really needed is technical expertise, not equipment. Specialists in technical areas can be retained as consultants or possibly hired full time to support specific patient-care or clinical specialties through clinical staff departmental support. This matter is explained further in Sec. 6.5. Such specialists acquired on a consulting basis should be looked upon as an available technological resource.

Computer applications in medicine form a large and growing area of technological application. In the clinical environment, however, caution is the byword. Some clinical engineering programs find that involvement with applications of computers to health care can cause a nearly all-consuming emphasis of the program on one or a few specialized areas. It is wise to treat computer applications in much the same manner as the acquisition of medical devices. If the device or computer programming has to be developed in-house, it must be recognized that traditionally this sort of development work requires a great deal of creative time and expense which is usually underestimated (see Sec. 5.8). If the computer applications are turnkey, then critical evaluations of what the computer hardware and software can and cannot do must be accomplished in advance with the results translated into clinical terminology for relevant evaluations. The end-users must acquire a clear understanding of what the computer system can and cannot do for them. Once done, the operational constraints and the technical specifications of such a system determine whether the system is really warranted.

In summary, the acquisition of technology through devices and personnel must be carried out carefully and without haste. It may be just as important to reject new technologies as to add them. Many of the existing modalities of health-care delivery can be very difficult to improve upon. New and modern is not always synonomous with improved or better. Be sure when acquiring new technological ability to use rigorous evaluation.

Technological performance assurance

Once a medical device, computer system, or technical staff is added, the assurance of the success of the technological application is generally coupled closely with the abilities of the users. Health-care staff education is a primary and fundamental concern. Within the clinical engineering program, there must be a training function including clinical engineering personnel who are competent educators. The teaching skills are, of course, above and beyond requisite understanding of the medical devices or applied technology. If the technical knowledge of a particular device or system is inadequate, the deficiencies must be made up through various methods. One of these can be the utilization of vendor-supplied in-service education. In this case, the education may have to take place at two levels. First, the in-house clinical engineering staff may have to be instructed to be declared competent by the vendor. Then as a team, the vendor and the clinical engineering staff can approach the health-care staff so that the vendor can assist with the first round of in-service education, after which he can transfer the continuing education processes to the in-house staff. This technique is particularly useful for complex devices such as the intra-aortic balloon pumps or computer-driven arrhythmia analysis systems.

Inasmuch as one of the major functions of the in-house program is to provide competent technical support of patient care, the users of the devices must feel comfortable in calling on the clinical engineering staff and they, in turn, must feel comfortable with their skills. In-service education should therefore be the first requirement for assurance of success for technically sound devices and methodologies.

Accurate records for equipment control are needed for technological performance assurance of medical devices. The purpose of the equipment-based records is to document the usage of devices and record servicing, preventive maintenance, and repair. Equipment control and record-keeping are discussed in detail in Chap. 9. Preventive maintenance and repair, discussed in Chap. 10, when combined with the educational responsibilities of Chap. 12, and the program management methods of Chap. 13 form the basis of technological performance assurance for devices.

Preventive maintenance and repair. An essential function of clinical engineering is to ensure that proper preventive maintenance and repair is accomplished on all medical devices. The clinical engineering program may have a division that accomplishes these functions or they may be done under warranty, under an outside contract, or even by another department within the hospital. In any case it is the responsibility of the clinical engineering program to ensure that all preventive maintenance and repair of clinical devices is properly carried out and documented. A documentation system to record all preventive maintenance and repair efforts must be established as discussed in Secs. 9.5 and 10.5. It is primarily through the documented preventive maintenance and repair program that technological performance assurance is accomplished.

There are three major classes of malfunctions associated with technology in health care. The first is a malfunction caused by medical device failure. Second, the problem may be caused by inadequate or improper medical in-service education. And third, unusual physiology or multiple-device interaction may result in confusing presentation or operation to either or both the medical and clinical engineering staff. The latter can be considered as a subset of inadequate or incomplete in-service education, but it is generally not possible to provide effective in-service education to cover all contingencies. Within the preventive maintenance and repair documentation program, there should be allowances for service calls related to in-service education. In this manner, through appropriate documentation, the clinical engineering program can not only determine the weak links with respect to preventive maintenance and repair regarding device failures, but can also provide in-service education. By communicating the results of this documentation through appropriate channels, very effective and localized in-service education and/or maintenance techniques can be applied. This sort of feedback system is invaluable for quickly locating and rectifying problem situations.

6.4 Clinical Engineering Program Structures

Although hospitals vary widely in administrative structure, size, and complexity (see Sec. 2.5), the basic structure of the clinical engineering program remains the same. A hospital may be in a community, university medical center, government (such as a Veterans Administration hospital), or a foreign country. For any of these situations the functional aspects of clinical engineering programs dictate a basic type of structure—a composite of various technical capabilities to meet a common need. Differences in existing structures of clinical engineering programs are caused by deviations from the ideal program function we describe. The structure may be modified by political considerations within the hospital and by the unique nature of individuals who comprise a clinical engineering department. The latter generally means that individuals can accomplish multiple functions for the department, and, therefore, it is difficult to draw an organizational chart showing all of these capacities. In any case, a general plan for clinical engineering departments is set forth here and modifications of this plan as it is to be established in large or small hospitals are a matter of professional judgment for clinical engineers.

Figure 6.1 shows a comprehensive clinical engineering program structure for a large hospital. Administratively, the clinical engineering program stands alone as a departmental entity with its own administrative reporting structure. Autonomy is therefore implied to develop and implement budgets for personnel, capital equipment, and operating expenses. Subsets of this structure are applicable for smaller, less complex hospitals where more than one basic function may be handled by an individual.

As the department structure becomes larger, the position of the director may tend to approximate that of an administrative engineer. In this case, engineering tasks must be allocated to associates, division heads, technical specialists, and consultants. If, on the other hand, administrative tasks are to be allocated instead of engineering respon-

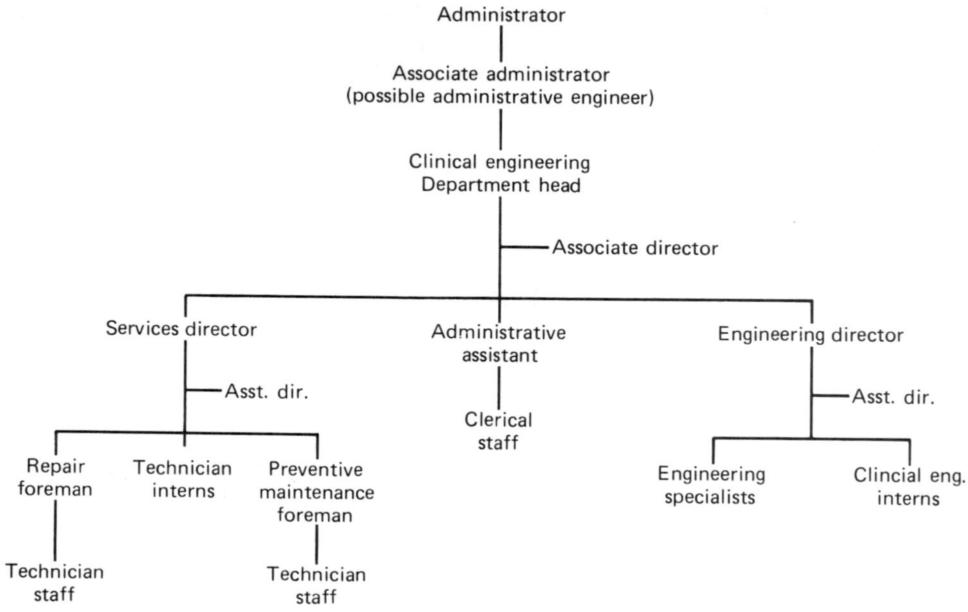

Figure 6.1 General organization structure of a clinical engineering department for a large hospital. Although the technician and engineering interns may be supervised by the services and engineering directors respectively, the interns will generally report to a fulltime staff member on an assignment basis (see also Figure 2.8).

sibilities, the director of clinical engineering will need an administrative aide(s) or will have to delegate administrative tasks to subordinate staff. The allocating of both technical and administrative tasks, which is generally the case, will result in a basic structure as shown in Fig. 6.1.

Essentially, a clinical engineering program structure follows a pattern similar to that of a corporation, in which there is an analogy between the board of directors (administration) with respect to the director of the department (president). Assistant directors (vice presidents) in various technical capacities must be added to form the nucleus of technical skills necessary while administrative assistants (secretary/treasurer) are also required to operate the entire program (business). In small departments, all of the skills and capacities required are the same, but with less depth and complexity. Therefore, fewer individuals may occupy one or more of the functional positions shown in Fig. 6.1.

It is the responsibility of the director of the program to recognize the strengths and weaknesses of the department in relation to the needs of the institution served. Positions must be justified, created, filled, monitored, or even closed out as part of the dynamic process of keeping in line with a cost-effective technical support role. This responsibility adds one other vital dimension to the skills necessary for the director of the program. That is, political strength with administration and medical staff is required in order to accomplish these administrative goals of creating and modifying the organization to meet changing needs. As it stands today, the development of clinical engineering programs in most hospitals is carved out—created by those directors who recognize the overall function of clinical engineering and who strive to attain the structure required. Each step and each position must be justified as part of a functional need, fulfilled and constantly reviewed.

There are three major divisions within the clinical engineering program structure. As the name implies, the service and maintenance division is responsible for almost all of the tasks associated with medical equipment once it has arrived and is in the hospital. This includes the incoming inspections for safety and performance, user education, implementation, preventive maintenance, and repair. The service and maintenance director should be a clinical engineer with considerable organizational skills (see Chap. 13). Within this division, responsibilities for repair are separated from preventive maintenance. Normally these task forces are led by biomedical equipment technicians with considerable technical and managerial expertise. It is important that preventive maintenance functions be separated from repair so that there are independent cross-checks on the quality of both preventive maintenance and repair. This will ensure that the preventive maintenance unit is not distracted to perform repairs, which have a tendency to supercede the preventive maintenance responsibilities. Avoidance of this pitfall will elevate preventive maintenance to a primary departmental function—the prevention of medical device failures.

The repair section should contain technician specialists to support various areas of the hospital, but with some personnel skill redundancy through training. Specialization is efficient, but the penalty of losing a specialist can be great for any repair support group. The object of a comprehensive repair program is to have a single phone number which is called for all repair requirements and the avoidance of requests for specific personnel. This can be achieved by having the call routed through the clerical support group so that the problem can be handled by any one of a number of people as designated by a working supervisor within the division. The details and requirements of both the repair and preventive maintenance programs are discussed in Chap. 10.

The second major division required is engineering which is responsible for finding solutions to longer range problems. This division is charged with such projects as new facilities planning, facilities renovation, medical device acquisitions, comprehensive in-service education programs, custom device development, computer applications, assistance in establishing or modifying hospital policies (such as equipment purchase protocols), support and review of the service and maintenance division, and, in general, working to solve problems more abstract in nature. In this area, specialty support covering the basic areas of engineering and science need to be acquired as needed to compromise an interdisciplinary support group. These should include expertise in electronics, mechanical engineering, chemical engineering, physics, mathematics, computer sciences, physiology, biochemistry, and medicine to name the major areas. Again, the number of personnel to fulfill this function does not change the overall basic structure, but the overall management system. To have a variety of expertise available usually requires consultants. Not all of the requisite scientists and engineers who support the engineering division can assume administrative responsibilities. The specialists should be chosen for their specialty skills in the area they represent and be encouraged to grow and increase these skills for clinical application. This requires a management system to reward technical growth, not just management growth (note this is also necessary for the technicians) (see Chap. 13). In smaller programs many of the engineering functions can be combined into a set of responsibilities for an individual, including working with consultants.

The third and functionally distinct division within the department is the administrative and clerical staff. In order to reduce the potential for duplicated records and files, one division is used for all clerical and administrative support. There should only be one equipment control file system, one technical manual library (which may have duplicate copies), one filing system, and one computer system with a common data base. Clerical help may be assigned to the various engineering or technician staff members as needed—both short- and long-term. Again, just as with the engineers and technicians, cross-training of the clerical staff with respect to the various functions within the department is necessary. The department can be crippled by the loss of one clerical staff member who may be the only one who knows the filing system. Although this management technique takes time to develop and is difficult to implement, it is, in the long-run, the best way. The clerical staff structure, just as the overall structure, should be set up in such a way that staff can move in and out of the structure without damaging the overall department integrity. As much as possible, the program structure should be functionally dependent, not personnel dependent.

Association with area educational institutions when possible is extremely useful. A program of clinical internships should be established which is of benefit to the hospital clinical engineering program. If done properly, this internship program will reduce the clinical engineering program operational costs while increasing capacity and throughput. The students serving an internship will obtain meaningful and useful experience. This technique and concept applies not only to the biomedical equipment technician, or clinical engineer, but also to the clerical staff.

Student intern programs should be directed by a full-time staff member as an administrative responsibility. Assignment of individual interns is made on the basis of departmental needs and intern skills. Interns must be closely managed in their respective assignments by full-time staff personnel. Since the intern personnel will be transient, in order to be useful the following guidelines should be observed: interns should work at least one academic year in a half-time appointment with scheduled hours; interns should be paid for their work and be held accountable as department employees; and all intern work should be documented so that other interns can complete as need be, projects that were started by other interns (Cook et al., 1973, 1978).

In-house large hospital clinical engineering program structure

The overall structure as indicated in Fig. 6.1 applies to the in-house large hospital. Alterations in this structure will occur to accommodate areas of emphasis that are found in private, university, or government hospitals. But the overall intent of the structure remains the same. The department must report to administration and have lines of communication open to the medical staff as indicated (Sec. 2.6). Additionally, fiscal responsibility must be a part of the overall structure and function (Sec. 13.5). Personnel profiles will tend to reflect the needs of that particular hospital. If it is a university medical center, for instance, and the medical faculty are involved in clinical research, within the clinical engineering division it is reasonable to have prototype capabilities in electronics and/or mechanics available through design engineers, researchers, and fabrication technicians.

In some large hospitals several in-house structures for clinical engineering within different departments may be formed. For instance, the cardiology unit could have a clinical engineering section to provide repair, preventive maintenance, and even some clinical engineering support. Surgery may have a similar group. This does not meet the overall intent of a clinical engineering program function for the hospital itself. Nucleated structures should be combined into one comprehensive organizational structure similar to that indicated in Fig. 6.1. This will lead overall to lower operational costs and increased efficiency for the hospital at large. Centralization prevents duplication of function and personnel. It becomes a professional obligation of various clinical engineers who find themselves in a large hospital exhibiting a decentralized structure to organize and pull the various structures together into one that shares common resources and exchanges knowledge for the overall benefit of the hospital.

In-house small hospital

For the purposes of definitions here, small hospitals are defined as those under 250 beds which do not offer highly technical services such as open-heart surgery. In hospitals of this type, obviously smaller programs are called for. Assuming that the hospital is large enough to merit a clinical engineer, we can talk about the same clinical engineering program structure as shown in Fig. 6.1. In the smaller hospital having intensive care, telemetry, cardiac catheterization, clinical laboratory, radiology, renal dialysis, and a busy emergency room, a clinical engineer is justified and may be able to cover all the clinical engineering program functions described. These include planning and development; medical-device acquisition; applications of technology; in-service education; and the direction of the preventive maintenance and repair program. If this is too much, delegation of administrative and technical responsibilities may be made to a secretary for the equipment control documentation and to the lead technician to manage the service and maintenance unit. It is best to keep in mind what the overall clinical engineering program function is, and then achieve as much of the function as possible through subsets of the structure outlined in Fig. 6.1. In this example, the small in-house program could be as illustrated in Fig. 6.2.

6.5 Clinical Engineering Program Personnel

The execution of the clinical engineering program function using the structure outlined is only possible through the selection and retention of qualified personnel. This is the most difficult aspect of establishing a program, since educational institutions vary considerably in

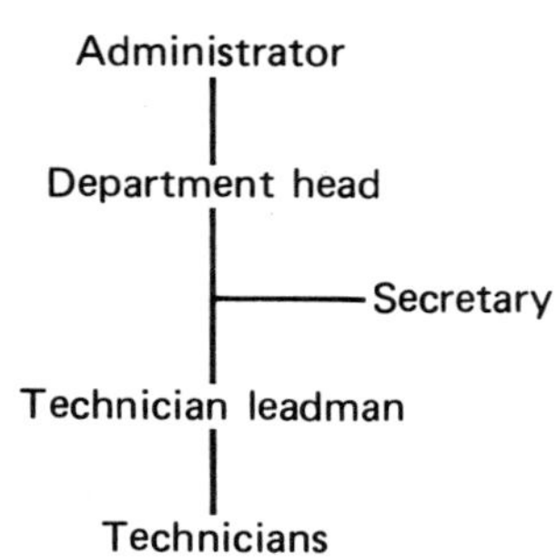

Figure 6.2 General clinical engineering program structure for a small hospital.

their ability to produce qualified clinical engineers, biomedical engineers, and biomedical equipment technicians. Standards for certification are unsettled and ill-defined (see Chap. 14). It is actually only through a proven performance record, that is, demonstrated successful experience, that satisfactory clinical engineers are clearly recognized. It is important to consider formal education as an alternative measure that can partially substitute for proven experience.

Biomedical and clinical engineers

In large hospitals the engineering functions required may include longer range development and engineering analysis typical of biomedical engineering. For this reason this section provides job descriptions for both clinical and biomedical engineers.

In job description development for biomedical and clinical engineers, it is important that the duties and responsibilities be kept relatively general so that the program director can utilize these interdisciplinary trained personnel in multiple capacities. Professional attitudes, behavior, and ethics for these personnel must be on the same professional plane expected of administrators, department heads, nurses, physicians, and other professional health-care workers. Hospitals are conservative institutions and this should be considered in the selection of personnel. As to qualifications, examples of job descriptions are included in Figs. 6.3 and 6.4 for the positions of clinical engineer in charge of a service and maintenance division and a biomedical engineer with specialist skills as they would fit into the structure described in Fig. 6.1. It is important to note that in regard to minimum qualifications, substitutions for experience and education are accomplished through a point system (Fig. 6.5). The reader is advised to use discretion in modifying this system to fit the needs of the particular institution served. The job descriptions presented are not meant to be all-inclusive or definitive, but to serve as examples of one approach.

Biomedical equipment technicians

Figure 6.6 gives a job description for an experienced biomedical equipment technician. Pacela and Forbes (1977) provide additional samples of job descriptions. The same general rules for acquiring clinical engineers apply to biomedical equipment technicians. However, the experience emphasis is perhaps even greater for a technician than it is for an engineer.

If the overall clinical engineering program deteriorates, the area in which the breakdown is likely to be most noticeable is that supported by the biomedical equipment technicians. Technicians work under a type of stress that is more immediate, and pressing

Clinical Engineer
Supervisor of Service and Maintenance Division

Job Description

General Statement:

Race, creed, color, national origin, sex, and age (except when specified by law) are not considerations for employment.

Under the direction of the Biomedical Engineering Department Head.

Under the direction of the Clinical Engineering Department Head, engineering as directed to include equipment evaluation, inspection, installation, user education, maintenance and preventive maintenance. Provide engineering expertise in accomplishment of departmental responsibilities as required.

Minimum Qualifications:

B.S. degree or equivalent plus combination of additional levels of education, licensure, certification, and experience per attached screening appendix.

Duties:

Coordinate a preventive maintenance and repair program for designated clinical equipment to meet current legal and accreditation standards as well as standards of safety and efficacy as required by current standards of engineering practice.
Control and coordinate initial incoming inspections of clinical equipment.
Implement as directed, a clinical equipment documentation and tagging system.
Hire and review the performance of biomedical equipment technicians.
Direct the 24-hour on-call emergency service.
Assist as required in specialized patient care involving technology.
Provide advice and technical input to new system installations and engineering projects as directed by department head.
Assist in the repair of nonroutine or particularly complex devices.
Assist, and at times coordinate, installation and user education for clinical devices or systems.
Review equipment hazard reports and provide notice to clinical departments and area supervisors; follow-up with appropriate corrective action.
Participate in incident investigations and reporting as related to clinical equipment failures to define causes and corrective actions needed.
Prepare yearly budgets for the service and maintenance division.
Carry out responsibilities as designated by department head.

Figure 6.3 Job description for clinical engineer in charge of a service and maintenance division.

Biomedical Engineer

Job Description

General Statement:

Race, creed, color, national origin, sex, and age (except when specified by law) are not considerations for employment.

Under the direction of the Clinical Engineering Department Head.

Carry out Biomedical Engineering responsibilities in: clinical device evaluations, implementation, user training, and maintenance; planning patient care facilities; clinical research; specialized patient care; and provide specialized expertise in an established scientific discipline.

Minimum Qualifications:

B.S. degree or equivalent plus combination of additional levels of education, licensure, certification, and experience per attached screening appendix.

Duties:

Supervise technical personnel as directed.
Perform as Project Director/Biomedical Engineer.
Develop, coordinate, and implement education programs.
Technically evaluate clinical equipment; assist with implementation or installation.
Carry out custom fabrication of funded research projects.
Assist as required in specialized patient care involving technology.
Help develop policies for BME relating to sphere of responsibility.
Communicate effectively and frequently with Clinical Engineering Supervisor.
Develop software for hospital computer systems.
Provide technical support to the 24-hour on-call emergency service as specifically required or directed.
Provide maintenance services as directed.
Otherwise carry out responsibilities as designated by department head.

Figure 6.4 A job description for a biomedical engineering specialist.

Biomedical/Clinical Engineer Qualification Summary

	Pt. Value	Minimum Required: Biomedical Engineer		Minimum Required: Clinical Engineer Super.	
I. Education, Licenses, Certification		50 pts	95 pts overall (I and II)	45 pts	95 pts overall (I and II)
1. Ph.D.—BME	50				
2. Ph.D.—other science	40				
3. M.S.—BME or other science	20				
4. B.S.—Physics, Engineering	40				
5. B.S.—BME, Life Science	30				
6. Assoc. in Science (no higher degree)	20				
7. Assoc. in Science (B.S. different major)	10				
8. Assoc. in Science (B.S. same major)	0				
9. Applicable college course work not part of attained degree 10pt/15 semester units grad. level 10pt/30 semester units undergraduate level 1pt/150 contact hours training schools					
10. Engineering Registration	40				
11. Engineer in Training Registration	10				
12. EIT with Engineering Registration	0				
13. Clinical Engineering certification	25				
II. Experience		35 pts		40 pts	
1. Engineering work as clinical or biomed. eng. (10pt 1st 2 yrs; 5pt/yr to 5 yrs; 2pt/yr thereafter)					
2. Supervision of technical personnel (10pt/yr for 1st year; 5pt/yr to 5 yrs; 2pt/yr thereafter)					
3. Engineering or research work in industry or university (5pt/yr for 1st 5 yrs; 2pt/yr thereafter)					
4. Upper level teaching in Engineering or Life Science in college or university (5pt/20 semester units)					
III. Professional Activities		5 pts		5 pts	
1. Publications in career related professional journals (2pt ea)					
2. Presentations in career related scientific meetings (1pt ea)					
3. Patents on career related devices (2pt ea)					
4. Continuing education while employed as engineer.(1pt/100 contact hrs and certificate(s) of completion)					
NOTE: Minimum total points must be met for each category as well as the total.			100 Minimum Total		100 Minimum Total

Figure 6.5 Minimum qualifications for the positions of biomedical engineer and clinical engineer supervisor using a point scheme to equate education and experience.

JOB TITLE BIOMEDICAL EQUIPMENT TECHNICIAN LEADMAN

JOB DESCRIPTION

General Statement:

Race, creed, color, national origin, sex and age (except when specified by law) are not considerations for employment.

Under the (specific) direction of (position): Clinical Engineering Supervisor

Describe position purpose and responsibilities: Repair and maintain clinical equipment in-shop and on the floors under the direction of the Clinical Engineering Supervisor.

Minimum Qualifications:

1. Graduation from High School required/xxxxrequired, with a A.S. Degree in Electronic Technology or equivalent experience.
2. Five years experience required/xxxxrequired, in electronic repair and troubleshooting including at least one year of troubleshooting and repair of medical equipment.
3. Must have the following minimum speeds: Typing _______, Shorthand_________, Other speeds required:_________
4. Must have (or be willing and able to be) certified by AAMI as a BMET.
5.
6.

Uniform: (if any)

Female:

Male:

Duties: (in descending order of importance or amount of time)

Under the direction of Clinical Engineering supervisor, the BMET leadman plans for and supervises the daily activities of a group of skilled BMET technicians performing electrical, electronic, and electromechanical services required for maintenance, preventive maintenance, and safety checking of clinical and laboratory equipment. In addition, this person will also:

1. Receive work requests from Maintenance dispatcher and accomplish appropriate troubleshooting and repair.
2. Report hazards involving clinical equipment to Clinical Engineering Supervisor and area supervisors.
3. Accomplish work reports to reflect activities in response to work requests.
4. Share 24-hour on-call with other BMETs and Clinical Engineering Supervisor.

Tools, equipment and business machines required on the job:

Oscilloscope, multimeter, soldering equipment, light mechanical shop tools, safety analyzer, basic calculator functions.

Figure 6.6 Job description for an experienced BMET capable of supervising a BMET work force.

than that for the clinical or biomedical engineer. The frequency of encountering stressful, emergency situations is greater in service and maintenance than in longer range problem-solving. Biomedical equipment technicians are often faced with on-the-floor clinical situations involving medical device failures that may be critical. The selection of quality biomedical equipment technicians is absolutely crucial. Biomedical equipment technicians should be selected for professional qualities equal to those of clinical or biomedical engineers. It is important to regard the difference between biomedical equipment technicians and clinical or biomedical engineers as a career choice; both are professionals. Overall, in the selection of biomedical or clinical engineers and biomedical equipment technicians there must be considerable judgment used with respect to education and experience versus qualities of professionalism. The degree of success for the clinical engineering program in filling out the structure as indicated in the previous section will be determined by qualitative judgment factors in the selection and management of personnel (Sec. 13.6).

Student programs

As previously mentioned, room should be made in the clinical engineering program for student participation (Cook et al., 1973). Primarily this student participation is for clinical or biomedical engineering and biomedical equipment technician students. Section 12.7 describes the educational aspects of clinical engineering internships. Except in a university-oriented hospital structure, it is generally not advisable to have students working within the clinical engineering program for academic credit. It is preferable that students in the program be there for a specific purpose to serve the overall function of the clinical engineering program. This work situation is excellent experience for the student; but designing work experience to meet academic goals for the students may cause them to serve two masters. This situation can cause separation of goals between academic institutions and the hospital. It is strongly recommended that any students who work within the hospital be paid for their work, be held accountable for the quantity and quality of the work, and be supervised by experienced staff personnel. Additionally, it is recommended that the faculty of the sponsoring institutions be thoroughly acquainted with the assistantship or internship program in its early phases. This may even require that the faculty consider serving part or all of an internship program if they have not had the benefit of previous clinical experience. Such an arrangement can be established with the university.

The activities of the student interns in the clinical engineering program provide general, overall support of the departmental function. Educational goals are considered secondary and adjunct to the experience of performing useful work in the clinical environment. Generally, the responsibilities and duties of the student are not in the mainstream with respect to clinical practice. That is, the students work within the department under supervision on projects and problems that do not directly affect the quality and quantity of patient care until the students have proven their abilities. It is not responsible, for instance, to allow students to repair a medical device without qualified supervision and thorough checks upon the work before the unit is released back into the clinical environment. For an engineering student, it is not prudent to allow the engineering student to design or participate in the design of a new facility without adequate control and supervision from an experienced clinical engineering staff member. However, in both capacities as mentioned, say the repair of the instruments or the design of a facility, the student can be of great use to the full-time staff member as an assistant.

It is recommended that regular meetings be scheduled between participating university faculty and clinical engineering department staff members to review the various institutional goals of both the hospital and the university or college to make sure that both have been met in a satisfactory manner.

6.6 Clinical Engineering Program Examples

In order to place the discussions of the clinical engineering program function, structure, and personnel into perspective, some examples of clinical engineering programs are given. However, the names of the hospitals are fictitious. These examples can be extrapolated to other types of institutions of various sizes and are presented here only to show how structure as discussed in Sec. 6.4 can be modified to meet particular needs and stages of growth.

Central City Hospital is comprised of two separate facilities separated by several miles. Each facility is an acute care hospital offering a wide range of services with the larger facility (400 beds) specializing in cardiovascular care and the smaller facility (250 beds) specializing in orthopedic and neurological surgery. Both hospitals have emergency facilities, dialysis units, intensive care, telemetry, pulmonary, cardiology services, and extensive radiological facilities. Overall, there are about 3000 medical devices.

The clinical engineering program is structured as shown in Fig. 6.7 and has been under development for six years. The department maintains control over all medical device technical evaluations, technical specifications, preventive maintenance and repair, custom project work, and computer applications for clinical work. Almost all outside preventive maintenance and repair contracts have been assumed in-house and those that remain are under departmental review

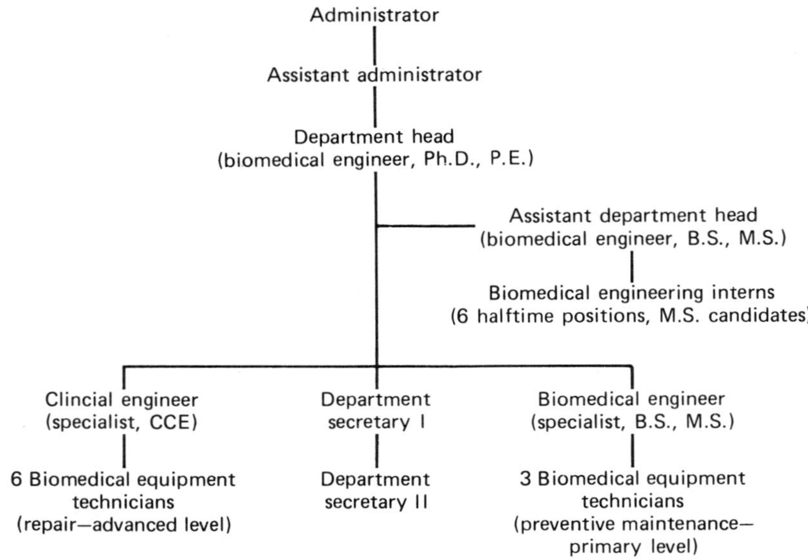

Figure 6.7 Central city hospital clinical engineering department as structured in 1977.

and control. A complete equipment control system has been established which is computer-based.

This structure is not fully mature. First, this department does not service radiology, a major clinical function. Second, as the preventive maintenance program becomes fully developed, it will be transferred from the medical engineer to the clinical engineer. The reason for the present separation is for program development (an engineering project). Third, medical staff input is absent. Plans are underway to create a permanent medical staff steering committee for the department.

Although not shown, the clinical engineer has additional skills in medical device modification/development and at times works as a resource engineer or specialist. When this occurs, a leadman for the BMET crew supervises the service and maintenance function.

The department head and assistant department head are working supervisors who both practice medical and clinical engineering. The student interns, all master's candidates in clinical engineering, report to the assistant department head and are assigned to any of the four engineers or technicians on a project basis.

By contrast, Memorial Hospital is a 300-bed institution offering most health-care services, but lacks dialysis and other specialties. In this hospital there is an administrative engineer in charge of physical plant operation and clinical engineering. The organization is shown in Fig. 6.8. Clinical engineering has the responsibility for the development and execution of budgets for personnel, equipment, and

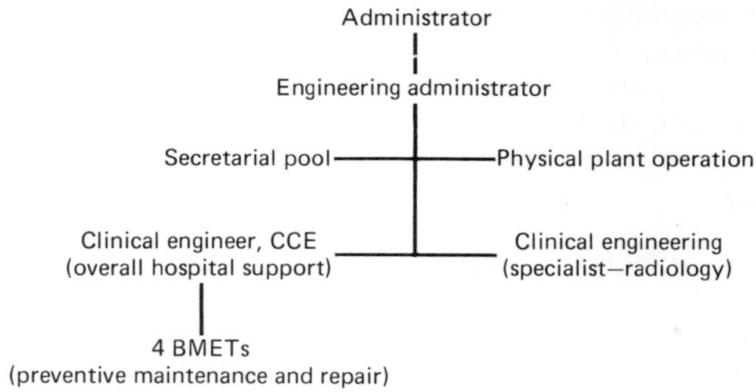

Figure 6.8 Memorial Hospital (300 beds) clinical engineering program structure.

operating expenses. The engineering expertise to cover all the program functions resides primarily with the administrative engineer and clinical engineering supervisor. The hospital is not large or complex enough to merit the acquisition of additional staff except for radiology which this hospital is beginning to handle in-house. There is a clinical engineer specifically to provide engineering support to radiology (a beginning project). The three engineers work as a team to solve problems and use common resources. This program has been under development for about four years.

It is important to note that the structure is a subset of that in Fig. 6.1 with the preventive maintenance and repair being prominent. The engineering group comprising the administrative engineer, the clinical engineering supervisor, and the clinical engineering specialist, work as a team to solve longer range problems. Once the preventive maintenance and repair sections were established, time became available for the clinical engineering supervisor to work on longer range problems. This provides for the present adequate engineering staffing for the support required. Both of these example programs began as one-person operations and were built by documenting needs (see Chap. 13).

6.7 Discussion and Summary

This chapter has provided an overview of clinical engineering program function, structure, and personnel in general terms. The details of each of these areas and specifics to accomplish requisite duties are the subject of subsequent chapters. The important thing to keep in mind is that the general function of clinical engineering is to apply technology to health-care delivery for the hospital served. This is a multidisciplinary, applied engineering function which, as programs become larger, cannot be accomplished by a single individual or by a single engineering discipline because of the depth of skills and educational background required. Management therefore becomes an important aspect of clinical engineering programs. The functional, structural, and personnel guidelines discussed here, when filled in by the specifics and details of the following chapters will provide useful guidelines for program development.

6.8 Study Questions

6.1 Define clinical engineering.

6.2 What is meant by a clinical engineering *program?*

6.3 What steps must a clinical engineer take to keep pace with changing technology in the hospital?

6.4 What are the duties and responsibilities of the clinical engineer in assuring safe and efficacious use of technological developments?

6.5 Summarize the roles of the clinical engineer in facilities planning.

6.6 How does planning for renovation differ from new facilities planning?

6.7 Describe the responsibilities of the clinical engineer in relation to codes and standards.

6.8 What advantages accrue to a hospital by involvement of clinical engineering in equipment acquisition and budget planning?

6.9 What does "technology acquisition" for the hospital encompass?

6.10 Describe the role of in-service training in technological performance assurance.

6.11 What is the purpose of equipment records in a clinical engineering program?

6.12 List the three major classes of medical-device malfunction.

6.13 Describe the function of each division shown in Fig. 6.1. What alteration in this structure might occur in a small hospital?

6.14 What are the advantages of separating the repair and preventive maintenance functions in the clinical engineering program?

6.15 What is the role of student interns in the clinical engineering program?

6.16 What considerations are important in selecting biomedical or clinical engigeers and biomedical equipment technicians?

6.17 Define the role and useage of consultants to clinical engineering programs.

6.9 References

Cook, A. M., B. H. Barkalow, W. M. Mason, and B. H. Hyndman (1973). Graduate bioengineering assistantships in a community hospital. *Eng. Educ.*, 64:194–197.

Cook, A. M., L. E. Rose, and B. H. Barkalow (1978). Cost effective clinical engineering internships in community hospitals. *J. Clin. Eng.*, 3(2): 173–178.

Oakes, J. B. (1975). Clinical engineering—the problems and promise. *Science*, 190:239–242.

Pacela, A. F., and E. G. Forbes (1977). Job descriptions and work objectives in clinical engineering: part 1. *J. Clin. Eng.*, 2(2): 157–164.

Thomas, E. L. (1974). Systems engineering and operations research. In C. D. Ray (ed.), Medical engineering. Chicago: Year Book, pp. 100–112.

7
Safety Program

JAMES J. FREEMAN, DAVID L. STONER,
JAMES B. SMATHERS, DAVID E. CLAPP,
and DEAN D. DUNCAN

7.1 Learning Objectives

Upon completing this chapter you should be able to:

1. Explain why a clinical engineering safety program is necessary.
2. Describe the major aspects of a clinical engineering safety program, and how they can be implemented.
3. Describe major hazards present in hospitals and the steps that can be taken to reduce them.
4. Perform electrical safety tests on medical instrumentation.
5. Perform tests for infection and radiation control.
6. Explain the role of the clinical engineer in hospital safety in relation to other safety officers.

7.2 Introduction

The hospital is an industrial facility similar to any factory staffed by healthy workers. The hospital harbors hazards that are life-threatening to the employees who must spend a working lifetime in these buildings as well as the patients who are in various stages of medical treatment. The clinical engineer must be aware of these hazards and be able to formulate programs to reduce these hazards below a predetermined level of acceptable risk.

The Joint Commission on Accreditation of Hospitals requires a multidisciplinary hospital safety committee as described in Sec. 4.7. Chapter 7 provides guidelines for the implementation of a clinical engineering safety program which satisfies these and other requirements.

7.3 Types of Hazards Present in Hospitals

The clinical engineer is a key figure in establishing a safe environment in the hospital. While physicians, nurses, and other medical personnel are busily serving patients under their care, the clinical engineer must focus on hazards that may disrupt or destroy the life-saving efforts of this medical staff. The hazards present in the hospital are widespread and complex.

Electrical hazards

Electrical hazards are insidious and pervasive. Electrical energy, if allowed to flow to sensitive body organs, can be life-threatening even at current levels below 20 μA. Aside from shock hazards, electrical energy is an ignition source for fires in the flammable hospital environment.

Mechanical hazards

Mechanical devices are necessary for many treatments in the modern hospital. These devices include mobility aids, transfer devices, prosthetic devices, mechanical-assist devices, and patient-support equipment. Each of these devices embodies numerous threats to the life and limb of patients as well as hospital staff. These devices must be subject to careful design review, failure indication, and the establishment of complete specifications for safe use.

Environmental hazards

The hospital is a subsystem existing within a larger, supporting ecosystem. The environment within the hospital is critical for life support and rapid patient recovery. Such elements as solid wastes, noise, utilities (natural gas, water, etc.), and structural aspects of the hospital building must be carefully controlled and managed to reduce hazards from the spread of infection as well as injury from physical objects. Beyond this, the hospital must exist in the community environment and solid wastes and air pollution are a significant output requiring the attention of the entire hospital staff with advice and recommendations from the clinical engineer.

Biological hazards

Infection control is obviously a major factor in hospital safety. Here, the term safety is used in its broadest sense since the threat to health from infection clearly threatens the safety of patients as well as staff members. Infection control is approached by a sampling and evaluation program to identify exposures followed by identification and elimination of sources. Sterilization, of course, is a primary tool in infection control, and it must be carefully evaluated in clinical engineering programs.

Radiation hazards

Finally, radiation and radioactive materials have assumed a broad role in medical treatment. The health hazard of these devices and substances requires the establishment of programs to control the use of diagnostic machines and therapy devices to prevent the unwanted release of ionizing radiation. Waste control is especially critical with radiological substances because of the extended periods during which these substances remain hazardous. The clinical engineer must have a firm grasp of the physics of ionizing radiation and its control to formulate and supervise control systems.

While the clinical engineer must shoulder the responsibility of ensuring a safe hospital environment for the patient, an efficient safety program requires a systematic cooperative interaction among all hospital personnel and the patients' physical environment. This chapter examines the role of the clinical engineer in the planning and execution of this safety program.

7.4 Electrical Safety Principles

The effects of electrical current on the human body are a function of the resistive paths of the body, the magnitude and frequency of the applied voltage or current, the current path, the length of time that the current flows through the body, and finally the overall physical condition of the person.

Macroshock

The reaction of the body to electrical current can take many forms. In the case of macroshock [currents greater than 0.5 mA passing through the body (Fig. 7.1)], large currents cause paralysis of the respiratory muscles due to continuous contraction of the chest muscles, respiratory failure due to damage of the central nervous system or neural inhibition, hemorrhage, and/or burns.

Figure 7.2 shows that smaller currents may exceed the threshold for perception or let-go current—that current that causes involuntary muscular contraction in the arms so the subject cannot let go. These mean limits are for a 60-Hz wave applied at the fingertips of the two hands of a 78-kg male (Olson, 1978).

The current thresholds for the various physiological reactions rises with frequency. As an example, the let-go current at 60 Hz is approximately 20 mA, while at 10 kHz it is about 90 mA (Dalziel, 1973).

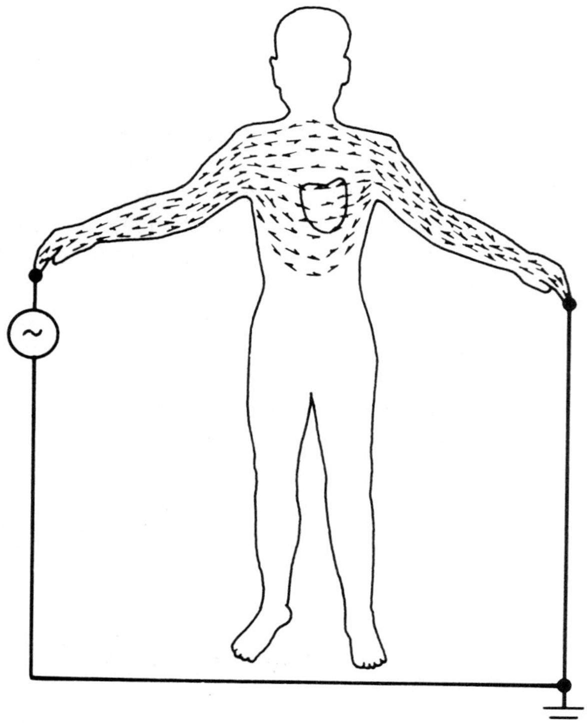

Figure 7.1 Macroshock current distribution. From F. J. Weibell, *Ann. Biomed. Eng.* 2:126–148, 1974.

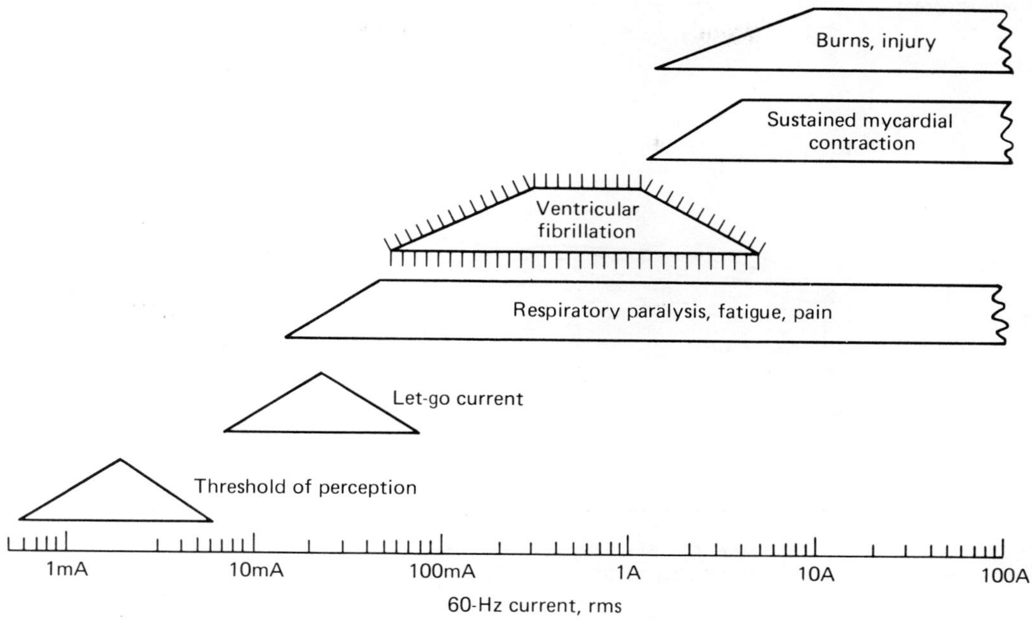

Figure 7.2 Physiological reactions to electricity. From J. G. Webster (ed.), *Medical instrumentation: application and design,* Houghton Mifflin, 1978, Reprinted by permission.

Microshock

The internal body resistance (about 500 Ω) is about 1 percent of the value of the skin resistance (about 50 kΩ). Therefore, if a potential is impressed within the body, a much larger current will result. Microshock is defined as currents normally less than 1 mA that enter the body at the heart through an invasive device. This is illustrated in Fig. 7.3. The primary reaction that can occur is circulatory failure due to ventricular fibrillation (unsynchronized contraction of the muscle fibers).

Procedures that make a patient susceptible to microshock include: insertion of a catheter from an external pacemaker, insertion of a fluid-filled catheter into the heart chambers, or insertion of an electrode into a heart chamber for intracardiac ECG measurement. There is little documentation on microshock-induced ventricular fibrillation, but studies have shown that currents as low as 20 μA have caused fibrillation in dogs (Weinberg et al., 1962; Geddes & Baker, 1971). Even less data exist on ventricular fibrillation thresholds in humans. Two major parameters involved in the experimentation have been the placement of the internal electrode and the

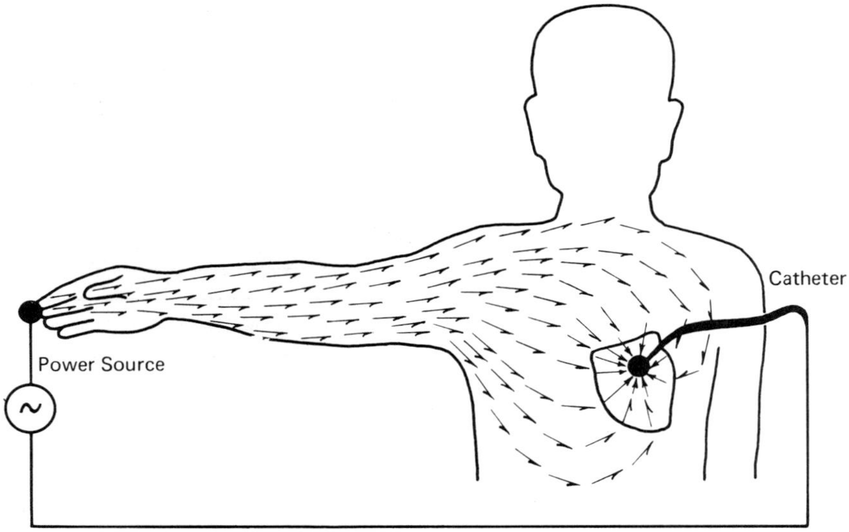

Figure 7.3 Microshock current distribution. From F. J. Weibell, *Ann. Biomed. Eng.* 2:126–148, 1974.

threshold itself. The most sensitive area for triggering fibrillation has been shown to be the inner right ventricular wall near the apex (Graystone & Ledsome, 1973). The threshold of fibrillation, with the catheter tip in this position, was reported to be 108 μA with a standard deviation of 35.8 μA. Graystone and Ledsome also reported that well below ventricular fibrillation thresholds (mean value of 21.1 μA, many below 20 μA) ventricular contraction was blocked while the current remained applied. Contraction resumed when the current was removed. Another study has reported the same phenomena but at higher thresholds (Green et al., 1972).

Classes of electrical risk

The hospital patient is more susceptible to electrical shock than the average person for two reasons. The patient at various times may have electrodes or fluid-filled catheters invasively applied to the cardiovascular system. Thus, small leakage currents may be large enough to produce microshock. The patient does not have the advantage offered by high skin resistance. Typical leakage currents of hospital equipment have been published (Norman, 1971). Since they were measured on the 230-V system in England, they are higher than those in the USA. They still, however, represent significant levels. These values are shown in Table 7.1 and were taken on normal operating equipment which was ungrounded.

Table 7.1. Leakage Current Ranges for Hospital Equipment (Norman, 1971)

Device	*Leakage Current (in μA)*
Suction pump	50 → 125
Volume respirator	85 → 175
Hypothermia unit	125 → 175
Electric bed	30 → 270
Patient monitor	20 → 350
Portable X-ray unit	30 → 600
Injection syringe	80 → 150
Electrosurgery unit	100 → 300
Portable operating room light	10 → 100
Track-mounted O.R. light	300 → 400

The other reason for the enhancement of the risk of electrical shock to hospital patients is their abnormal physical condition. For instance, a patient on medication may have a lowered threshold of perception to electrical current. A patient with reduced circulation may not notice the warming sensation of current being conducted through spilled liquids, dressings or water. Drug imbalance in the body may also lower the threshold of critical electrical shock. These added risks have been recognized by the NFPA and other standards-making organizations. They therefore have published levels of risk. These are shown in Sec. 4.4.

7.5 The Electrical Safety Program

Every effective electrical safety program must integrate the environment with hospital personnel, the patient, and administrative authority. It is comparatively simple to institute a number of equipment tests that detect hazards. It is more difficult to institute reliable standards that reduce these hazards. It is even more difficult to institute a program that commits the hospital to a program of continuing electrical safety.

Program parameters

The parameters of any electrical safety program may be summarized as follows:

1. Administrative commitment: The hospital administration must commit resources in the form of personnel, budget allocations, and authority to the program.

2. Training: Health-care delivery personnel must be trained to think and react to electrical safety hazards. This training will take different forms for the various groups of nurses, medical staff, and maintenance personnel.
3. Safety committee: A safety committee is an effective way of determining possible hazards and providing advisory and management authority in the various hospital departments.
4. Technical expertise: External technical consultation is usually warranted because of the complex and interrelated systems found in modern hospitals. A certain amount of in-house expertise is necessary to determine hazards. Shared service companies, medical researchers, and consulting engineers can expedite the diagnosis and treatment of electrical hazards.
5. Management and control: The clinical engineer or safety director must be given the authority to set up controls of personnel and environment. This authority includes a program that encompasses safety surveys, equipment inspection, specifying and testing of all new patient instrumentation.

It is not possible to initiate a program of this magnitude immediately. An effective priority list is necessary to implement any program in stages. An example of such a program (Kilpatrick and Kilpatrick, 1971) with accompanying priorities follows:

1. Protecting the electrically sensitive patient in existing facilities as quickly as possible, through reduction of unknowns and unknowing utilization of electrical equipment, a measurement and tutorial program.
2. Determination of specific optimal methodology for providing a safe, reliable, hospitalwide electrical environment. This investigation should include, as a minimum, these factors:
   (a) Determination of the changes in accepted practices necessary in the application, maintenance, and acquisition of instrumentation and appliances.
   (b) Degree of protection from macroshock and microshock hazards.
   (c) Reliability in the hospital environment.
   (d) Delay before implementation (extension of time of risk).
   (e) Obsolescence of equipment.
   (f) Degree of immunity from the consequences of human failure in operation or maintenance.

(g) Economic factors—costs of acquisition, installation, operation, maintenance, and training personnel.
(h) A detailed review of current and planned renovation and construction.

3. Protection for general patients and personnel in existing facilities as quickly as possible (hospitalwide, see item 1.) through a tutorial and survey program.
4. Development of formal preventive maintenance programs in each hospital, providing continuity of safety and reliability (the medical device that fails to function as expected during a medical emergency is just as dangerous as one that is a macroshock hazard).

The specifications for the electrical safety program must also take into consideration the "risk" classifications. It is not economically feasible to protect all patients with the same level of electrical safety. These parameters must be dictated by the classes and standards set forth in Chap. 4.

7.6 Implementation of the Electrical Safety Program

Any electrical safety program must entail more than just equipment. The program must include departmental organization and responsibilities, official policies, safety and performance testing, and test schedules (Chaps. 6, 9, and 13).

The organizational structure of the clinical engineering program must be such as to allow it to provide the necessary services and exercise the responsibility inherent to these services. The electrical safety aspect of this program includes the areas shown in Fig. 7.4. The program utilizing this block diagram must, of course, conform to the codes, standards, and regulations in Chap. 4.

The clinical engineering staff is a service facility that not only checks for leakage current hazards but furnishes other services directly involved with safety. One major aspect of this is the performance evaluation of instrumentation. The amplification, frequency response, etc. of an instrument can be checked at the same time as the safety check, and is inherently related to the safe use of any diagnostic or therapeutic instrument. A typical outline of these services is:

1. Safety testing
   (a) Complete evaluation of receptacle wiring, polarity, tension, and equipotential grounding in general-care areas on a yearly basis.

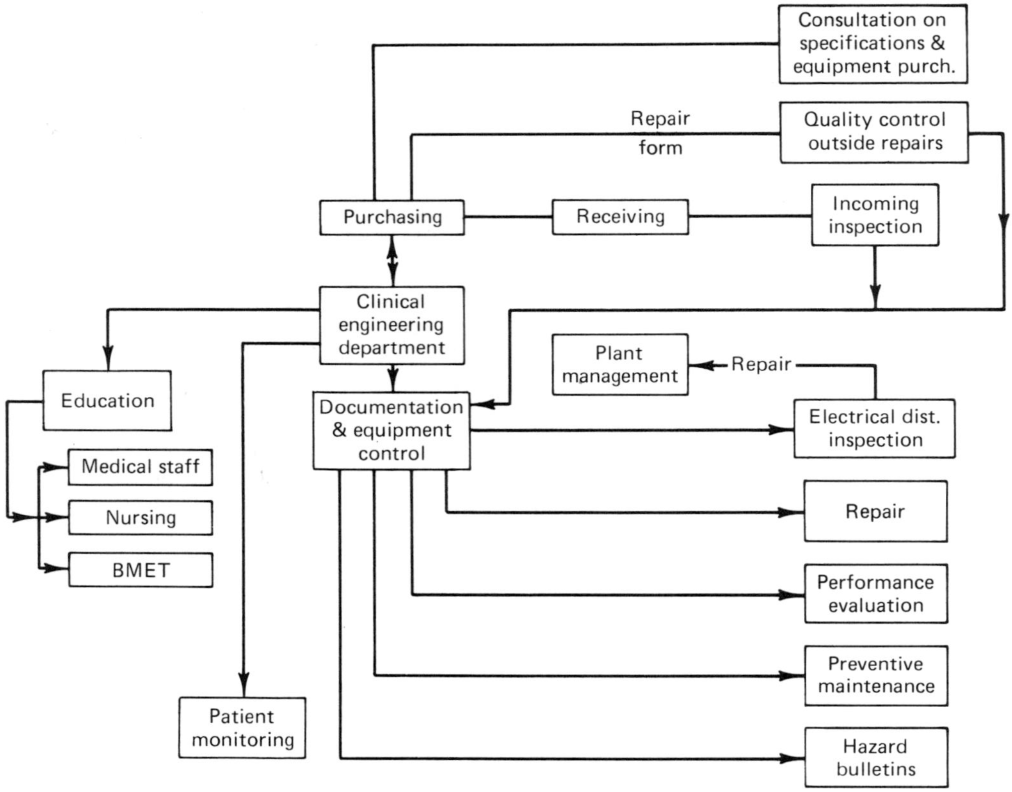

Figure 7.4 Areas considered in electrical safety.

(b) Evaluation of receptacle wiring, polarity, and tension in special-care areas on a bimonthly basis.

(c) Electrical safety evaluation of monitoring and patient-care equipment in special-care areas on a bimonthly basis.

(d) Calibration of defibrillator outputs on a monthly basis.

2. Performance testing

   (a) All monitoring and patient-care equipment in special-care areas are checked for proper performance and calibration on a bimonthly basis.

   (b) External pacemakers are tested after each use.

   (c) Pressure transducers used in special-care areas are tested after each use.

3. Repair: In-house facilities are maintained for repair of patient-monitoring equipment. Outside repairs are coordinated through the department (Chap. 10).

4. Emergency on-call service: Technicians are available 24 hours a day, 7 days a week to respond to emergency situations related to patient safety.
5. Incoming inspection: All patient-care equipment is tested for electrical safety and proper performance before being delivered to clinical areas for use (Chap. 9).
6. Education: Lectures and demonstrations on electrical safety, operation, and care of equipment are given to house staff and nurses (Chap. 12).
7. Equipment acquisition and selection: Assistance is provided in defining equipment to suit clinical needs, in evaluating specifications of available equipment, and in performing on-site trials (Chap. 8).
8. Purchase agreements: The department assists the purchasing department in preparing purchasing agreements under which equipment is purchased by the hospital (Chap. 8).

Safety tests

The remaining part of this section will be devoted to the equipment necessary for electrical safety testing and some examples of tests primarily designed for power distribution systems and patient instrumentation. Various aspects of electrical safety testing in hospitals are well documented. (Roveti, 1973b; Hewlett-Packard, 1972; Page & Aronson, 1973).

Some of the major devices necessary for safety and performance testing are:

1. Receptacle tester: Tests for neutral and ground resistance, high neutral voltage, ground faults, and mechanical tension.
2. Defibrillation tester: Measures the defibrillation energy output.
3. Electrosurgery tester: Measures output current of electrosurgical unit.
4. Electrical safety analyzer: Measures leakage current.
5. Performance testing: To test the performance of most instrumentation, a function generator, oscilloscope, and microvoltmeter are normally needed.

Electrical power distribution

The maximum potentials allowed between any two exposed conductive surfaces in the patient vicinity as specified by NEC ($f < 1$ kHz across 1 kΩ) are:

1. General-care areas: 500 mV under normal operation.
2. Critical-care areas: 100 mV under normal operation.

In order to achieve the above specifications for a general-care area, conventionally grounded electric wiring with grounding conductors may be used. A typical grounding system of this type is shown in Fig. 7.5.

A grounded installation with an equipotential grounding system supplements the conventional system and reduces any remaining potential difference by adding bonding surfaces to reduce the ground-wire impedances. This system, however, has the same fault conditions problems as the conventional grounding system. A typical grounding system of this type is given in Fig. 7.6.

In order to prevent the flow of large ground-fault currents, an ungrounded-isolated power distribution can be used. A typical isolated system, with a LIM (line-isolation monitor or fault detector), is shown in Fig. 7.7.

Instrumentation and electrical safety

An electrically safe environment for the patient means that the power-distribution system must have approved attachment plugs,

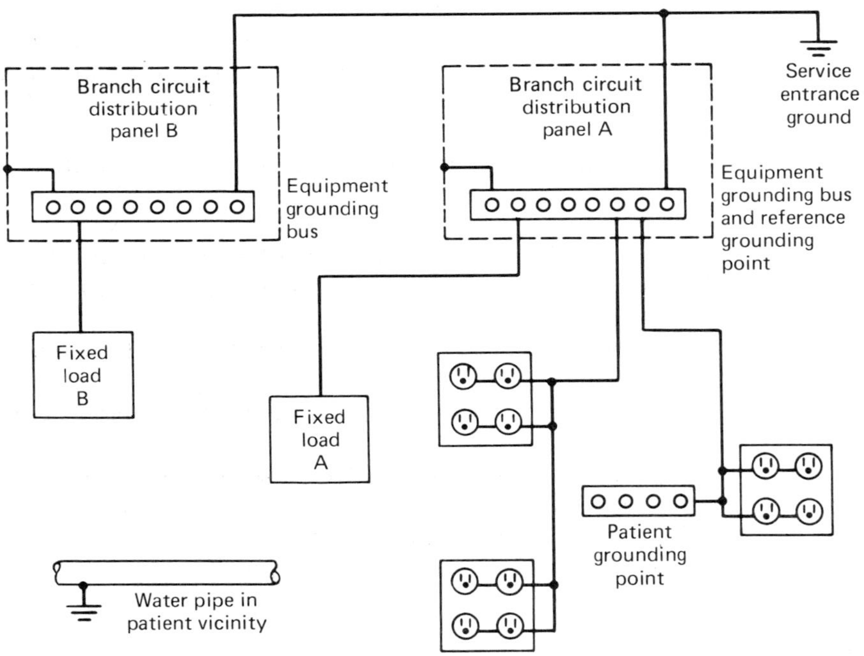

Figure 7.5 Typical grounding system for general patient location. Reproduced with permission from NFPA 76B-T (Tentative Standard for the Safe Use of Electricity in Patient Care Facilities), Copyright 1973, National Fire Protection Association, Boston, MA. NOTE: Tentative means that the Standard has not finally been adopted by the Association.

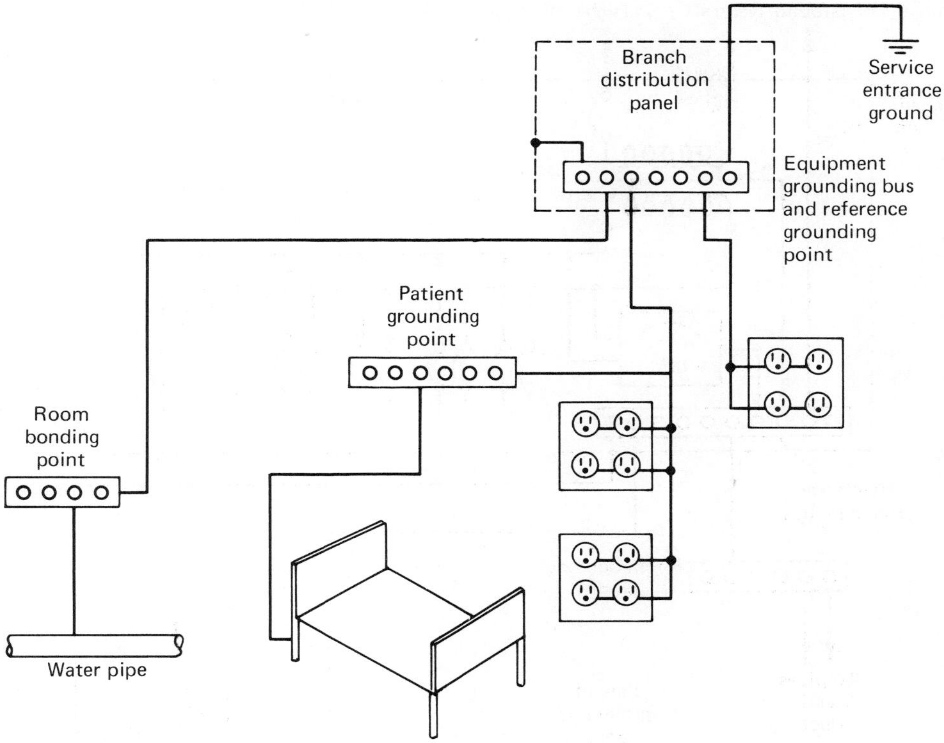

Figure 7.6 Typical grounding system with equipotential bonding conductors. Reproduced with permission from NFPA 76B-T (Tentative Standard for the Safe Use of Electricity in Patient Care Facilities), Copyright 1973, National Fire Protection Association, Boston, MA. NOTE: Tentative means that the Standard has not finally been adopted by the Association.

power cords, protection of wiring within the instrumentation power-control switches, overcurrent protection, insulation, etc., in order to limit leakage currents. These requirements are published by the National Fire Protection Association (1973b).

The tests for leakage current limits are normally preceded by measurement of the integrity of the grounding circuit and the measurement of the resistance to ground for each wire of the plug. There are three recommended leakage current tests for instrumentation. These are performed with all switches in all positions.

1. Leakage current from exposed metal to ground. Current measurements are taken with the ground plug disconnected and the power switch first off, then on. Then the test is repeated with the polarity reversed. This is shown in Fig. 7.8.
2. Leakage current between patient leads and ground. The test is made with the patient leads active, the power

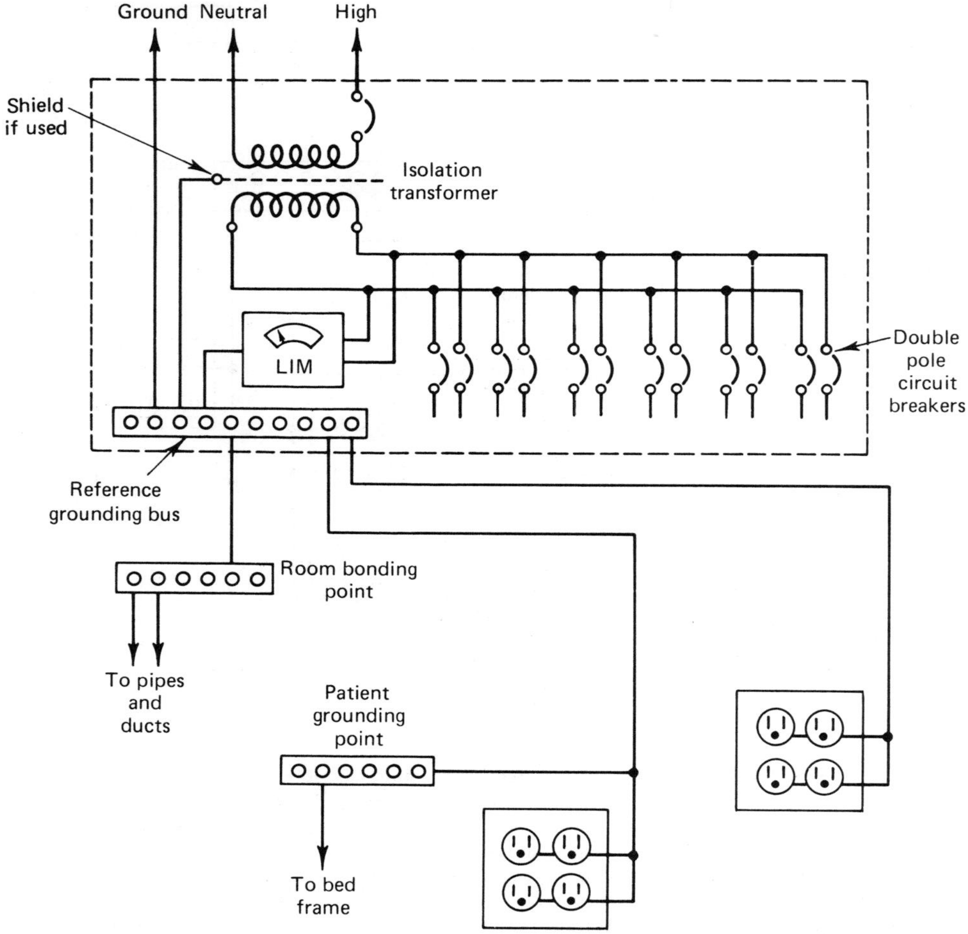

Figure 7.7 Typical grounding and bonding system for isolated power distribution system. Reproduced with permission from NFPA 76B-T (Tentative Standard for the Safe Use of Electricity in Patient Care Facilities), Copyright 1973, National Fire Protection Association, Boston, MA. NOTE: Tentative means that the Standard has not finally been adopted by the Association.

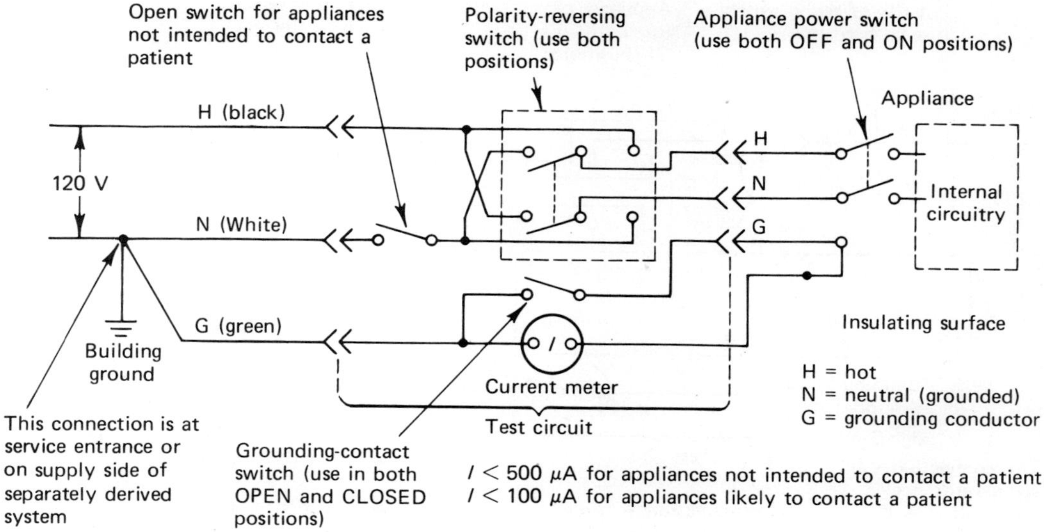

(a)

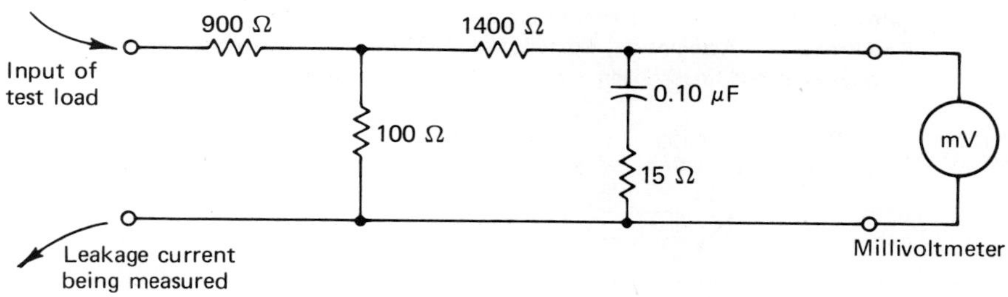

(b)

Figure 7.8 (a) Chassis leakage-current test. (b) Current-meter circuit to be used for measuring leakage current. It has an input impedance of 1 kΩ and a frequency characteristic that is flat to 1 kHz, drops at the rate of 20 dB/decade to 100 kHz, and then remains flat to 1 MHz or higher. Reproduced with permission from NFPA 76B-T (Tentative Standard for the Safe Use of Electricity in Patient Care Facilities), Copyright 1973, National Fire Protection Association, Boston, MA. NOTE: Tentative means that the Standard has not finally been adopted by the Association.

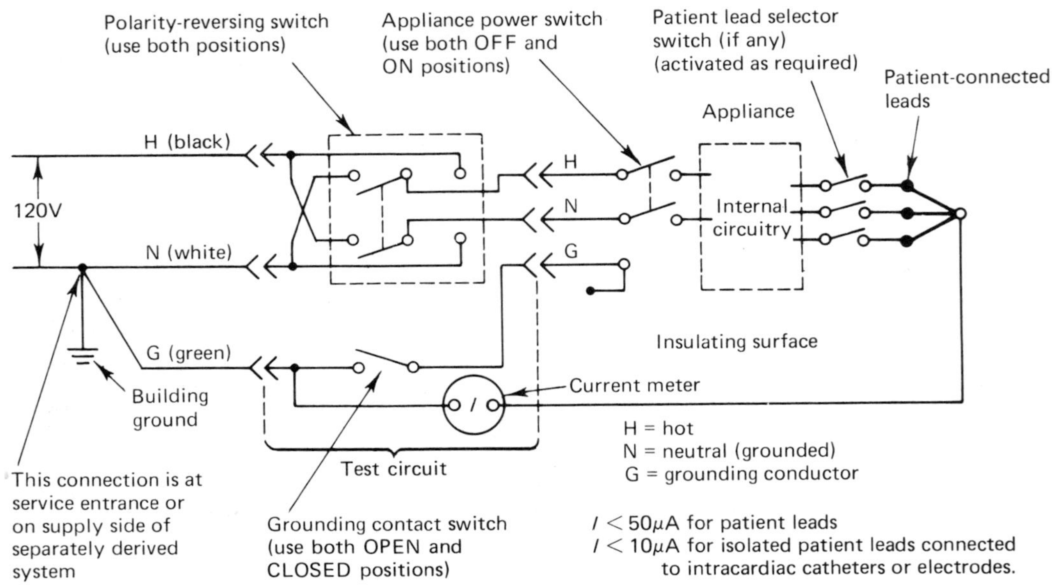

Figure 7.9 Test for leakage current from patient leads to ground. Reproduced with permission from NFPA 76B-T (Tentative Standard for the Safe Use of Electricity in Patient Care Facilities), Copyright 1973, National Fire Protection Association, Boston, MA. NOTE: Tentative means that the Standard has not finally been adopted by the Association.

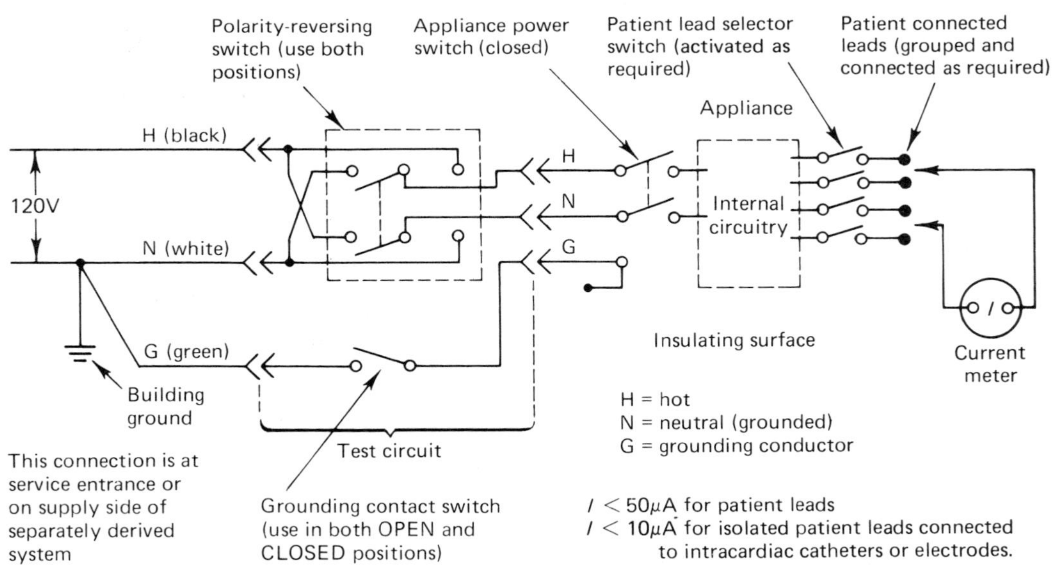

Figure 7.10 Test for leakage current between patient leads. Reproduced with permission from NFPA 76B-T (Tentative Standard for the Safe Use of Electricity in Patient Care Facilities), Copyright 1973, National Fire Protection Association, Boston, MA. NOTE: Tentative means that the Standard has not finally been adopted by the Association.

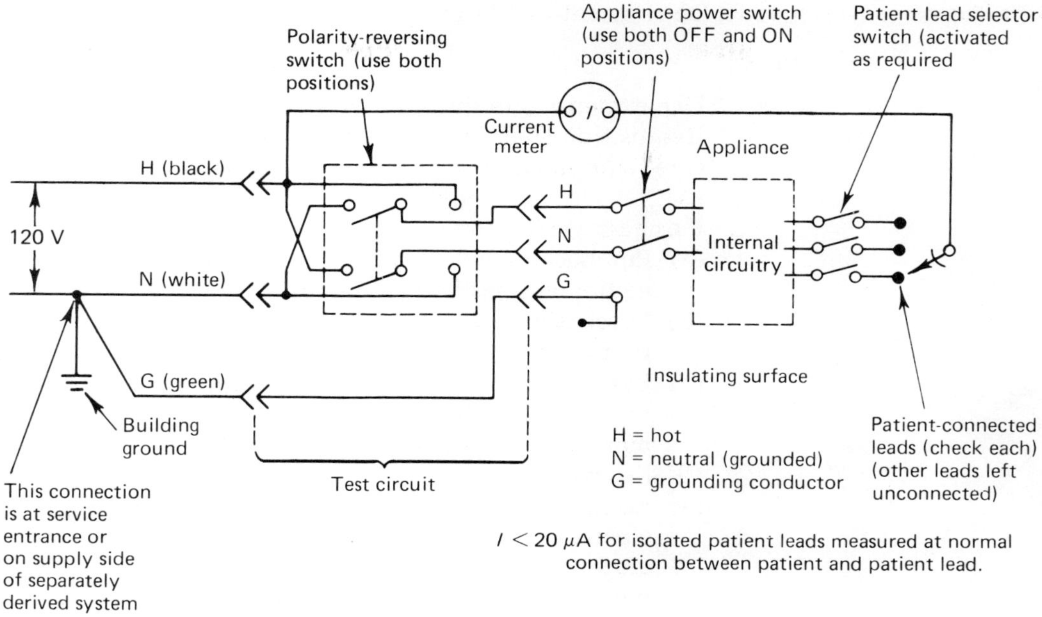

Figure 7.11 Test for A.C. isolation current. Reproduced with permission from NFPA 76B-T (Tentative Standard for the Safe Use of Electricity in Patient Care Facilities), Copyright 1973, National Fire Protection Association, Boston, MA. NOTE: Tentative means that the Standard has not finally been adopted by the Association.

switch off, then on and with the polarity reversed. This is shown in Fig. 7.9.

3. Leakage between leads. This is shown in Fig. 7.10.
4. Leakage between isolated patient leads and ground (risk current). This is shown in Fig. 7.11.

Instrumentation example of safety and performance testing

Every instrument used in the patient area must be initially checked and then rechecked for safety and performance on a regular basis. A weekly, monthly, and an extensive testing procedure every three or six months is normal for isolated power areas. Other areas may be extensively tested every six months or yearly. Following is an example of an initial safety and performance test procedure for an ECG instrument.

Initial equipment testing

Policy:

All medical electrical equipment designed for connection to patients, or for use in a patient's room, will be inspected by the Clinical Engineering (CE) Department prior to being placed in service.

Procedure:

1. Departments ordering equipment.
   Requisition shall include the phrase, "Electrical patient care equipment—subject to initial inspection before use."
2. Purchasing department.
   - (a) Purchasing will notify the CE Department that equipment has been ordered.
   - (b) Purchasing Department will notify Receiving Department to deliver units to the CE Department.
   - (c) Purchasing will notify Accounting that payment will not be made until advised by the CE Department that units have passed incoming inspection.
3. CE department.
   - (a) Equipment shall be tested to insure that it meets electrical safety requirements. If necessary, hospital grade plugs will be installed on the equipment. Tags indicating the unit has passed inspection will be placed on the unit.
   - (b) CE Department will notify Accounting that payment is authorized and that the equipment is ready to be tagged with a property control number.
   - (c) A file will be initiated for each piece of equipment for complete documentation of test results.
   - (d) When the equipment is tagged, the CE Department will notify the department ordering the equipment that it is ready for use.
4. Accounting department.
   - (a) Accounting will dispatch personnel to tag the equipment.
   - (b) Accounting will withhold payment until notified that the equipment has passed incoming inspection.

Electrical safety test protocol

Purpose:

To measure grounding resistance and leakage current of electronic equipment used in patient care, as well as examining the physical condition of line cords, plugs, strain reliefs, and the general condition of the equipment.

Frequency:

Instrumentation in all intensive care areas, operating rooms, and special procedures rooms shall be inspected at bimonthly intervals. Other equipment shall be tested quarterly.

Procedure:

1. Equipment without patient leads.
   (a) Measure resistance of line cord from ground pin to chassis.
   (b) Measure leakage current from chassis by attaching clip lead to chassis and connecting power cord of instrument to leakage current tester. Leakage current is to be measured with the equipment power off and on and in the following wiring configurations: grounded with normal polarity and ungrounded with reverse polarity.
   (c) Examine condition of power cord all along its length for cuts or splits. Examine strain reliefs and type of plug used. Check fuse and note rating.
   (d) Operate all controls and check for smooth operations. Verify that meters or other indicators are functioning.
   (e) Measure isolation of power cord by checking resistance between hot and ground and neutral and ground.
2. Equipment with patient leads.
   Perform all tests indicated above as well as the following:
   (a) Resistance from reference electrode to ground.
   (b) Leakage current from each lead under all wiring conditions—record highest value and lead involved.
   (c) Leakage current between leads under all wiring conditions—record highest value and leads involved.
   (d) Measure isolation of patient leads by applying 115 V ac to leads.

When tests are completed, the equipment is to be labeled with either the red and white label indicating the equipment is suitable for connection to patients with low resistance connections to the heart or with the yellow and black label indicating the instrument is suitable for general patient use only.

The following standards shall be used for determining the category of usage:

	Patient Contact	*Isolated Input*
Resistance of ground wire	0.5 Ω	0.15 Ω
Chassis leakage current	100 μA	100 μA
Lead leakage current	50 μA	10 μA

7.7 Mechanical Safety

Introduction

In this section we are concerned with the nonelectrical design considerations that make an apparatus safe for patient use. We begin with a discussion of general design considerations applicable to medical equipment intended to be used for diagnostic, therapeutic, and monitoring purposes within the health-care facility. Next we will discuss proper support of the bedridden and wheelchair-bound patient, who if placed on an improperly designed mattress or wheelchair cushion could develop severe tissue breakdown problems (decubitus ulcers). Finally, we will analyze the possible hazards associated with patient mobility devices and consider the special problems architectural barriers can present to the handicapped individual.

General design considerations

The clinical engineer must first realize that the primary concern of the physician, nurse, or technician who will be using medical instrumentation, is not for the care and safety of the instrument but for the care and safety of the patient. Therefore, it is not uncommon for medical instrumentation to suffer a given amount of abuse during routine use. For example it may be dropped or subjected to liquid spillage. Obviously abuse of this type can lead to both mechanical and electrical malfunctions, thereby creating a safety hazard to operator and patient alike.

One answer to the problem of equipment abuse is staff education in the form of short courses or seminars on medical equipment (Chap. 12). The courses, conducted by the clinical engineer, can be designed to cover the specific needs of each unit in the hospital.

An additional approach to the inevitable problem of equipment abuse, and in some cases misuse, is for the clinical engineer to ensure that all medical equipment purchased incorporates the principles of fail-safe design. Fail-safe design (Chap. 8) ensures that the occurrence of a failure will leave the system unaffected or convert it to a state in which no injury or damage can result. In some instances, this action may cause inactivation of the system. In any case, the fundamental principle is that a fail-safe system will first protect the operator and the patient; second, prevent damage to the equipment; and last, prevent loss of function or degraded operation. Fail-safe designs can be categorized into three types:

1. Fail-passive arrangements reduce the system to its lowest energy level. The system will not operate until corrective action is taken, but no further damage will result

from the hazard causing the inactivation. Circuit breakers and fuses for protection of electrical circuits are fail-passive devices. The circuit breaker or fuse opens when the system is overloaded or a short circuit occurs; the system is deenergized and safe.

2. Fail-active design maintains an energized condition that keeps the system in a safe mode until corrective or overriding action occurs or activates an alternate system to eliminate the possibility of an accident. An example of such an arrangement is a monitor system that activates a visual indicator if a failure or adverse condition occurs in a critical monitoring operation. Fail-active systems can also incorporate features by which a malfunction in the warning system itself is indicated by a continuous, blinking, different color, or auxiliary light. In this way, there would be a very high degree of certainty that the monitoring and warning systems are operative.
3. Fail-operational arrangements allow system functions to continue safely until corrective action is possible. This type of design is the most preferable since, unlike the other two types, there is no loss of function. However, there is the disadvantage that there may be an increased danger compared to the other two types. An isolated power system and its associated line-isolation monitor provide an example of a fail-operational arrangement. Should a fault occur between one of the current carrying lines and ground, the system is no longer isolated from ground and the line-isolation monitor will alarm. However, all equipment connected to the isolated power transformer will continue to function normally while the operator takes steps to locate and remove the defective device.

As mentioned earlier, medical instrumentation can receive fairly rough treatment during routine use. This is especially true in areas where the personnel may be called upon to perform exhaustive work under extreme pressures, such as intensive-care units, operating rooms, emergency rooms, and delivery rooms. In addition to the unique electrical design requirements that are necessary to eliminate the microshock hazard, special consideration must also be given to the mechanical design of medical instrumentation. Mechanical design criteria have been established which, if met, will enable medical instrumentation to withstand vibration and mechanical shock, will provide protection against liquid spillage, corrosion, sterilization and disinfection procedures, and moreover will minimize the effects of excessive humidity and temperature on equipment performance.

Several studies have proven that 40 to 50 percent of existing devices in hospital use today, are defective for two basic reasons: (1) They did not meet manufacturer's specifications; (2) They were hazardous either to the patient or the operator (Simmons, 1972). Hospital injuries have been reported that range from burns to contusions, lacerations, and fractures. Therefore, there is certainly a need for meaningful evaluation techniques of medical equipment (Chap. 8, 9).

The Emergency Care Research Institute (ECRI) provides medical device evaluations which are published in *Health Devices* (Sec. 15.6).

MET standards (MET Electrical Testing Co., 1976) are tests for potential mechanical problems such as water protection, physical abuse, and fail-safe control, along with electrical safety evaluation tests such as leakage, grounding, and insulation. The standard is based on existing standards, some of which are mandatory (OSHA, FDA, and NEC) and others that are guidelines by voluntary groups (NFPA, AAMI, UL, and AHA). Portions of the MET-501 standard are in Sec. 4.9.

Tissue trauma

Tissue breakdown, in the form of pressure sores (decubitus ulcers) or abrasion of the skin surface, can occur as a result of pressure, shear, or torsion, continuously or intermittently applied to the skin surface (Kosiak, 1959; Schell and Wolcott, 1966; Dinsdale, 1974). Common troublesome areas are the heels, trochanters, sacrum, scapula, and iliac crest. For many patients with a potential breakdown problem, a skin trauma relief material, formed into either a pad or insert, can be provided to aid in force and energy distribution. However, given the variety of materials and systems now available and the limitations in fundamental knowledge, it has become difficult to select appropriate protection for a given patient.

The basis for protection offered by a material to the patient can be seen in Fig. 7.12 (Rogers, 1973), which provides a guideline for time-to-breakdown as a function of the applied pressure. When considering a material for the totally immobile patient (i.e., coma or aged) or the immobile and insensitive patient (i.e., paraplegia or quadriplegia), it must be realized that this patient will be subjected to loading for long periods of time but preferably at controlled pressures. This would place him in the right-lower portion of Fig. 7.12. As a result of this sustained loading, local ischemia can be a contributing factor to the development of a pressure sore. Therefore, the pressure-relief material must be capable of controlling the pressure profile to the patient to maintain minimum pressures. On the other hand, a patient who is mobile but insensitive (i.e., diabetes, nerve injury, leprosy, or amputee) will probably encounter breakdown

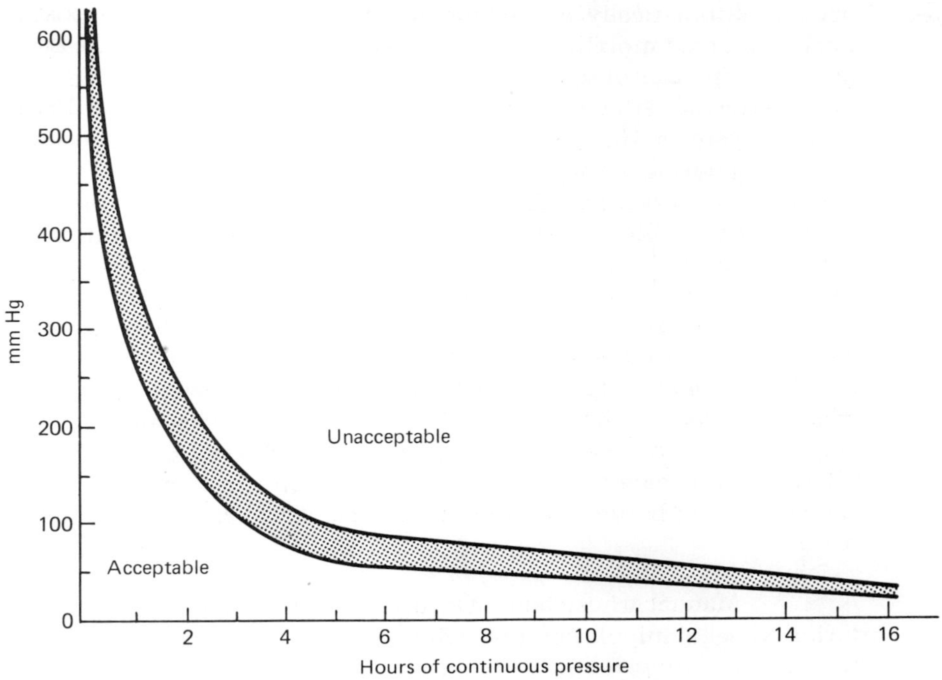

Figure 7.12 Tissue tolerance guidelines. Courtesy of John Rogers, MS. Director of Tissue Trauma, Rancho Los Amigos Hospital, Downey, CA.

problems due to intermittent impact or shear loading during walking. At high pressures this places him in the left-upper portion of Fig. 7.12. Since high loads of short duration are encountered by this mobile but insensitive patient, mechanical damage rather than local ischemia is probably the main factor contributing to tissue breakdown. As a consequence the material must have excellent energy-absorbing properties and stand up well to cyclic loading.

In selecting the appropriate material or system to prevent tissue breakdown, the patient's pathological state must also be considered. Local factors such as skeletal irregularities or loss of compliant tissue due to scarring, and cellular factors, such as inflammatory reactions to stress or infection, can cause a shift in the pressure versus time-to-breakdown curve. Additionally, tissue breakdown can be accelerated by the presence of excessive heat or moisture. Finally, any material or system designed for patient use must be considered acceptable by the patient.

Patient mobility devices and barriers

Electric wheelchairs. The "standard" electric wheelchair is basically a manual wheelchair with the additional components necessary for automatically propelling and steering the device. Other electrically operated mobility devices range from a powered bed operated by a puff-sip control to a golf-cart type "wheelchair" controlled by a tiller-foot-pedal system. There are hazards common to all these mobility systems. Hazards associated with batteries include acid spills or splashes during servicing, hydrogen gas generation during charging, and electrical shock hazards. Since much of the servicing of the electrical system would be done by the occupant or an attendant, these persons need to be aware of the hazards and actions to be taken in case of accident. Rubber gloves, aprons, and eye protection should be worn during servicing, which should not be done in the vicinity of open flames. While the training period to use electric wheelchairs may be reasonably short, subsequent use presents some hazards to bystanders due to the occasional jerky and erratic motions of the wheelchair. Also, the heavy weight of the chairs (approximately 110 lb) dictates care on the part of the attendants in moving or lifting the chair or its batteries to prevent muscle strain and pinched hands or fingers.

Manual wheelchairs. Occupant hazards include hand injury during wheel manipulation through narrow doors, falling forward out of the chair (depending on physical condition), flipping the chair backward, and muscle strain or overexertion while propelling the chair. These hazards can generally be avoided by careful operation by the occupant. Hazards exist also to attendants, especially in folding and lifting the wheelchairs. Finger/hand injuries can occur as well as back strain caused by lifting the chair in and out of a car.

Other mobility aids. Crutches, canes, and four-legged walkers have relatively few hazards and these can usually be eliminated through training and careful use. Some common hazards are rubber "feet" such as crutch tips slipping on wet floors, patient imbalance causing falls, and inadvertent folding of a four-legged walker also causing falls. Use of such devices in a crowded environment may likewise result in a fall.

Automotive mobility aids. In general, paraplegics have adequate upper body strength and function to drive a sedan-type automobile with the aid of hand controls for functions normally performed by the feet. Paraplegics using a manual wheelchair may have adequate strength to independently transfer from their wheelchair to the driver's seat of a two-door sedan, then fold the wheelchair and place it behind the driver's seat. Quadriplegics, depending on level of cervical lesion, may not have the upper body strength, function, or balance required to drive a conventional car with manual controls. In recent years, however, many quads have become more mobile through the use of electric wheelchairs and a van-type vehicle equipped with a wheelchair lift, restraints, and power-assist devices for driving. All handicapped drivers, with the assistance of adaptive equipment, are normally required to pass the same licensing tests as able-bodied drivers.

Manual driving aids. These include devices for operating brake and accelerator, gear selector, turn signals, headlight controls, parking brakes, etc. The Veterans Administration has a standards document (Veterans Administration, 1975) concerning design and test criteria for these devices. Included are requirements for strength, safe operation, electrical and mechanical components, and durability.

This special equipment (approved under the above standard) should be installed by a manufacturer's representative or a qualified mechanic. Maintenance is usually minimal, but the item should be checked periodically for proper operation. A driver training program is recommended for new drivers using assistive devices.

Wheelchair lifts. These devices, shown in Fig. 7.13, are installed in the side or rear door of vans or buses and provide vehicle entry/exit for the wheelchair-bound driver (or passenger). The platform on which the wheelchair rests is typically 28 in. × 40 in. and is raised, lowered, and folded by one or a combination of mechanisms: cables, chains, hydraulics, motors, linkage, etc. They must be safe for use by the para or quad population as well as for attendants and family of the disabled user. As for hand controls, the VA has developed a standard for wheelchair lifts to be purchased for veteran beneficiaries (Veterans Administration, 1977).

Ramps (for van entry/exit). Ramps can be purchased as an alternative method of wheelchair entry/exit for vans. They require an attendant to push or pull the wheelchair along the ramp. Hazards associated with ramps are mostly attendant-related: unsure footing while pulling or pushing, back strain, and perhaps hand/finger injury in close-quarters maneuvering. Obviously, if the attendant loses control or drops the wheelchair, the occupant may be injured. Pulling or pushing the wheelchair up the ramp at a ramp angle of greater than approximately 20° is very difficult, especially with an adult wheelchair occupant. This situation can be avoided by using longer ramps with smaller angles of incline.

Wheelchair restraints. As of mid-1976 there were a number of commercially available wheelchair restraints for vans on the market. These devices, shown in Fig. 7.14, included floor-mounted pedestals with clips that attach to the wheels, T-bars that span the lower frame members and bolt to the van floor, floor-mounted seat-

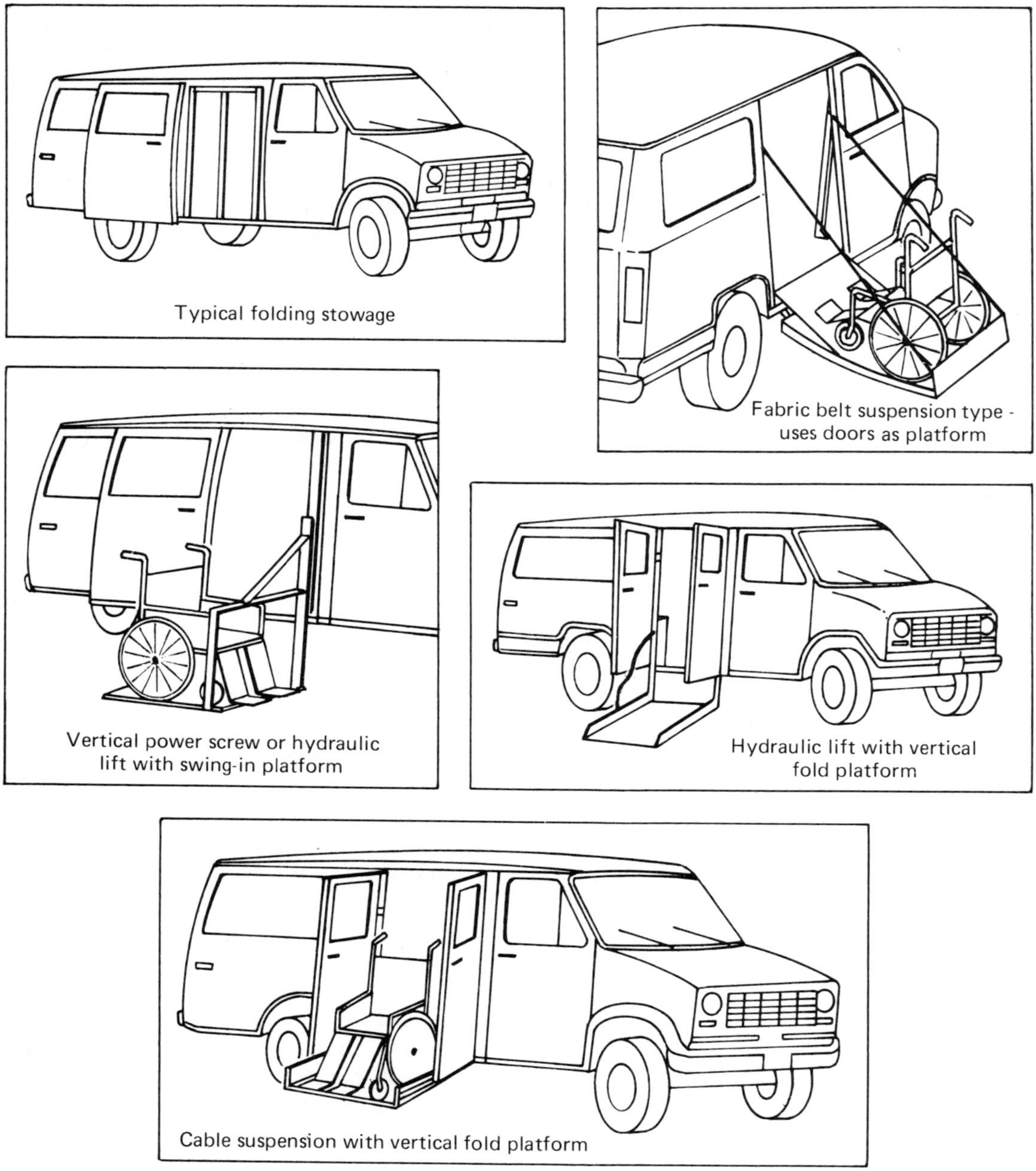

Figure 7.13 Basic lift types.

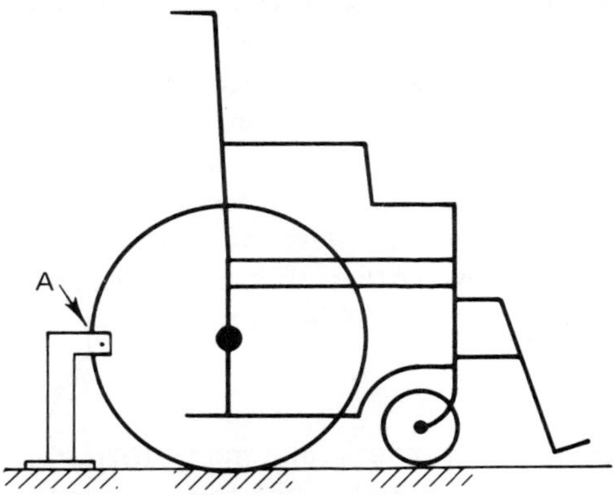

Pin or clamp attaches to each wheel rim at Point A (floor or wall-mounted)

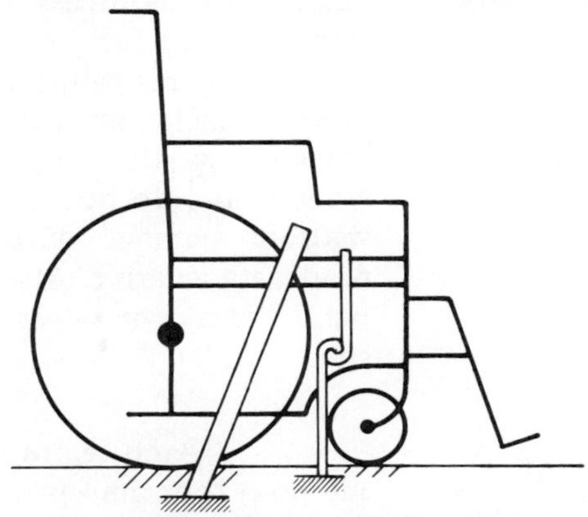

Occupant lap belt and vertically-acting over-center toggle on each side

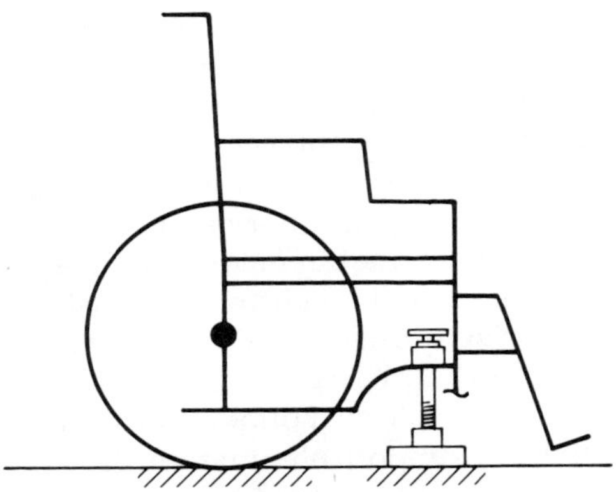

Lateral bar across lower frame members, attached with a single vertical wing-bolt to floor fitting

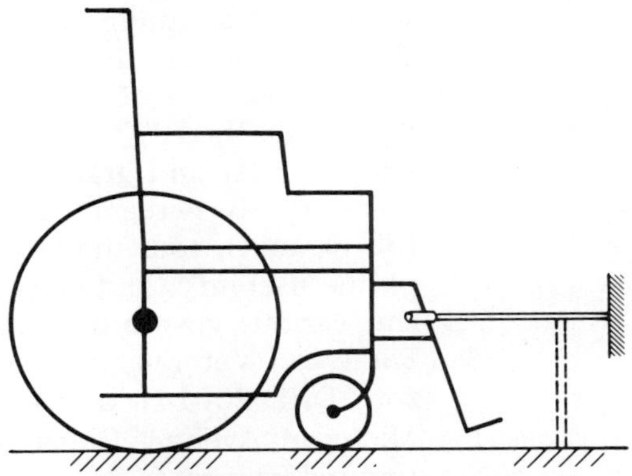

Dash, engine cover, firewall, or floor-mounted clamp, attaches to one footrest frame

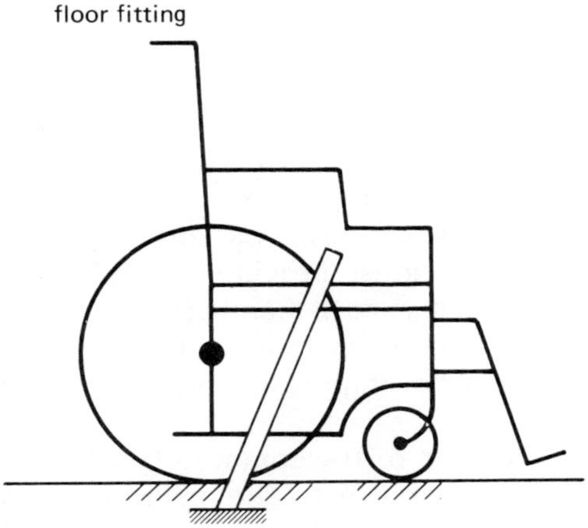

Occupant lap belt only

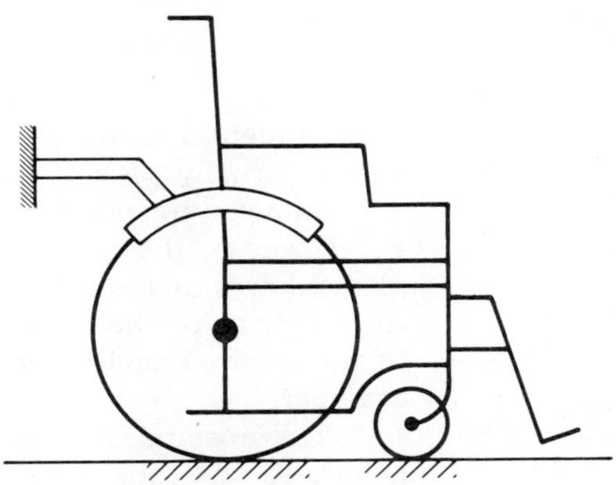

Tightly-fitting "fender" over one or two wheels

Figure 7.14 Typical restraint types.

belts, dash-mounted clips that attach to the forward wheelchair frame, etc.

These restraints are designed under the assumption that "standard" wheelchairs will be used by the van driver or passengers. Basic research studies (Rider et al., 1976) have shown that "standard" wheelchairs do not have the inherent design strength to be restrained without damage during an automobile accident of even low to moderate severity. Also, these studies have shown that most restraints bend, break, or release during crash loading of less than 5 *g* (a low-speed crash).

Architectural barriers. These barriers take many forms and all pose the same problem: preventing (or delaying) the wheelchair-bound person from entering buildings, doors, rooms, sidewalks, etc. Barrier-free living has become a subject of considerable activity in the wheelchair community in recent years and with much success. National, state, and other governmental agencies have enacted statutes for the purpose of eliminating barriers to mobility. The ANSI standard (ANSI, 1971) gives guidance on such subjects as: sidewalk width (4 ft) and grade (5 percent), parking lot accessibility, ramp grade (8.33 percent) and other dimensions, door-opening width (32 in. min), stair dimensions, toilet-room requirements, public telephone usability, and elevator requirements. In addition to ANSI and the various government agencies, private organizations such as the Paralyzed Veterans of America, Inc. (7315 Wisconsin Ave., Washington, D.C. 20014), and the National Paraplegia Foundation (333 N. Michigan Avenue, Chicago, IL 60601) support elimination of architectural barriers.

7.8 Environmental Factors

Environmental concerns affect the hospital in two ways: internally and externally. First, the hospital requires a specialized internal environment to permit effective treatment and patient recovery. Furthermore, the internal environment must provide healthful and pleasant working conditions for the many staff members who spend the majority of their working lives in that environment. Both aspects of the internal environment are of special importance to the clinical engineer.

The hospital, however, is a member of the community where it is located and thus is a significant contributor to the environmental load of that community. Second, then, the clinical engineer must consider such external environmental factors as increased traffic, sources and treatment of water and other utilities, sewage treatment and disposal, solid waste disposal, and possible air pollution due to power plants, incineration, and chemicals used in hospital operations.

This section, then, deals with the many aspects of the hospital internal/external environment. It is impossible, of course, to provide a total treatment of this broad topic in these few pages. Here, the goal is to touch on each major area, to provide an overview for the clinical engineer, and to motivate consideration of these topics supported by study in more complete sources.

Waste disposal

Waste of all types produced by hospitals is increasing at an alarming rate. One estimate (Jacobsen, 1969) suggests that the average hospital generates 15 pounds of solid waste per day per bed or a ton and a half of waste per day for a 200-bed hospital. Much of this waste poses significant and specialized problems for the clinical engineer. Many patient-treatment items are disposable (e.g., syringes, trays, meal-service items, etc.) for reasons of hygiene and convenience. These items present a substantial problem when disposal must be undertaken.

In addition to the volume of hospital wastes, certain items pose a health or safety hazard and must be specially handled. Radiological and infectious wastes represent obvious hazards. Radiological wastes are specifically discussed in Sec. 7.10. These wastes clearly require specialized handling which must be carefully engineered and maintained on a controlled basis. Infectious wastes are deserving of similarly careful control measures. Most hospitals have established programs for such wastes, including sealed disposal containers with final disposal by incineration. In all cases, the clinical engineer should study such disposal schemes to ensure that those practices are adequate.

The majority of hospital wastes are bulk materials which must be incorporated into the waste stream of the community. Many hospitals rely on unprocessed bulk systems for disposal (e.g., "Dempster Dumpster") which do not modify the state of the waste. Others use incineration which reduces solid waste volume significantly but with possible air-pollution problems. Wet grinding has become more widely used in hospitals (Weintraub and Kenn, 1970) with disposal into the sanitary sewer system. At the same time, compaction (Jacobsen, 1969) appears as a valuable method to reduce the volume of solid waste prior to disposal in a sanitary landfill. In either case of grinding or compaction, incineration is required for disposal of pathological specimens.

Liquid and airborne effluents are produced by all hospital facilities. Recent legislation and the general environmental concerns of the nation dictate standards concerning these disposal routes. Federal standards as well as state and local requirements establish the acceptable limits of contamination. The role of the clinical engineer, in partnership with hospital management, is to ensure compliance with these standards.

Noise

Environmental noise pollution, sometimes referred to as "ear pollution" is ubiquitous. The hospital harbors significant noise sources, and in some cases, the health and recovery of patients may be affected by exposure to noise. In all cases, patients and hospital staff are subject to annoyance, distraction, and disruption of communication due to noise exposure.

Noise measurement is complicated by units of measure used to quantify exposures. The decibel (dB), coined in honor of Alexander Graham Bell, is used to describe sound pressure levels and is defined as:

$$\text{dB} = 10 \log \left(\frac{P}{P_0}\right)^2$$

where P is the measured sound pressure (N/m^2) and P_0 is a reference sound pressure (2×10^{-5} N/m^2) accepted as the minimum threshold of human hearing. Thus, the dB is a dimensionless ratio distributed over a logarithmic range from 0 dB (threshold of hearing) to 140 dB (threshold of pain). Importantly, the numerical spread of these ranges (0–140 dB) is actually a range of seven orders of magnitude (10^7) in absolute sound pressures. Thus slight differences in dB readings on a sound-level meter imply large changes in sound pressure impinging on the ear (e.g., a 6-dB increase in sound pressure level is a doubling of the absolute sound pressure).

Sound pressure level (SPL) measurement is further complicated by the adoption of A weighting. This weighting corrects measured SPL values to correspond to the actual frequency response of the human ear. At low frequencies, for example, the A weighting subtracts up to 40 dB from the observed SPL, whereas at high frequencies (near 4 kHz) the weighting causes an addition to the observed SPL. The ear is relatively unresponsive to low frequencies while resonances in the hearing mechanism cause the ear to be far more sensitive (and subject to damage) from higher frequencies. SPL measurements and exposure standards are normally stated as dB(A) values.

Noise levels in hospitals may be expected to fall in the range of 50 to 85 dB(A). A recent study (Falk and Woods, 1973) revealed an average of 57 dB(A) in recovery rooms and acute-care units with peaks to 86 dB(A). This same study showed noise in incubators in the range of 58–61 dB(A). These values do not appear excessive when compared with industrial exposures which commonly fall in the range of 90 to 100 dB(A). The current standard for industrial exposures (OSHA) is established as 90 dB(A) continuous exposure for an 8-hour workday. At present, however, there is some consideration by federal agencies to reduce this standard to 85 dB(A) for an 8-hour workday.

Noise exposures in hospitals are related to other influences not found in the industrial environment. There is some evidence that noise has an adverse effect on the pituitary-adrenal axis (Henkin and Knigge, 1963) as well as the cardiovascular system (Geber, 1967). Beyond this, however, there is evidence of deleterious effects of noise in the presence of certain drugs. One study demonstrated that patients under treatment with aminoglycosidic antibiotics are exposed to higher risk of hearing loss in the presence of otherwise equivalent noise exposures (Falk, 1972). In all cases, excessive noise is an unpleasant stress on recovering patients and certainly interferes with the rest that is required in the recovery process.

No specific standards for noise exposures in hospitals presently exist. The American Academy of Ophthalmology and Otolaryngology (Committee on Conservation of Hearing, 1969) recommends 80 dB(A) as the upper limit for no loss of hearing due to noise exposure. Certainly, levels below 70 dB(A) are preferred, considering the essential benefit of quiet in any recovery process.

The clinical engineer should measure noise levels in his area of responsibility on a periodic basis to reduce patient exposures to harmful and annoying sound levels. A type-2 sound-level meter (e.g., GR Model 1565) is compact and suitable for survey-type studies. These meters are reasonably accurate (±3 dB) and can reveal overall sound pressure levels of noise sources so that the clinical engineer can make simple suggestions to reduce noise (e.g., identification and positioning of noisy equipment, identification of high-noise areas in the hospital, evaluation of exposures of patients unable to communicate overexposure including the aged and newborn). An excellent reference for basic measurements and surveys may be found in the General Radio Handbook of Noise Measurements (Peterson and Gross, 1974) which is readily available and inexpensive.

Water quality

Water is a necessary utility which the hospital must obtain from outside sources. This utility is of special interest to the clinical engineer since it may serve as a medium for the spread of infection from outside sources to the hospital as well as the spread of infection within the hospital. Water of acceptable microbiological quality is essential not only for drinking purposes but also for solutions used in patient treatment and laboratory procedures.

Water typically is obtained from public systems although some hospitals have private, internal water systems. In either case, the hospital engineering staff is responsible for ensuring the quality and treatment of the water before it is delivered to patient-treatment areas. Typically, a complete chemical analysis of water is required to establish quality levels. These analyses may be performed by local public health agencies or by hospital staff.

The typical treatment of the hospital water supply includes deionization, distillation, and sterilization. In addition, boiler fed water must be chemically treated although this system is separate from the potable water system. Deionization is a routine process which the clinical engineer should check occasionally for potential hazards. Deionization, of course, does not remove all inorganics or organics or bacteria. Bacteria are a special problem since the resin beds used for deionization provide an ideal growth medium for these organisms. Distillation provides a purer form of water although it is not a sterile supply. Distillation will remove inorganics and most organics but bacteria may be inadvertently reintroduced by a final filtration process. The reverse osmosis method which is receiving wider acceptance appears to be a useful method for providing deionized, distilled water in hospitals.

Sterilization of water provides the most acceptable form for treatment purposes. Since sterilization is achieved by processing in autoclaves, the supply is limited to the capacities of these units. The quantities produced in most hospitals are sufficient for treatment purposes. As with any sterile substance, handling and storage can eliminate the sterile properties of water.

The hospital must have an uninterrupted supply of water for medical treatment and other uses. The specialized treatment of water demands the serious attention of the clinical engineer. Even plumbing in hospitals is specialized (National Standard Plumbing Code ASA–A–40) and requires unique designs. Most hospitals, for example, use a "looped" system which permits a continuous flow of water in the event one source of supply is disrupted. Backflow and backsiphonage are especially serious threats that can spread infection in the hospital. Fixed air gaps, backflow preventers, and booster pumps are typically specified to reduce or eliminate these hazards.

General building safety

Introduction. The physical structure of the hospital may pose hazards to the patient and worker alike. In this sense, a considerable amount of information can be learned from industrial experience. The National Safety Council (1976) lists the following most common causes of compensable work injuries in industry as a whole:

1. Manual material handling (lifting, etc.).
2. Falls.
3. Struck by falling, moving objects.

The clinical engineer can readily translate these industrial causes as likely areas for injury in the hospital environment. Lifting of patients and other heavy objects are likely causes of staff injury while patients obviously should not be permitted to lift objects of any substantial size or weight during the recovery process. The only approach to avoiding injuries due to lifting is common sense and an effective education/training program.

Falls or physical injury due to collision of mobile equipment are not an uncommon accident cause in the hospital. Modern hospitals require considerable hallway traffic with patients transported to and from treatment and diagnostic facilities as well as treatment devices transported to patients. All of this activity threatens staff and patients who use the hallways as routes of travel. Accidents and injury due to falls and collision can be minimized through a simple prevention program including:

1. Establishing a traffic flow pattern.
2. Ensuring clean, dry, unobstructed, and slip-resistant surfaces.
3. Maintaining railings and stair surfaces in good condition.
4. Supervising patients to minimize unnecessary travel.

The general area of hospital safety is well covered in the publication *Safety Guide for Health Care Institutions* published jointly by the American Hospital Association and the National Safety Council (1972). This book contains useful checklists which should help the clinical engineer to insure safe practices throughout the hospital.

Gas cylinder safety. Gas cylinders and distribution systems are usually considered an integral part of the hospital structure. Gases are a vital element in patient treatment as well as laboratory procedures. Most hospitals have a permanently installed gas distribution system which provides for orderly distribution of gases to the many points of use throughout the building. In these systems, gas cylinders may be stored in a centralized location separate and apart from other hospital functions. With a single-point gas cylinder facility, a maximum of control over storage, gas delivery, and cylinder disposal is possible. When the gas cylinders are distributed widely throughout the hospital, a standardized program of cylinder safety is essential. Any gas cylinder should be accorded the following elements of care (Weiss, 1976):

1. Identify cylinders by written name in addition to color coding.
2. Secure cylinders to wall or within enclosures.
3. Cap all cylinders not in immediate use.
4. Protect cylinders from temperature extremes.
5. Never use oil or grease on oxygen or nitrous oxide equipment.
6. Never permit smoking or use flames near oxygen or nitrous oxide equipment.
7. Check gas regulators and do not use a regulator for oxygen or nitrous oxide after it has been used with any other gas.

Nonionizing radiation hazards. Nonionizing radiation is a significant health hazard in all hospitals. This type of radiation includes ultraviolet, microwave, and laser radiation. In most cases, the eye is the primary organ at risk although other organs may be affected by some forms of nonionizing radiation at certain intensity levels.

Ultraviolet radiation, frequently used in sterilization procedures, produces primarily erythema (reddening) of the skin. The effect is most common, of course, in the painful sunburn produced by overexposure to solar radiation. In certain cases, UV radiation can cause keratitis and skin carcinogenesis. UV exposures are best controlled by limiting exposures as a function of energy (J/em^2).

Microwave radiation is commonly used in hospital diathermy treatment; and, of course, it is widespread in microwave oven applications. The biological effects of this radiation are largely thermal with most hazard to the eye (cataractogenesis), although some Soviet investigators claim effects in the central nervous system (International Commission on Radiation Units, 1977). The thermal effect of microwave radiation, of course, is the desired therapy in microtherapy and diathermy treatment. As with UV radiation, microwave radiation is controlled by limiting exposure, and microwave sources should be periodically surveyed with measurement equipment to ensure exposures are below predetermined allowable levels.

Lasers are receiving an increasing role in medical treatment. The very organ most commonly treated, the eye, is the organ most vulnerable to damage from laser exposures. Lasers are especially hazardous due to the long potential travel of the beam and the possible reflections of the beam from surfaces in rooms where the beam is operated. Eyewear is the most common method of protection although a careful program of area restriction is advisable to prevent inadvertent injury to unwary persons.

Skin burns from surgical preparation. Agents used in the preparation of the patient's skin before surgery have two purposes: mechanical cleansing and sterilization. Freon TF, which has been classified by Underwriter's Laboratories, Inc., as nonflammable, is now used in place of ether as a defatting agent. It has been shown (Klein and Leape, 1976) that Freon, when combined with 70 percent isopropyl alcohol, produces a chemical irritative effect on the skin which is enhanced markedly by pressure and mild heat. Therefore, if care is not taken to prevent Freon and alcohol from collecting under the patient following surgical prep, skin burns can occur. Should a burn occur in surgery, the clinical engineer will undoubtedly be called to investigate. Since electrical equipment, such as electrosurgery, the warming blanket, or monitoring equipment is always first implicated by the surgical team, it is important for the engineer to be aware of other causes of burns.

7.9 Infection Control

Infection control in hospitals has improved dramatically in recent years. Control has been achieved by recognizing the means of growth, reproduction, and transmission of pathogenic microorganisms. Microorganisms, of course, are essential to human food production and bodily functions. Under certain conditions, however, these microorganisms pose a serious threat to debilitated patients in a susceptible state.

Nosocomial (hospital-acquired) infections are spread by people, food, fomites (bedding, clothes, etc.), animals, and arthropods. Recognizing and controlling these sources is the key to infection control. The common forms of infection control include eliminating or controlling the reservoirs, killing by sterilization, reducing numbers of organisms by disinfection or sanitation, creating an unfavorable environment for growth, and disturbing or eliminating the means of transmission.

The clinical engineer is an important member of the infection control team. In this section, some aspects of his likely duties are outlined. For an infection control program to be effective it must be composed of three key steps: recognition, evaluation, and control. First, the sources and types of pathogenic microorganisms must be known; next an instrumented evaluation must be established to determine the quantity, type, and source of microorganisms at a given time and place in the hospital; and finally, a control program must be established to reduce or eliminate the identified sources and their paths of transmission to the site of infection.

Modes of disease transfer

All transmittable diseases start with an infected host. In order for the disease to spread to a new host, a certain chain of events must occur. Interruption of the chain at any point will stop the transmitting process. Thus, once it has been determined that an infection problem exists, an effort must be made to determine the pathway from one host to another. Although no two diseases are identical, and the variations in transfer mechanisms are endless, certain similarities do exist. An understanding of these similarities can be helpful to the clinical engineer who might be called in to help investigate infection outbreaks in the hospital or clinic.

Figure 7.15 summarizes the disease transmission chain. First, there must be a reservoir which is an infected host who may be either symptomatic or asymptomatic. Next in the chain is the primary vehicle, which is anything the host puts out from his body. Spread of the infection from the primary vehicle to the new host can take place in three modes. The most obvious mode is direct contact between the new host and the primary vehicle. However, transmission can also occur by a secondary vehicle or a vector. Examples of secondary vehicles are flies, food, fluids, dust particles, and fingers. Vectors are intermediate hosts that become infected with the disease, but do not actually develop the disease symptoms. Finally, there must be a portal of entry by which the infection enters the new host.

As an example of this transmission chain, let us examine the spread of infectious hepatitis, a disease that attacks the liver (USAF School of Aerospace Medicine, 1972). The cause of infectious hepatitus is a virus about which little is known. There are two types of viruses. The SH virus is present only in the blood and never in the stool, while the IH virus is present in blood and stools (primary vehicles). Therefore, the SH virus is only spread by means of plasma, other blood products, or inadequately sterilized syringes, and the IH virus is spread through contaminated water and food (secondary vehicles). As a result, the portal of entry is either the mouth or blood stream. The reservoir for infectious hepatitis is a human carrier who is free from symptoms; there are no known vectors or intermediate hosts. It is possible to interrupt the infectious hepatitis transmission chain by assuring that water supplies are free of contamination, properly sterilizing needles and syringes, and avoiding multiple injections with the same needle. In a few special instances, the use of a gamma globulin immunization provides some protection against the disease caused by the IH virus.

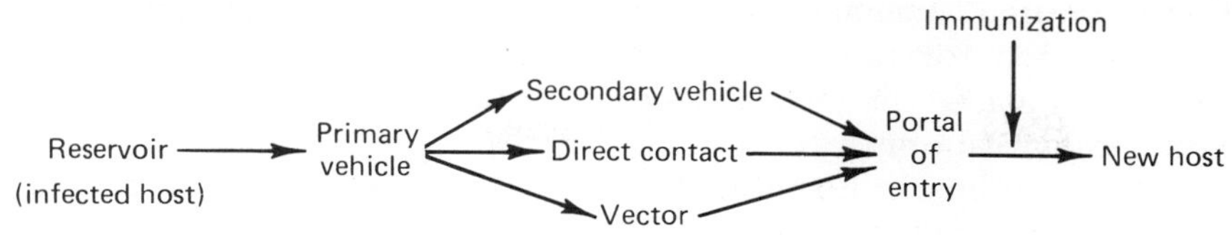

Figure 7.15 Transmission of disease chain.

Microbial sampling programs

Most airborne and surface contamination settles by gravity to floors where it is transferred from room to room by shoes or mobile equipment. There are no standards at present for the microbiological cleanliness of floors and thus it is not possible to evaluate surfaces as part of a program of routine sampling. Random sampling programs, however, are useful for establishing a record of microorganism counts for reference. In the event of an epidemic, sampling must be immediately and extensively established to locate and eliminate sources and routes of contamination.

Some microorganisms are water borne (e.g., salmonella, shigella) and periodic sampling of water systems must be performed by hospital staff or an outside health agency. Samples may be obtained from water taps, drinking fountains, and other points in the general hospital water system. It is important, of course, to ensure the use of sterile containers and procedures in obtaining and transmitting the sample to the analysis laboratory. In addition, sterilized water supplies should be occasionally sampled to determine the presence of live pathogenic microorganisms. As with all microbial samples, testing is normally conducted in the microbiological laboratory.

Surface sampling should be conducted on a periodic basis. The American Hospital Association (1968) recommends a minimum of 10 samples from any surface, e.g., floors, apparatus and equipment including furniture, masks, tubes, and solutions. A sampling scheme could be established, for example, to evaluate housekeeping methods with 10 samples before and 10 samples after cleaning. A much larger number of samples, say 60, could increase statistical confidence but also increases the cost and effort of collecting and evaluating the results.

Surface samples are normally taken with cotton swabs moistened in sterile water. The swabbing technique is not standardized although some effort should be made to standardize this procedure among individuals collecting samples to attempt to achieve some consistency in method. In some cases, a large swabbed area may be desired to obtain a quantity of microorganisms, while if reproducibility of results is desired, a smaller, more consistent area for sampling should be established for each sample obtained. After sampling, the swabs should be placed in closed, sterile containers for delivery to the microbiology laboratory.

Air sampling in hospitals is possibly the most common type of test for pathogenic organisms. This may be due to the fact that airborne microorganisms are most likely to threaten the patient by entry through the respiratory system except in areas where wounds are open to the atmosphere. Furthermore, air sampling is a far more convenient means of evaluation and analysis than surface samples where the surface is difficult to swab such as blankets or bedding.

Air sampling generally may be divided into three groups: (1) liquid medium entrapment (impingement), (2) surface entrapment, and (3) filter entrapment. All three methods may be used in a given hospital although surface entrapment (slit sampler) appears as the most widely used. In all cases of air sampling, it is necessary to know the air volume drawn through the sample during the sampling period so that the final microorganism count can be stated as a function of volume of air (e.g., colonies per cubic meter). Liquid impingement methods involve bubbling contaminated air through a liquid. This approach may be undesirable since it breaks down clumps of bacteria by a washing action and causes a higher count than might be obtained by surface methods. Filtration methods (e.g., membrane filters) may be used but this approach may also cause problems in obtaining particle growth or when washing particles from the filter.

Solid-media-type filters appear to be the most widely used air samples in hospitals today. These samples are of two types: sieve or slit-type samplers. Slit samplers are available from several manufacturers (e.g. Casella) and collect microorganisms through a slit opening. The collected particles are uniformly deposited in a capture substance as a dish rotates slowly under the slit for several revolutions during the sampling period. A waxy, agar substance is placed in the dish to capture microorganisms and maintain them in a viable state. Sieve-type samplers cause the captured flow of air and suspended particles to pass through succeeding stages of sieves of smaller and smaller sizes. Particles are thus sized as they are captured on dishes (with agar surface) under each sieve stage. Thus, with the sieve samplers, total bacteria concentration is obtained with an additional particle size distribution.

Sampling programs, regardless of method, should be established on a random basis. A series should be taken throughout the day, for example, in a surgical suite, delivery room, and patient room. The selection of specific room, as well as time of day, should be randomly determined. Further detailed information on microbial sampling may be obtained from the National Communicable Disease Center, U.S. Public Health Service, Atlanta, Georgia 30333.

Sterilization techniques

Sterilization is a key step in establishing a program of infection control. It is important to distinguish sterilization from disinfection. The process of disinfection may reduce the numbers of potentially hazardous microorganisms but sterilization is intended to totally kill all microorganisms and bacterial spores. All objects used in surgical procedures must be sterilized including linens, paper, nylon, glass, and metal objects. Normally, objects may be presterilized and packaged for later use although all such items have a limited shelf life (usually up to one year).

Sterilization is achieved through three principal methods: heat, chemical, or radiation sterilization. Heat sterilization, either by steam or dry heat, seeks to raise the temperature sufficiently to kill all microorganisms on the objects processed. The temperature and time required are functions of the process employed: autoclaving (steam) requires less time and temperature but requires an evacuated chamber; dry heat methods require longer sterilization periods and higher temperatures but operate at ambient pressures. Some sharp instruments are sensitive to high temperatures and thus dry heat should not be used for these items. The autoclave provides the best all-around performance in heat sterilization but the device must be checked periodically by use of test strips (spore tests), regular maintenance, and mechanical testing, to ensure it is operating properly.

Chemical sterilization (liquid or gas) offers the advantage of avoiding the need for high heat in sterilizing sensitive objects. Compounds such as ethylene oxide, formalin, or glutaric aldehyde may be used for these purposes. Most of these substances must be handled carefully since they may represent a health hazard to users (poisonous, flammable, or other hazard). Such devices must be carefully monitored to minimize hazards to hospital staff and patients both during sterilization and during storage when some remaining trace quantities of gas or liquid may leak from the objects into the surrounding environment.

Radiation sterilization is a promising method of sterilization with a capacity for high throughput. Although irradiation is typically a lengthy process (24 hours) it can be designed into a conveyor-type system which provide continuous flow of objects through the sterilizer. Radiation sterilization offers a clean, dry process which ensures full exposure of objects since they are irradiated from all directions in the process. Ionizing radiation, of course, poses a threat to humans and such systems must be carefully operated by qualified personnel.

Sterile field techniques

The surgical suite, because of higher threat of infection, requires specialized care to minimize risk to patients in a debilitated state. In some cases, such infections can result in death of a patient who might otherwise survive, or the loss of a limb which has been painstakingly reconnected. In all cases, the highest source of microorganisms in the surgical suite is shedding from members of the surgical team. Normal scrubdown procedures for walls, floors, and other surfaces before and after surgery minimize the survival of microorganisms on these surfaces. All the sterile field techniques adopted in the surgical suite are directed at controlling the reservoir of microorganisms introduced by the entering surgical team as well as the patient himself.

The surgical suite is normally divided into zones of increasing antisepsis. The outermost area is devoted to scrub rooms, changing areas, and patient preparation areas while within the operating room a boundary is maintained between sterile and unsterile areas. The peripheral members of the surgical team remain in the unsterile area away from the patient (e.g., anesthetists, orderlies, etc.) while the surgical team maintains the highest level of sterility. Even in the unsterile area, sterile masks, caps, and gowns are worn to prevent contamination of the sterile field by airborne particles from the skin as well as droplets from the nose and mouth. The surgical team who are directly working in the sterile field wear sterile gowns, caps, and gloves and for some procedures will double some garments (e.g., two gowns, two pairs of gloves, etc.) to further minimize risks. A sterile field is created by draping the patient with sterile drapes open only in the area of the proposed incision. During the surgical procedures, sterile instruments and other materials are rigidly protected from contamination, and if contamination is suspected, these instruments are segregated from uncontaminated items. In some extremely sensitive surgical procedures, the surgical team may wear space-suit type garments with a separate supply of breathing air to totally isolate them from the sterile field although these garments are the exception rather than the rule in present-day hospitals.

In many modern hospitals, paper gowns and other garments have been adopted in preference to linen items. This trend is partly due to the easy disposal and one-time use of these garments. Beyond this, however, paper garments pose far less hazard from the airborne fibers that are released from laundered linen items. Fibrous particles, of course, may be a vehicle for potentially hazardous microorganisms in the operating room air.

7.10 Radiation Safety

The general topic of regulations has been treated in Sec. 4.8. This section addresses the implementation of a safety program to meet those regulations.

External sources. Therapeutic radiation sources are generally calibrated and verified for proper operation by either a staff or consulting medical radiation physicist, since the efficacy of the treatment is critically dependent on the delivery of the proper radiation dose. However, if verification of proper operation on a daily basis is the responsibility of the clinical engineer, the following areas should be considered.

The equipment manuals for cobalt 60 units should be studied to determine the proper method to manually close the shutter (terminate treatment) in case of power failure or equipment malfunction. Several different shutter designs are shown in Fig. 7.16.

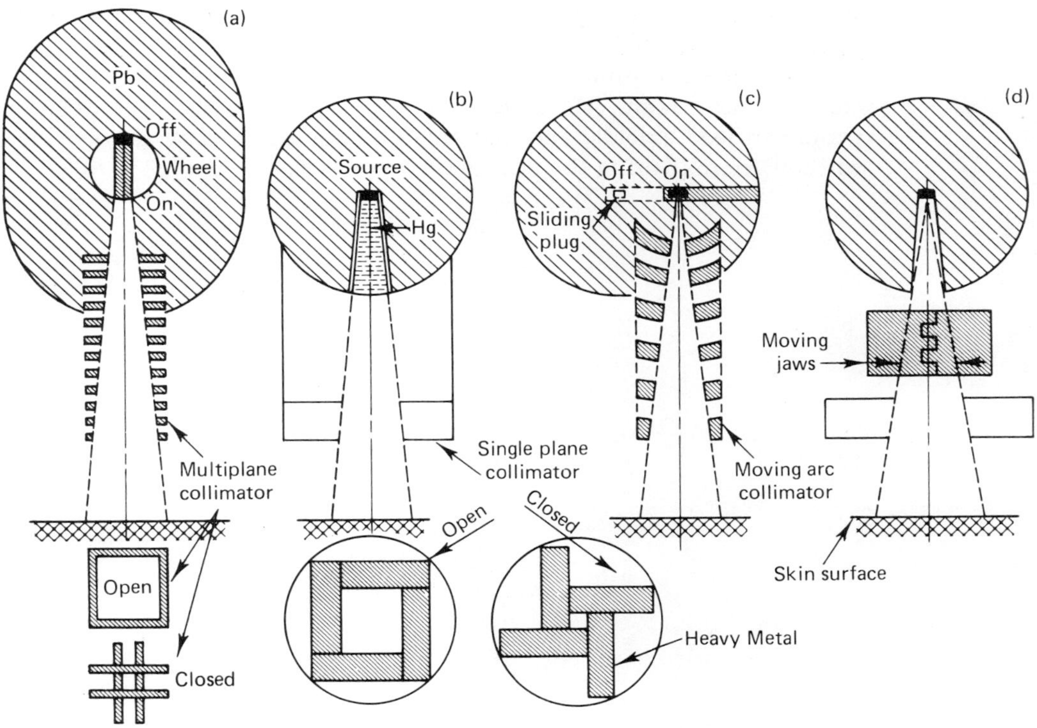

Figure 7.16 Schematic representation of 4 different types of isotope machines. (a) Unit using rotating wheels to carry the source from the "off" to the "on" position and a multiplane collimator to control the size of the beam. (b) Mercury type shutter with single plane arc collimator. (c) Sliding plug-type mechanism with moving arc collimator. (d) Moving jaw mechamism and single plane collimator. From H. E. Johns and J. R. Cunningham, The physics of radiology, 1969. Courtesy of Charles C. Thomas, Publisher, Springfield, Illinois.

Periodic dose checks can be accomplished using LiF solid-state dosimeters supplied thru the mail by the Radiation Physics Center, M. D. Anderson Hospital, Houston, Texas 77025. The cost of the service is nominal and it provides a third-party check on system calibration. An excellent practical manual on calibration of radiotherapy sources is available from the International Atomic Energy Agency (Massey, 1970).

Internal sources. Radiation therapy from permanently implanted radioactive seeds (class C sources), or brachytherapy with cesium or radium removal sources (class D sources) necessitates protection for other patients, nursing staff, and visitors during the therapy period. Generally the patient is placed in a private room and a lead shield, similar in design to a blackboard on a stand, is placed beside the bed to shield the nursing staff attending the patient from stray radiation. A typical gynecological holder for brachytherapy sources is illustrated in Fig. 7.17.

To prevent unnecessary exposure to other patients, two precautions are taken. All sources are transported within the hospital proper in lead shields, and, patients undergoing therapy are confined to their rooms during the course of treatment.

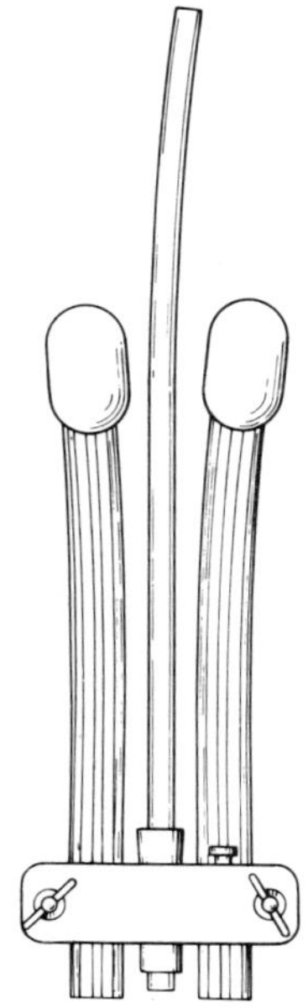

Figure 7.17 Brachytherapy applicator assembly. Sources are loaded in the tubes through the base which remains outside the patient.

Freeman, Stoner, Smathers, Clapp, and Duncan

Nursing care. Only those patients undergoing internal source therapy require any special precautions. The nursing staff should be advised to stand behind any shielding available when possible and remain several feet from the bedside when just conversing with the patient. All soiled linens, bed pans, dressings, waste receptacles, etc. should be checked for the presence of a radioactive source prior to removal from the room.

Patient discharge. When removable brachytherapy sources are used, the patient and his bed area should be checked after completion of treatment for any sources which might have become separated from the holder.

Diagnostic radiology

Diagnostic X-ray machines. The safety considerations involved in diagnostic radiology are primarily those of proper shielding of the X-ray tube and protection of the X-ray technician from multiple exposures to the low radiation levels. Figure 7.18, a sectional diagram of a simplified X-ray tube, illustrates with a heavy black line the integral shield about the tube which limits the solid angle of X-ray emission to that allowed by the thin window in the shield. The metal shield is fabricated from several pieces and must be periodically checked for leakage of X-rays through cracks at the joints. Film or

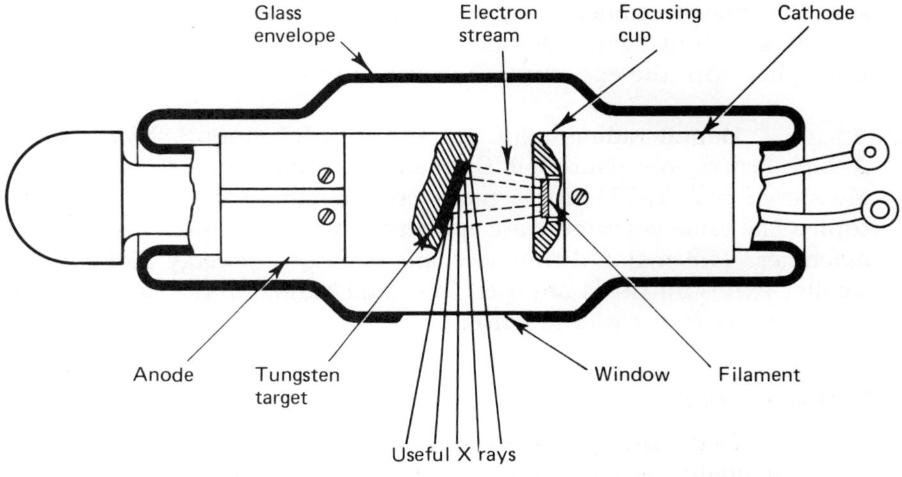

Figure 7.18 A simplified X-ray tube with a stationary anode and a heat filament. From W. Bloom, J. Hollenbach, and J. Morgan, *Medical radiographic technic, 3rd ed.,* 1965. Courtesy of Charles C. Thomas, Publisher, Springfield, Illinois.

LiF dosimeters placed over the joints and left there for a series of exposures is one method of checking for radiation leakage.

To protect the technician against multiple exposures to low levels of scattered radiation the control system is positioned such that during system operation the X-ray technician is shielded against scattered radiation. Generally 3 to 6 mm of lead is sufficient shielding and a lead glass window in the shield wall allows visualization of the patient during the procedure. Under no conditions should extensions of the controls be installed which would allow the technician to stand outside of the shielded area and operate the X-ray system. The walls of the room at which the X-ray machine may be directed should offer sufficient shielding such that there is no exposure to personnel in adjoining offices to transmitted X rays. NCRP Report #49 (National Council on Radiation Protection and Measurements, 1977) treats this topic in detail and should be consulted for detailed information.

Fluoroscopy and angiography. Generally the greatest exposure to personnel involved in diagnostic radiology occurs during fluoroscopy, angiography, or other special studies. As the procedures require that some staff members remain in the room during the examination, the safety effort is directed toward minimizing the staff exposure. Lead drapes, protective lead panels, lead impregnated aprons, gloves, etc. should be used whenever appropriate. Personnel in the room should be restricted to only those absolutely essential for the examination and they should place themselves behind the lead drapes and stand back from the examination area whenever possible. To prevent unnecessary exposure to the patient and the staff a "deadman" foot switch should be used for the X-ray tube's operation. This stops the exposure whenever the foot is removed.

Dental radiography. Refer to NCRP Report #35, *Dental X-Ray Protection* (National Council on Radiation Protection and Measurements, 1977), for a more comprehensive treatment of the topic. The same comments are applicable here as for diagnostic X-ray machines. The major differences are that lower-energy X rays and smaller exposure fields are used in dental radiography. The shielding problems are thus easier to solve.

Nuclear medicine

Diagnostic procedures. The hazards in the administration of small quantities of radioactive material to a patient as a gas or a liquid are quite minimal. In the case of gases we should be certain that the expired radioactive gases are exhausted through a duct system which terminates at the top of the building and at a safe distance from the building air intake. The exhaust duct system should also be shielded during its passage through the building to minimize staff exposure.

For liquid sources, the administration materials that come in contact with the isotope, such as syringes, should be disposed of separately as radioactive waste. The patient's urine or fecal material however, is disposed of through the sanitary sewer system and no concern need exist about the patient's perspiration.

Therapeutic procedures. For class A sources, those administered by mouth, the precautions vary depending on the amount administered. Assuming a maximum amount has been administered, as for control of cancer of the thyroid, the patient can be a hazardous source of radiation to others and should be confined to his room. The urine may be collected for the first 24 to 48 hours and if collected, it should be disposed of as a radioactive waste. If not collected, normal sanitary facilities may be used for disposal.

Bed linens and other soiled materials should be monitored for radioactive contamination prior to removal from the room. If contaminated they should be sealed in plastic and segregated from the normal laundry. Subsequent treatment would depend on the half life of the radioisotope but could range from temporary storage to allow for decay of the radioactive isotope to disposal of the material as radioactive waste.

Class B sources, radioactive materials injected into body cavities, can pose more of a material contamination problem since this normally occurs as a postoperative procedure. There are consequently dressings to change and soiled bed linens which quite often are contaminated with low levels of radioactive material. Along with the materials used to cleanse the surgical area, these items must be segregated and treated as radioactive waste unless proven otherwise by a careful radioactive survey.

Patient excrement may be disposed of through the sanitary sewer system; however, the container should not leave the room for cleaning or use elsewhere without monitoring for radioactive contamination.

To minimize staff exposure, the patients should be encouraged to care for themselves to the extent their medical condition will allow. The room and the patient's chart should be posted with notices that the patient has received a radioactive isotope and the necessary precautions are to be observed.

For a more detailed discussion of this topic the reader is refered to NCRP Reports #48 and 37 (National Council on Radiation Protection and Measurements, 1977).

Radiopharmaceutical laboratories. If radioactive procedures are used within the hospital, the assumption is made that the hospital pharmacist or one of his staff have received training in the proper handling of radioactive isotopes. Details on proper isotope handling techniques within the pharmacy are thereby omitted. Refer to Chase and Rabinowitz (1967) for additional information in this area.

The monitoring of isotope containers that are to leave the pharmacy for surface contamination which could be removed on handling the vial requires that the container be wiped with a damp piece of filter paper or gauze pad and this wipe then be monitored for radioactivity at some distance from the vial. This prevents source radiation which penetrates through the shield from affecting the wipe measurement.

Waste disposal was treated in more detail in Sec. 7.8. However, the guideline for the radiopharmaceutical laboratory is that all radioactive waste, liquid and solid, must be collected for proper disposal.

Nursing care. Without jeopardizing proper nursing care, there are simple precautions that can minimize the radiation exposure of the medical staff. Where appropriate, bedside shields should be used and the staff directed to stand behind them when possible. Simple conversations can be held with the staff member 6 to 8 feet from the bed, thus minimizing exposure. Pregnant staff members should not care for patients undergoing radioisotope treatment. The staff should be advised that total radiation dose received is a function of time, distance, and shielding. More shielding, greater distances, and shorter times all tend to minimize radiation exposure.

Patient discharge. Guidelines for appropriate radioactive levels that may exist in the patient at the time of discharge are given in NCRP Report #37 (National Council on Radiation and Measurements, 1977). The recommended levels vary considerably from one radioisotope to another and space does not permit specific numbers to be given here. Note that this situation is one of many in which the published guidelines have the effect of regulations or accepted codes of good practice.

Morgue. In the case of bodies that contain sealed sources, class C or D sources, the sources should be removed from the body, returned to the isotope storage area for cleaning, and disposal or reuse as appropriate. After complete source removal, a radiation hazard no longer exists.

For those bodies that contain solutions or dispersions of radioactive material, class A or B sources, the question arises as to whether an autopsy is to be performed or not. If performed, it should be done within the hospital with all fluids and sections removed considered radioactive.

Fluids may be discharged to the drain with copious quantities of water used to flush them down. Samples retained for subsequent analysis should be stored in shielded containers with the amount of shielding dependent on the radioactive levels present within the samples.

The body itself may be released to an undertaker for burial. However, if it is to be embalmed this should occur in the autopsy room and immediately after autopsy if one was performed. Again all body fluids removed may be discharged down the drain.

All instruments, materials, and clothing that come in contact with the body shall be considered radioactive and treated accordingly. After a separate thorough washing, instruments should be monitored and if free from contamination returned to normal service. After pathological study, all specimens should be disposed of as radioactive waste.

Staff members should be monitored at the exit of the autopsy room for any contamination, i.e. hands, soles of shoes, front of clothing. All contaminated materials remain within the room until properly packaged for safe transport.

Once embalmed, the body requires no further special treatment as prolonged exposure at close distances to mortuary staff members is quite unlikely.

Waste disposal

Gases. Gaseous radioactive isotopes should be used within a well-ventilated room and exhausted into a hood that discharges through a separate duct at some elevation above the roof of the hospital. This exhaust point should be quite removed from the air intake for the hospital ventilation system to prevent recirculation of the gases throughout the hospital. Simple smoke bombs released in the hood can verify the maintenance of this condition.

Human excreta. Present guidelines allow the disposal of human excreta through the sanitary sewer system. In general, the water volume usage within a hospital is sufficient to dilute the releases to levels that are nonhazardous to the general public.

Solids. All solid waste should be sealed in heavy plastic bags and placed in containers for transfer to a radioactive waste disposal company. High-level radioactive waste may have to be placed in a lead or concrete shield in addition to meet transportation regulations. Consult your state regulations and the disposal company for guidance in this area.

For materials that are contaminated with short half-life isotopes, it may be more economical to store the waste in a shielded area until it no longer contains detectable levels of radiation and then dispose of it as normal trash. This must be decided on an individual basis depending on the availability of an adequate storage area and fire regulations.

Liquids. The transportation of liquid waste in fluid form is generally prevented by local, state, or national regulations. Several mechanisms exist to solidify it and among these are incorporation in a vermiculite-concrete mixture, a simple concrete mixture, or a polyester resin. The weight percentage of water that can be incorporated and the shielding offered by the solidifying agent varies considerably among the options. Generally, the waste disposal company will have a prefered method which they will request to be used.

Again the option to store short half-life waste, allow radioactive decay to take place, and finally to dispose of it down the drain exists. The economic tradeoff of space vs. disposal cost must be evaluated on a case-by-case basis.

7.11 Study Questions

7.1 What are the major types of hazards found in hospitals?

7.2 Define microshock and macroshock.

7.3 How can microshock currents be introduced?

7.4 Why is a hospital patient more susceptible to microshock hazards than an average person?

7.5 What are the priorities for implementation of an electrical safety program?

7.6 List the five parameters of an electrical safety program, and describe the relationships among them.

7.7 List the eight elements of an electrical safety program.

7.8 List the major pieces of equipment necessary for implementation of an electrical safety program.

7.9 What are the advantages of an equipotential grounding system for hospital critical-care areas?

7.10 When is an LIM useful in protecting a patient from electrical shock?

7.11 What types of leakage current tests are typically carried out as part of electrical safety program?

7.12 What are the three types of fail-safe designs, and what is a medical example of each type?

7.13 What are the main causes of tissue trauma problems in patients. What are the bases for protection of tissue from trauma?

7.14 List at least three hazards resulting from the use of electrically powered wheelchairs. Repeat for manual wheelchairs.

7.15 What are the major functions provided by manual driving aids?

7.16 Describe the hazards associated with wheelchair restraints installed in vans.

7.17 What are the two major aspects of environmental control in hospitals?

7.18 Describe the special considerations that are important in disposal of solid wastes by hospitals.

7.19 What is meant by "A weighting" in SPL measurements?
7.20 Describe several noise exposure problems that are unique to the hospital environment.
7.21 Why is the water supply system of particular interest in developing a hospital safety program?
7.22 What procedures constitute the typical treatment of hospital water supplies?
7.23 What are some unique aspects of hospital plumbing systems?
7.24 List the most common causes of compensable work injuries in industry, and give an example of each type that may occur in a hospital.
7.25 Describe the elements of safe handling and use of gas cylinders.
7.26 What are the major types of nonionizing radiation present in hospitals?
7.27 Describe the primary risks of importance in nonionizing radiation.
7.28 List the common modes of infection control.
7.29 What are the three key steps in an infection control program?
7.30 Describe the disease transmission chain.
7.31 Describe the major sampling procedures for pathogenic organisms.
7.32 What is the difference between sterilization and disinfection?
7.33 List the major methods of sterilization.
7.34 Describe the procedure employed in the surgical suite to insure asepsis.
7.35 What steps should be taken to ensure the safety of external therapeutic radiation sources?
7.36 What are the major concerns and precautions associated with internal radiation sources?
7.37 Describe the primary safety considerations associated with diagnostic radiology.
7.38 During which diagnostic radiology procedures does the greatest exposure to personnel occur?
7.39 What are the procedures used for class A and class B radiation sources in nuclear medicine?
7.40 Summarize procedures for safe disposal of radioactive wastes.

7.12 References

Agnew, G. H. (1969). Special considerations in design of the operating area. *Anesthesiology*, 31(2): 116.

American Hospital Association (1968). Infection control. Chicago.

American Hospital Association and National Safety Council (1972). Safety guide for health care institutions. Chicago.

ANSI (1971). ANSI standard A 117.1-1971. New York: American National Standards Institute.

Bruce, D. L., K. A. Eide, H. W. Linde, and J. E. Eckenhoff (1968). Causes of death among anesthesiologists: a 20-year survey. *Anesthesiology*, 29(3): 565.

Chase, G. D., and J. L. Rabinowitz (1967). Principles of radioisotope methodology. Minneapolis: Burgess.

Code of federal regulations (1976). Dept. of Transportation 49 CFR 170–179, Coast Guard 46 CFR 146, Federal Aviation Agency 14 CFR 103. Washington, DC: U.S. Government Printing Office.

Code of federal regulations (1977). U.S. Title 10 and Title 42, Washington, DC: U.S. Government Printing Office.

Cohen, E. N., J. W. Bellville, and B. W. Brown (1971). Anesthesia, pregnancy, and miscarriage: a study of operating room nurses and anesthetists. *Anesthesiology*, 35(4):343.

Committee on Conservation of Hearing (1969). Guide for conservation of hearing in noise. American Academy of Ophthalmology and Otolaryngology.

Dalziel, C. F. (1972). Electric shock hazard. *IEEE Spectrum*, 9(2):41–50.

Dinsdale, S. M. (1974). Decubitus ulcers: role of pressure and friction in causation. *Arch. Phys. Med. Rehab.*, 55:147.

Falk, S. A. (1972). Combined effects of noise and ototoxic drugs. *Envir. Health Prospect*, 2:5–22.

Falk, S. A., and N. F. Woods (1973). Hospital noise-levels and potential hazards. *N. Engl. J. Med.*, 289(15):774–780.

Food and Drug Administration (1977). Bureau of Radiological Health Technical Publications. Rockville, MD.

Friedlander, G. D. (1971). Electricity in hospitals. *IEEE Spectrum*, 8(9):40–51.

Geber, W. F., and T. A. Anderson (1967). Cardiac hypertrophy due to chronic audiogenic stress in the rat. *Comp. Biochem. Physiol.*, 21:273–277.

Geddes, L. A., and L. E. Baker (1971). Response to passage of electric current through the body. *J. AAMI*, 5:13–18.

Graystone, P., and J. R. Ledsome, (1973). Microshock hazards in the hospital. *Proc. Int. Conf. Med. Biol. Eng.*,:159

Green, H. L., E. B. Raftery, and I. C. Gregory (1972). Ventricular fibrillation of healthy dogs to 50 Hz current in relation to earth leakage currents of electromedical equipment. *Biomed. Eng.*, 7:408–414.

Hallen, B., H. Ehrner-Samuel, and M. Thomason (1970). Measurements of halothane in the atmosphere of an operating theatre and in expired air and blood of the personnel during routine anaesthetic work. *Acta. Anaesth. Scandinav.*, 14:17–27.

Henkin, R. I., and K. M. Knigge (1963). Effect of sound in the hypothalmic–pituitary adrenal axis. *Am. J. Physiol.*, 204:710–714.

Hewlett-Packard (1972). Patient safety. Application Note AN718. Waltham, MA.

International Commission on Radiation Protection Reports (1977). New York: Pergamon Press.

International Commission on Radiation Units (1977). ICRU Reports, Washington, DC.

Jacobsen, T. L. (1969). Compaction system reduces disposal hazards. *Hospitals*, 43:89–90.

Joint Commission on Accreditation of Hospitals (1976). Accreditation manual for hospitals. Chicago.

Kilpatrick, D. C., and L. B. Kilpatrick (1971). Electrical safety standards in the health care delivery system. *Crit. Rev. Bioeng.*, 1(3):289–331.

Klein, B. R., and L. L. Leape (1976). Skin burn from freon preparation. *Surgery*, 79(1):122.

Kosiak, M. (1959). Etiology and pathology of ischemic ulcers. *Arch. Phys. Med.*, 40:62.

Linde, H. W. and D. L. Bruce (1969). Occupational exposure of anesthetists to halothane, nitrous oxide and radiation. *Anesthesiology*, 30(4):363.

March, R. C. (1972). Laminar air flow concepts for the hospital operating room. Traumatology convention. Gottingen, Federal Republic of Germany.

Massey, J. B. (1970). Manual of dosimetry in radiotherapy. Vienna: International Atomic Energy Agency.

MET Electrical Testing Co. (1976). MET standard 501: standard for performance and safety certification of medical equipment. Baltimore, MD.

National Council on Radiation Protection and Measurements (1977). NCRP Reports. Washington, D.C.

National Fire Protection Association (1973a). Standard for the use of inhalation anesthetics. NFPA 56A. Boston.

National Fire Protection Association (1973b). Electricity in patient care facilities. NFPA 76B-T (tentative). Boston.

National Fire Protection Association (1976a). Fire protection handbook. Boston.

National Fire Protection Association (1976b). Sprinkler systems. NFPA Code 13. Boston.

National Fire Protection Association (1976c). Life safety code. NFPA Code 101. Boston.

National Safety Council (1976). Accident facts. Chicago.

Norman, J. (1971). Electrical safety in hospitals. *Lancet*, 1(7707):1020–1021.

Olson, W. H. (1978). Electrical safety. In J. G. Webster (ed.), Medical instrumentation: application and design. Boston: Houghton-Mifflin.

Page, C. F., and M. H. Aronson (1973). Electrical safety. *Med. Electron. Data*, Series running from 1973 through 1975.

Peterson, A. P. G., and E. E. Gross (1974). Handbook of noise measurement. Concord, MA: General Radio Company.

Pfeiffer, E. A. (1968). Electrical stimulation of sensory nerves with skin electrodes for research, diagnosis, communication and behavioral conditioning. *Med. Biol. Eng.*, 6:637-651.

Pfeiffer, E. A. (1972). Microshock hazard from fluid-filled cardiac catheters, *Med. Electron. Data*, 3(6):51–59.

Rider, M. J., M. McDermott Jr., and J. G. H. Thompson, (1976). Use of wheelchairs as vehicle seats: current practices and recommendations. TR-060276. New York: VA Prosthetics Center.

Rogers, J. (1973). Downey, CA: Rancho Los Amigos Hospital.

Roveti, D. (1973a). Conductivity tests in the operating room. *Med. Electron. Data*, 4(3):96–98.

Roveti, D. (1973b). Electrical safety test procedures for hospitals. *Med. Electron. Data*, 5-part series, July 1973 to May 1974.

Schell, V. C. and L. E. Wolcott (1966). The etiology, prevention, and management of decubitus ulcers. *Missouri Med.*, 63:109.

Simmons, D. A. (1972). Medical and hospital control systems: the critical difference. Boston: Little, Brown.

United States Nuclear Regulatory Commission Regulatory Guides (1977). Washington, DC.

USAF School of Aerospace Medicine. (1972). Introduction to intestinal diseases and parasites affecting man. Brooks AFB, TX.

Veterans Administration (1975). VA standard design and test criteria for safety and quality of special automotive driving aids for standard passenger automobiles. VAPC-A-7505-8. New York: VA Prosthetics Center.

Veterans Administration (1977). VA standard design and test criteria for safety and quality of automatic wheelchair lift systems for passenger motor vehicles. VAPC-A-7703-2T. New York: VA Prosthetics Center.

Weibell, F. J. (1974). Electrical safety in the hospital. *Ann. Biomed. Eng.*, 2:126–148.

Weinberg, D. I., J. L. Artley, R. E. Whalen, and H. D. McIntosh (1962). Electric shock hazards in cardiac catherization. *Circ. Res.*, 11:1004-1009.

Weintraub, B. S., and H. D. Kenn, (1970). Use of wet grinding units for disposal of hospital solid wastes. *Hospitals*, 44:28–37.

Weiss, W. H., (1976). Gas cylinder safety. *Hospitals*, 50:91–93.

Wessell, L. (1971). "Freon" as a solvent. *AORN Journal*, 14(4):88.

Whitcher, C. E., E. N. Cohen and J. R. Trudell (1971). Chronic exposure to anesthetic gases in the operating room. *Anesthesiology*, 35(4):348.

Whitcher, C., R. Piziali, R. Sher, and R. J. Moffat (1975). Development and evaluation of methods for the elimination of waste anesthetic gases and vapors in hospitals. HEW Publication No. (NIOSH) 75:137.

8
Equipment Acquisition

J. B. OAKES and D. H. GORDON

8.1 Learning Objectives

Upon completing this chapter you should be able to:

1. Develop specifications for medical equipment based on clinical need.
2. Carry out a detailed prepurchase evaluation of medical equipment.
3. Draft purchase and acceptance specifications for medical equipment.
4. Describe and explain the equipment acquisition process.
5. Identify vendors that supply specified equipment.
6. Evaluate buy/build options.

8.2 Introduction

The care of today's patient is increasingly supported by high-technology electronic and electromechanical equipment. The practicing clinical engineer will find that a major portion of his time is devoted to the acquisition, maintenance, and control of such patient-care devices and systems. Therefore, this chapter, and several subsequent ones, will deal with various aspects of this involvement in considerable detail.

8.3 Overview

This chapter deals specifically with several facets of the equipment acquisition process. It describes a logical sequence of steps which can be taken to assure a successful acquisition program. Referring to the flow chart of Fig. 8.1, the first three of these steps involve proper definition of the medical requirements, assessment of environmental conditions, and a survey of commercially available equipment. This is the information-gathering part of the process, and assures that the medical needs are correctly translated into detailed, realistic, and quantitative engineering requirements. This requires a collaborative effort among people of differing backgrounds. Thus, as suggested by Fig. 8.1, an iterative process of coordination and review is normally required to assure that medical and/or engineering enthusiasm do not outrun fiscal constraints or the technical state of the art.

With this information in-hand, a decision can be made either to purchase the equipment "off-the-shelf" (i.e., by specifying a vendor's catalog number) or to purchase via a more detailed "specification-bid–contract" route. The latter procedure is normally followed if a system of equipment with specially designed interfacing and system control features is required. Assuming this is the case, the steps to be taken include generation of a system specification, solicitation of proposals, proposal evaluation, vendor selection, and contractual acquisition of the system. Here again, iteration among the hospital staff will undoubtedly be required, but the potential vendors may well be included in the iterative loop. If off-the-shelf purchase is the decision, a purchase order is substituted for the more complex procedure just described.

After a purchase order or contract has been issued, the procedures to be used for equipment acceptance must be prepared. Upon acceptance, the equipment is properly identified for entry into the equipment control program (Sec. 9.3), and the acceptance data for

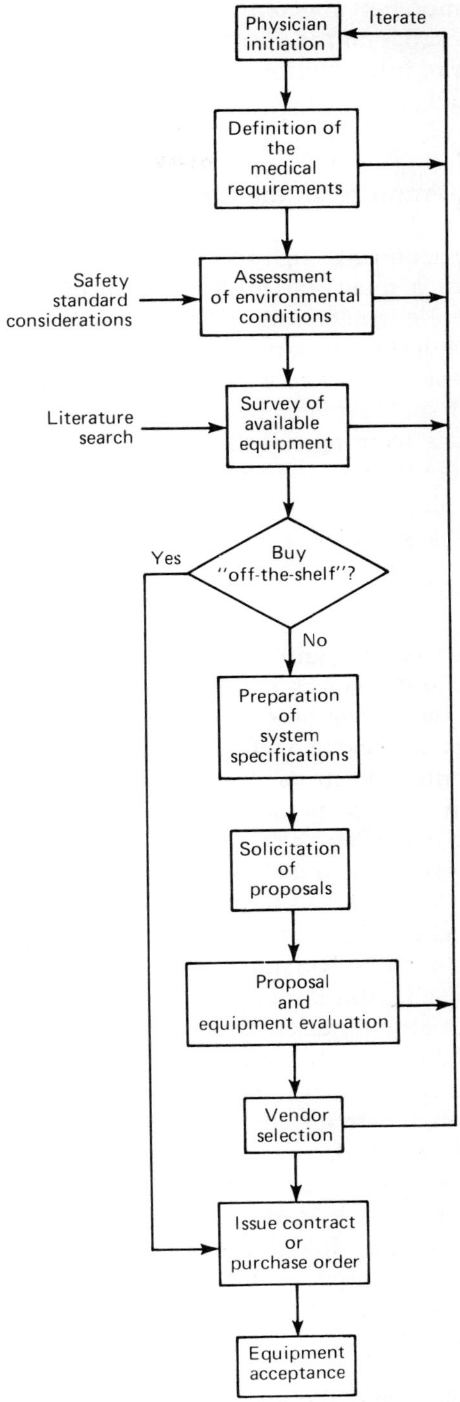

Figure 8.1 Equipment acquisition process.

each system component become initial input into the hospital's quality assurance program for patient-care equipment (Chap. 10).

With this brief introduction, let us look at each of these steps in much greater detail.

8.4 Definition of the Clinical Requirements

The process of acquiring high-technology patient-care equipment should begin with a quantitative description of the clinical requirements. The clinical engineer can contribute effectively in such quantifying efforts. He or she can serve as a consultant to the clinical staff, a member of a team tasked to solve the problem at hand, or a leader or coordinator of such a team. Regardless of which role the clinical engineer assumes, a team approach to defining clinical requirements, and to defining the ultimate solution, is recommended. This approach best assures that all key people have the same clear understanding of the problem to be solved, and that the solution will be optimal.

The team

Medical, nursing, and clinical engineering representation on the "problem-solving" team is recommended. The team should be as small as practical, with members chosen based on motivation to solve the problem at hand, and a willingness to understand the needs of the other represented interests. While the rationale for this sort of team composition may seem obvious, it is worth stating explicitly. The medical staff (physicians and surgeons) has a strong interest in the machine or system output. The output may be a single number, like heart rate, or a complex picture, perhaps in the form of a radiograph. The medical staff is somewhat less interested in the operation of the machine or system. Nursing, on the other hand, is generally intimately involved with the knobs and dials used to obtain the result in the form the doctor requires. Clinical engineering is the technical representative on the team. It serves as technical liaison with vendors, and once a system is procured it assures that the system is safe to use. Through clinical engineering's contribution to user education, system effectiveness is also assured.

Each member of the team must query the others and members of their respective staffs in order to (1) define the general clinical goals, (2) specify the physiological variables to be measured or controlled, (3) characterize the likely range of these variables, and (4) determine any restrictions of a medical nature. Let us look at these four areas in more detail.

Define the general clinical goals. The purpose of this step is to write a statement of the clinical objectives related to the problem at hand. These objectives should be stated in a precise, succinct manner, in a language that is clear to all interested parties. This task is not always as easy as it sounds, inasmuch as doctors, nurses, and engineers frequently do not speak the same language. The resulting statement of objectives will serve as a basis for choosing and evaluating equipment that contributes to the desired solution, and so it must be developed with care.

Caution should be taken to assure that the statement of goals does not take the form of a "wish list," particularly in the case of systems of equipment. If the list of goals and objectives becomes unmanageable, or if it appears that some of the goals are desires rather than requirements, it may be useful to list the goals in order of priority.

Specify the physiological variables. With the clinical goals determined, the physiological variables to be measured or stimulated by the equipment must be defined in order to achieve these goals. It is obviously desirable, in the interest of simplicity and economy, that these variables be directly measurable. If that is not the case, there may be some choice as to the indirect measurements to be employed, and a combination of engineering and medical judgment may be required to select the best instrumentation approach.

Characterize the range of physiological variables. The range over which each of the physiological variables that is to be measured or stimulated must be described. These data will eventually be used to determine the type of transducer to be employed, its dynamic range, and the type of signal conditioning or excitation equipment that will be required. In equipment for clinical use, it will usually be easy to describe the physiological signal ranges to be encountered. For equipment of a research nature, this may be difficult or impossible, and may require the specification of more flexible, adaptable, and expensive equipment to assure experimental success.

Restrictions of a medical nature. In the course of defining the medical requirements for the equipment to be acquired, serious attention must be paid to any medical restrictions imposed. For example, is the patient at risk from infection such that invasive measurements cannot be employed? How rapidly should a change in a monitored variable be made known to the medical staff? Are there certain formats of output data which are not acceptable? The question of such restrictions should be raised directly with the users during discussions of the medical requirements, since these can have a profound impact on the equipment to be acquired.

With the information derived from the four steps listed above, two important documents can be generated. The first, of course, is a written definition of the medical requirements, which can be circulated, criticized, modified, and eventually adopted by the users. This information can become input data for a specification or purchase order for the equipment, as we will see in a subsequent section. The second document which can be generated is a rather detailed block diagram of the equipment to be acquired. This diagram, of particular use to the clinical engineer, will also be of great value in the acquisition process.

8.5 Environmental Survey

A step which holds equal importance to the definition of medical requirements is the gathering of information concerning the environment in which the equipment will operate. The clinical engineer must be aware of any potential restrictions that the environment places on the use of the equipment to be acquired. A number of possible restrictions are reviewed below.

Space

One generally conceives of space as a restriction applying only to large systems of equipment, and often this is true. However, the question of available space should be raised with the medical staff for all equipment to be acquired, to assure that any potential problems are recognized early. Negotiations for additional space often take many months and may involve the highest echelons of the organization. On a detailed level, and particularly for large systems, a preliminary environmental survey should include considerations of the space that will be required for staff and patient access, for maintenance (including access to equipment cabinets for maintenance personnel and their equipment) and for power cables, signal cables, and environmental control (e.g., raised floors for computer installations). For large equipment, dimensions of the access route to the proposed location for the equipment and its transporters should be determined. For individual patient-care equipment, space within the patient room should be studied, remembering that bedside space is at a premium, especially in patients at risk.

Power

The initial environmental survey should next include a listing of the present availability of power in the proposed area. With the assistance of the plant engineering group, a survey of the local ac power system should be made. This will include the details of the feed system to the room, including voltage, number of phases, available current per phase, and the presence and type of emergency power. The availability of compressed air, a form of power sometimes required in patient-care equipment, should be noted, as should existing access to central oxygen, anesthesia, suction, water, and waste disposal services.

Weight

The medical staff may want to locate individual instruments or small systems on existing wall or floor-mounted support structures. The clinical engineer should assess the strength and stability of such structures. For large systems with heavy system components, the maximum safe floor loading of the area to be employed, and of the access route for equipment being transported to that area, should be checked.

Temperature range

Most electronic and electromechanical patient-care equipment operates satisfactorily over the controlled temperature range normally provided for staff and patients. In areas that are not occupied on a 24-hour basis, however, environmental controls may be disabled at night and on weekends. In such instances, a recording thermometer is an invaluable aid in judging the long-term temperature environment. If large systems or instruments that purposely generate heat as a part of their operating cycle are visualized, the existing air-conditioning capacity in the space the equipment will occupy should also be determined.

Vibration and shock

The characteristics of any unusual mechanical environment to which the equipment will be subjected should be determined. As in the case of temperature, the environment of any system components that will be located outside areas normally occupied by patients and staff should receive particular attention.

Electric and magnetic fields

Some patient-care equipment is particularly susceptible to the presence of static or intermittent electric and magnetic fields. Various physiological monitoring systems fall into this category, although proper design can do much to reduce this sensitivity. Digital computers are also subject to disruption in high fields. In any case, we should be particularly careful to note the presence or proximity of diathermy and electrocautery devices, radio paging antennas, and large electric motors. Special shielding, or relocation of equipment, may be required to cure severe interference problems, and a field survey may be cost effective if a potential for such interference exists.

Explosive and flammable environment

If the equipment to be acquired is to be used in an operating room or an area where explosive material may be employed or stored, special equipment designed and powered in such a way as to eliminate the danger of explosion must be used. If the equipment is to be located in an area where an oxygen-enriched atmosphere may be present, this fact should also be noted.

Humidity and moisture

Much patient-care equipment will operate in a humidity-controlled environment. If this is not the case, or if an apparatus or component must operate in especially high or low relative humidity, this fact should be noted. If the humidity environment is severe or unknown, a recording humidity meter will be useful.

Occasionally, some equipment may have to operate in an environment containing water droplets or standing water. Such a requirement should be noted, since equipment can be designed to be splash or drip-proof if required (Sec. 4.9). The presence of automatic sprinkler heads should also be noted. Such systems tend to have a deleterious effect on expensive electronic equipment. In certain instances, fire codes may allow these systems to be replaced by a nonaqueous automatic extinguisher system which does no permanent harm to the equipment. In any situation where a potential exists for a water spray, accidental or otherwise, the presence of a workable floor drain to prevent flooding is particularly important.

Applicable standards

A detailed review of codes, standards, and regulations applying to the health-care process is the subject of Chap. 4. It will be particularly useful for the clinical engineer to make a list of the specific standards that apply to the equipment he is acquiring, and to the space it will occupy. He should likewise consult the organization's internal standards — which he has probably helped develop.

8.6 Survey of Available Equipment

With the clinical needs defined and the environment known, the next step in the acquisition process is a search for specific information on equipment that may meet these requirements. In carrying out this survey, it will be helpful to have a one-page summary, or checklist, of the most significant characteristics of the desired equipment for ready reference. The survey process itself is an important one, and should be carried out thoroughly regardless of the level of experience of the clinical engineer or the level of complexity of the system to be acquired. The hospital will benefit in many ways from the time spent in this phase.

The literature search

In many cases, you will have reason to expect that the system requirements can be met, more or less fully, by commercially available equipment. In these instances, the literature search might begin among the advertisements carried by the various publications relating to the intersection of the medical or biological fields and engineering or by the current industrial periodicals (see Secs. 15.4 and 15.6). From advertising or editorial materials in these periodicals, a list of potential vendors and specific product information applying to your application can be generated. Other sources of information on commercial products are the newsletters and information services listed in Sec. 15.7. The latter are subscription services, and are valuable because they often contain comparative evaluation of competing products. Additional sources of technical information are those publications which list manufacturers by product categories. Section 15.4 lists a number of such publications. With good fortune, the result of this effort will be a listing of potential vendors of commercial equipment that meet all, or most, of your clinical requirements. If you have also made some notes as to the typical performance and environmental specifications of that equipment, you can compare reality with desire by using the checklist.

A less fortunate circumstance is one in which the search so far has been unsuccessful, or only partially successful. In this case, the implication is that parts of the equipment system may have to be specially designed, or that a different approach to system implementation may be required. At this juncture, a deeper search into the current technical and scientific literature is desirable and valuable. Here, the objective is to determine who is carrying out development work in your particular area of medical interest, and what instrumentation they are using in this effort. The desired output in this case is a list of researchers, and a collection of abstracts and reprints that detail their efforts. Perhaps the most rapid way of accessing this literature is through use of one of the medical indexing systems identified in Sec. 15.5. Most medical libraries and many public library systems have several of these index systems available. Certain of these are also accessible by remote computer terminal, and a key-word system is used to narrow the printout to your particular specialty.

Vendor contacts

The process above has provided a list of potential vendors for some, or all, of the equipment you wish to acquire. The local sales representative of each of these manufacturers should be contacted. There will be specific equipment specification sheets you will require, and these should be requested. In addition, it is often useful to describe the particular application you have in mind, in case a related instrument in the manufacturer's product line might have some value to you, or in instances where an instrument presently in development might meet your needs and time scale. Experienced sales representatives are an excellent source of information of benefit to you and your organization, and developing a suitable rapport will be valuable to both participants. One important end result of your vendor contacts is the broadening of your own files. With increasing experience, you will frequently find the answers to instrumentation problems at your fingertips. Care in assuring that you are on the right mailing lists will help keep these files up to date.

Running cost estimate

As the data collection process proceeds, it will be useful to establish some idea of the cost of the components, or system of components, to be acquired. A useful approach is that of the "running cost estimate." As the name implies, this is a cost estimate which is accumulated and modified as one gathers technical information about the equipment to be acquired. As your view of the equipment needed improves, and a listing of system components, or a block diagram, develops, approximate costs can be assigned and a running total established. You can be sure that this information will be of great interest to the medical staff and that you, as the clinical engineer, will be asked for the "bottom line" figure many times during your deliberations. Where applicable, remember to include installation costs in this estimate, as these are frequently a significant component of the total cost.

Iteration

Let us review the situation that now exists. You have a well-thought-out list of the clinical requirements to be met, in written form. You have made a complete environmental survey, and are aware of the various governmental and internal standards that may constrain the equipment. You have a file of manufacturer's data, and possibly of research papers, which indicates that equipment can be acquired, or designed, to meet the medical requirements. You also have some idea of the cost of those system components that are commercially available, and of the cost of installation.

With this information, a most important process of iteration can begin. You must make your findings available in a succinct but understandable form to the users. A formal presentation, with handout material covering the important points, has been found to be a useful approach. You should recapitulate the user needs, summarize the environmental situation, and provide the cost information you have derived. You should then point out any instances where the stated user needs and the environmental constraints have a strong impact on cost. If development will be required to meet some of the needs, this should be pointed out, and the uncertainty of cost estimates in such a situation should be made apparent. If you have recommendations concerning modification of the user needs or changes in the environment (e.g., a different location for the equipment) which could have significant impact on cost, present them.

Your objective is to foster discussion and interaction among the committee members. This is the "moment of truth," and the iterative process is the best way, and perhaps the only way, of examining tradeoffs between user needs and cost. It is important that this process of maximizing the system "cost effectiveness" be carried out at the user level, and that out of the process comes a rationale to support acquisition of the system. This rationale will be necessary when funds are requested from the hospital administration, who, lacking realistic justification, may make a decision based on cost alone, rather than on cost effectiveness.

Decision

The iterative process just described should result in a final list of user needs and environmental constraints representing a consensus, and having the support of the team. At this point, a decision as to the approach to be used for acquisition can be made. There are four major directions that acquisition can take.

Direct purchase. If the user needs can be met by the acquisition of instruments or systems that are commercially available, a simple purchase order may be the solution. In this case, evaluation of the equipment of competing vendors (see Sec. 8.10) and selection of a vendor from those qualified (see Sec. 8.11) should precede the actual generation of the purchase order (see Sec. 8.13).

Contract purchase. In larger systems, it is often the case that the user needs may be met only by assembling a number of commercially available components, properly interfacing these components, and designing in the appropriate system control features. Many manufacturers of individual medical instruments are qualified to carry out such jobs and are interested in accepting such system design work under contract. If this is the chosen route, the next step involves generation of a specifications for the system. This step, and the process of soliciting proposals responsive to the specification, are the subject of Secs. 8.7 and 8.8. Equipment evaluation, vendor selection, and the contractual phase follow (Secs. 8.9 through 8.12).

Combination. An alternative to system procurement from a single vendor is for the clinical engineering department to carry out the component interfacing itself. In this case, commercial system components are purchased directly. It would be wise to generate a detailed interface description for every commercial component so that your in-house design effort is itself completely defined. A caveat is in order if this course is followed. The responsibility for proper system operation now is divided between the vendor(s) of the commercial components and the clinical engineering group. In case of trouble during the system checkout phase, disagreement can result as to the source of the problem, and troubleshooting may be replaced by "finger pointing." You may be sure, however, that your organization will hold you responsible for getting the system debugged and working.

State-of-the-art development. Despite the iteration process, it may be that the users needs cannot be met by commercially available components or systems. Such a situation may well arise in a teaching institution carrying out medical research. In such instances, state-of-the-art development may be required. Such development may be carried out internally, on a collaborative basis between the clinical engineering group and the user (who will in this case be a member of the medical research staff). Funds for such development may be derived by grants from the public or private sector. The information on user needs will, in this case, be of great value in preparing grant request proposals. Alternatively, you may choose to approach several commercial instrumentation vendors to carry out this development. In this instance, the route outlined in Secs. 8.7 through 8.12 should be followed. Be aware that when there are no commercially available answers to your clinical problems, you will interface with a hospital group concerned with human experimentation. You must collaborate with the medical staff in drafting the protocol to which this committee must react before developmental work can proceed. The details of specification preparation will be discussed below, and will be illustrated by pages excerpted from a specification for a physiological monitoring system.

8.7 Specifications and Interface Control

A formal specification is a necessity in the procurement of all but the most standard medical devices or systems. The specification must be carefully written so that selected vendors may submit meaningful proposals. The specification, coupled with the winning vendor's proposal, will also become a part of the defining contractual documents. Because of its importance, the generation of a specification will be given generous attention in this section, and examples from a typical specification will be used to illustrate the discussion.

The format suggested here is indicated in Table 8.1, and parallels that which has been employed for military procurement for many years. Following such a format assures that every item is included that may have a bearing on the system being specified. Sections that do not pertain may be omitted.

Scope

This section of the specification contains a brief description of the areas that the specification will cover. One generally defines the user organization, and establishes any abbreviations that will be employed in the text of the specification. Table 8.2 illustrates a typical statement of scope.

Applicable documents

This section lists any documents that will be of assistance to vendors who respond to the solicitation of proposals. The list should contain only pertinent drawings; it is normal to make these drawings a part of the specification to the extent that they are refer-

Table 8.1. Specification Format

1.0	*Scope*
2.0	*Applicable Documents*
	Precedence
	Specification Changes
3.0	*Requirements*
	Design Objectives
	General System Requirements
	Specific System Requirements
4.0	*System Testing and Acceptance*
	Vendor Test Requirements
	System Acceptance Testing
	Acceptance Criteria
5.0	*General*
	Documentation Requirements
	Training
	Warranty
	Maintenance

Table 8.2. Specification for Cancer Center Patient Monitoring Instrumentation

1.0 *Scope*

This specification covers the design, installation, and acceptance testing of patient monitoring equipment for the Regional Oncology Research Center ("Cancer Center") at the Central City Hospital ("CCH").

2.0 *Applicable Documents*

The following documents form a part of this specification to the extent specified herein:

2.1 *CCH*

Construction Drawing E-22, entitled "Physiological Monitoring System," dated May 13, 1974, and including two addenda dated August 12, 1974.

Sketches CB155 and 155A, entitled "Nurse's Station Casework," dated April 5, 1976.

2.2 *Precedence*

In case of conflict between documents referred to herein and the requirements of this specification, the requirements of this specification shall govern.

2.3 *Specification Changes*

Requests for changes to this specification subsequent to its issuance shall be made only in writing and shall be subject to written agreement between the Central City Hospital and the vendor.

---

enced in the text of the specification. This section also addresses the problem of precedence of documents referenced in the specification, and defines the route to be followed for the incorporation of changes in the specification following its initial issuance. Table 8.2 illustrates this discussion.

Requirements

The requirements section is the core of the technical portion of the specification. Generally, this section becomes more detailed as it progresses. The first part of the section sets the stage by describing to the vendors the motivation for the system, and provides, in narrative form, a description of the system design objectives. Table 8.3 illustrates a typical description of this sort. You will note that the words are carefully chosen to provide a succinct description. Details are left until later.

The second part of the requirements section describes the general requirements that the system to be proposed must meet. In our example, illustrated in Table 8.4, these are requirements that are imposed by the philosophy of the medical user, by the presence of an existing electrical cable system, and by the desire for a rugged mechanical assembly. If the system or any of its components must perform under unusual environmental conditions, such information should also be included in Sec. 3.2.

Table 8.3.

3.0 *Requirements*

3.1 *System Design Objectives*

The monitoring instrumentation described herein has as its purpose the acquisition of physiological variables from selected Cancer Center inpatients. Certain of these variables will be selected for display in the patient rooms and for display and recording at central nursing stations that serve the inpatient areas. Additionally, alarms that indicate out-of-limit values for these variables must also be available at the nurse's station and in the patient rooms.

More specifically, the patient monitoring equipment will be required to:

(a) Acquire physiological variables in a safe manner from ill patients.
(b) Generate suitable alarm signals when these variables exceed adjustable thresholds.
(c) Display electrical replicas of the physiological signals, plus any alarm signals generated, in the patient room, and also transmit these signals to a central nurse's station.
(d) Display selected variables, either visually or graphically, at the nurse's station.
(e) Display alarm signals visually and aurally at the nurse's station.
(f) Provide maximum modularity to allow easy exchange of malfunctioning and spare subassemblies.

Table 8.4.

3.2 *General System Requirements*

3.2.1 *Monitoring Philosophy*

In consonance with Sec. 3.1, the following general patient monitoring philosophy has been adopted by the Cancer Center. Each of the two inpatient floors of the center will be served by two nurse's stations. Each nurse's station will serve a maximum of 16 rooms, and will contain a console on which physiological signals brought from each patient room may be displayed and recorded, and on which the appropriate alarms may be displayed. Provisions are also included for transmitting the monitoring signals for up to six patient rooms from one nurse's station to another on the same floor, and for transmitting selected signals to a remote minicomputer for analysis. This will be accomplished through manual patching.

Equipment for signal conditioning and alarm generation will be provided in portable form through the use of patient monitoring carts. Two types of carts will be employed, an acute level cart and an intensive level cart. Several of each type cart will be available for use on each inpatient floor, and will be assigned to a patient room as the need arises.

3.2.2 *Cabling*

The interroom and intraroom cabling required to support the patient monitoring system will be provided as part of the Cancer Center building itself. (Refer to Drawing E-22). Located on the headwall in each patient room will be a patient monitoring module. This module will contain four insulated phone jacks and four 6-pin MS-type receptacles. Phone jacks #1, 2, & 3 will allow temperature sensing via thermistor probes, and shielded 3-conductor cables lead from these jacks directly back to the nurse's station. Phone jack #4 will be reserved for transmitting an infusion pump alarm signal directly from bedside to nurse's station. The four MS receptacles are wired via low-noise, shielded, 2-conductor cables to the patient monitoring relay module, located away from the area of the patient bed so

Table 8.4. (Continued)

that the patient monitoring cart will not occupy space at the bedside. Thus, the MS-type receptacle #9 provides all input signals for the patient monitoring cart.

Two receptacles are provided for forwarding conditioned physiological signals and alarms to the nurse's station; these are labeled #10 and #11 in Drawing E-22. It is planned that one of these cables be reserved for analog physiological signals, and the other for alarms and control signals (dc levels, relay closures, etc.) Thus, the MS-type receptacles #10 and #11 collect all output signals from the patient monitoring cart.

Starting with the third part of the requirements section, we address the specific requirements of the system. The format here will depend on the system being defined. If the system can be broken down into several identifiable subsystems for purposes of system description, it will be best to do so. This will reduce the need for a cumbersome hierarchical numbering system in identifying the various sections of the specification. In our example, the system conveniently breaks down into two segments. Patient monitoring carts of two types are addressed in Sec. 3.3 (see Table 8.5), and the nurse's station monitoring console is defined in Sec. 3.4 (see Table 8.6).

In studying Tables 8.5 and 8.6, you will notice the absence of numbers strictly defining module size, the range of physiological variables, and the like. It is generally true that, whenever possible, a specification should define *what* is to be done, rather than specifically *how* it is to be done. This approach allows a vendor to respond by proposing a system that, insofar as possible, uses instruments from his standard commercial product line, and reduces the possibility that he might have to design or modify an instrument to meet some unnecessarily restrictive mechanical, electrical, or physiologic constraint. In our example, we have imposed only two restrictions as a result of pre-existing conditions. The nurse's station console envelope is defined, and the connector types used in a cable network already in place are described.

In cases where a custom-designed instrument or system is being specified, it may be desirable to provide some rather detailed guidance to potential vendors. In such cases, the above comments do not apply, and the requirements section can, and should, become very specific indeed.

System testing and acceptance

This section of the specification notifies the vendor how you propose to test the system he delivers, and describes the criteria you will use in determining its acceptability. It is generally useful to request the vendor's inplant acceptance test data as a starting point for your own tests. Table 8.7 includes rather nonspecific statements in these areas, and such an approach is appropriate in the case of our

Table 8.5.

3.3 *Patient Monitoring Carts*

3.3.1 *Mechanical Design*

Each patient monitoring cart shall provide a mechanically stable, tip-proof wheeled cabinet to house the modular amplifier, alarm, display, and control circuitry appropriate to its purpose. Each cart shall contain a drawer and a shelf suitable for holding additional instruments, and shall be appropriately cabled to match the connectors of the patient monitoring relay module. Corner bumpers shall be provided to protect the cart and patient room walls from mechanical damage caused by collision.

3.3.2 *Acute-Care Patient Monitoring Cart*

The acute-care patient monitoring cart shall be designed to provide information concerning a patient's pulse and respiration rates. The cart shall:

1. Display pulse and respiration rates.
2. Display the waveforms employed in deriving these rates (e.g., ECG and respiration waveforms).
3. Provide adjustable high- and low-rate alarms for each rate signal, with audible and visual status indicators. The audible alarms on the cart shall be capable of defeating.
4. Transmit the rate, waveform, and alarm signals to the nurse's station monitoring console.

It is not mandatory that the modular concept of Sec. 3.1(f) be employed in the design of this cart.

3.3.3 *Intensive-Care Patient Monitoring Cart*

The intensive-care patient monitoring cart shall be designed to provide four channels of patient physiological data. It shall house the waveform display module of Sec. 3.3.3.8, the ECG module of Sec. 3.3.3.1 and any combination of three of the other modules listed in Secs. 3.3.3.2 thru 3.3.3.7.

The ECG module shall be keyed in such a way that it can be inserted in only one of the module receptacles on the cart, and connector #5 of the patient monitoring module shall be dedicated and keyed as the only allowable input connector for this channel. All other modules described below shall be mechanically interchangeable in the remaining cart receptacles, and their electrical output connections shall be so designed that they excite the appropriate nurse's station displays and alarms regardless of the cart receptacle into which they are inserted.

Table 8.6.

3.4 *Nurse's Station Monitoring Console*

3.4.1 *Mechanical*

The equipment described in this section shall be housed in a vendor supplied console. This console shall fit within the following above-counter envelope:

5' 0" long
1' 2" high
2'11½" deep

The counter top itself can have appropriate openings to allow the required signal cables to access the console. Alternately, the console itself may be recessed through the counter top, and protrude moderately into the space below, to allow these connections, as long as the above-counter envelope is observed. In either case, since some cable-patching

Table 8.6. (Continued)

capability is to be provided on the console (see Sec. 3.4.2 below), the console design shall be such as to make this patching as simple as possible.

3.4.2 *Waveform Display and Recording Capability*

Each nurse's station monitoring console shall contain one 4-channel nonfade CRT-type display for simultaneously monitoring the waveforms received from any of its 22 locations (i.e., 16 of its assigned rooms plus six from the other nurse's station on the same floor). Unused channels on this display shall be capable of being blanked.

Each nurse's station monitoring console shall contain a single-channel ECG strip chart recorder which can be assigned to any of the 22 locations. With this recorder, it shall be possible to record either a real-time or a delayed ECG waveform.

Each nurse's station monitoring console shall contain a 4-channel strip-chart recorder which can be assigned to record all the waveforms from any of the 22 locations. With this recorder, it shall be possible to record either real time or delayed waveforms.

3.4.3 *ECG Capability*

In addition to the general waveform display and recording capability of Sec. 3.4.2 above, a single heart-rate display shall be provided, with manual switching which allows its assignment to any of the 22 locations.

Individual visual high- and low-rate alarm indications shall be available for each of the 22 locations. A single audible alarm shall be provided for the totality of locations.

Table 8.7.

4.0 *System Testing and Acceptance*

4.1 *Vendor Test Procedures and Data*

The vendor shall submit a written copy of test procedures and performance certification for the system hardware. These procedures and supporting individual data sheets shall verify that the vendor specifications and CCH requirements have been met. These documents shall be received by CCH before shipment of hardware.

4.2 *CCH Acceptance Testing*

The CCH acceptance test procedure will include complete verification of functional operation. It will include tests to verify acceptable quality of signals under clinical conditions. For these tests, the ECG signals will be derived from a human volunteer and the transducer amplifiers will be connected to the pressure transducers under pressure from a manometer. The Central City Hospital also may choose to verify the published specifications of the vendor's catalog components and to verify the performance specifications for noncatalog or modified devices. This testing will be completed within 30 days of the receipt of all system components.

4.3 *Acceptance Criteria*

Should the system not meet the requirements of this specification and the vendor's device data sheets prior to acceptance by CCH, it shall be the vendor's responsibility to localize and correct the fault or to notify the CCH if the fault is external to the vendor's system. Vendor test data substantiating the adequacy of the system, or localizing a fault, shall be made immediately available to CCH personnel. Should signal quality tests made in accordance with 4.2 not be acceptable, it shall be *the joint responsibility of the vendor and Central City Hospital* to localize the problem. Interference arising from the leads from the patient to the vendor-supplied patient cables shall be the responsibility of Central City Hospital, assuming all device specifications have been met.

example, where most of the equipment is expected to be made up of the vendor's standard modules. This section, of course, can be and should be much more extensive and detailed if the equipment is to be custom-designed. In the case of computer systems, you would normally require inplant satisfaction of a benchmark test program, prior to shipment and acceptance at your facility.

General

This section of the specification is available to cover those details that do not readily fit the other sections. In our example (Table 8.8), it includes a description of the documentation that a vendor is expected to supply with his equipment or system. Notice that acceptance of the system is contingent on the adequacy of this documentation. In-service training is also a contractual requirement.

Table 8.8.

5.0 *General*

5.1 *Documentation*

Acceptance requires that each module or subsystem shall be adequately documented in the following categories:

(a) Specifications.
(b) Installation instructions.
(c) Operation manual.
(d) Input/output connection identifications.
(e) Functional description.
(f) Theory of operation.
(g) Maintenance and calibration procedures.
(h) Detailed functional block, circuit, and mechanical diagrams.
(i) Replacement parts list (using standard commercial designations).

These items should appear together to aid in understanding and maintenance of equipment.

5.2 *Training*

On-site training of CCH personnel shall be provided by the vendor. Operator training for each unit shall commence immediately after acceptance. All three shifts of operating personnel shall be accommodated during this training period, with scheduling and session length to be mutually agreed upon.

5.3 *Warranty*

The vendor's standard warranty is expected to supply parts and service at no charge to CCH for the first year of equipment operation. Custom designed units or first production run units shall be updated with all applicable engineering changes by the end of the first year of service at no cost to CCH.

5.4 *Maintenance*

Facilities of the nearest vendor service organization shall be available to repair and return the equipment to operation.

Our expectations insofar as warranty period and availability of maintenance facilities are also included.

8.8 Solicitation of Proposals

We now have in hand a specification that describes, in as much detail as necessary, the characteristics of the equipment to be acquired. This specification should be circulated for technical approval with an attached signature/cover sheet. As a minimum, approval should be obtained from the medical, nursing, and engineering members of the original team. Possibly more signatures are required by protocol in your organization.

While this approval is being obtained, a request for proposal (RFP) should be drafted. Normally, this document takes the form of a cover letter to which the specification, and a copy of all applicable documents listed in the specification, are attached. It is best that you, as the clinical engineer, generate this draft, rather than leaving the task to your contracts department, since you now know more about the system to be acquired than anyone else. An example of a draft RFP is included in the appendix of this chapter; this example is, in fact, the RFP letter that accompanied the specification discussed in Sec. 8.7.

The request for proposal should list the enclosures, and then contain a very brief discussion of the reason for the request. The core of the request is contained in a section entitled "Guidelines for Proposal Preparation." In this particular case, the vendor is asked to bid his commercial modules insofar as possible in meeting the system requirements. He is encouraged to propose alternative solutions that meet the intent of the specification, and those items are listed that his proposal should specifically address. Since the system is expected to be modular, the vendor is asked to provide a detailed cost breakout so that the decision as to how many of each type of module to procure can be more easily made.

The final section of the RFP indicates the approach that will be used to evaluate his proposal, and includes a disclaimer allowing rejection of any or all bids as unsuitable. Directions for proposal submission, (when, how many copies, and to whom) are also included.

This draft cover letter may be submitted to your contracts department, along with a list of potential vendors and a request that you be allowed to review any revisions of the letter and vendor list before it is sent. On the matter of potential vendors, do not include on the list any vendor from whom you would not purchase the equipment, for whatever reason. An unacceptable vendor often becomes a low bidder, and you may be stuck with him.

8.9 Preliminary Proposal Review

All vendors should respond to your RFP by the appointed day, either with a proposal or a "no-bid" letter. To assure this responsiveness, it is an accepted practice to have your contracts department call each vendor a week or so before the due date, and inquire of his status. This may result in one or more requests for a deadline extension, and it will be your decision to grant such a request, or not. If an extension is granted, it must be granted to all vendors.

Assuming that you have accumulated a stack of technical proposals, the first step in their evaluation is to determine whether each one is responsive to the RFP. A well-organized technical proposal will key its response directly to your RFP letter, with a section devoted to each item of that letter. (See Sec. 4 of the draft RFP letter.) If any proposal is missing specific information that you required in the RFP, make a note of this.

Next, review the responsiveness of the proposals to the specification itself. Note any lack of responsiveness or alternate technical proposals. Note any instances in which your specification has been misinterpreted by the vendor. If this misinterpretation is due to lack of clarity in the specification, you may want to issue a clarifying specification addendum to all vendors.

A final step in the preliminary proposal review is to verify that each cost proposal submitted is broken out as the RFP required. Note any deviations here.

With this proposal and cost information in hand, you are in a position to request from each vendor such information as to make his proposal complete. This request should be made in writing, through your contracts department, and is generally a worthwhile iteration in order to assure you a maximum number of qualified bidders. Any specification addenda should also be submitted at this time to all vendors.

At this point, significant exceptions to the specification should be reviewed by the entire evaluation team, where a decision as to feasibility of these exceptions can be made. It is poor policy to continue negotiation with a vendor when it is known that the exception he takes to the specification will likely disqualify him.

At the completion of this process, the engineering and clinical evaluation of each vendor's equipment can begin. It is often the case that these evaluations can be carried out in parallel.

8.10 Equipment Evaluation

Your request for proposal has notified each vendor that evaluation of certain of the equipment he proposes to supply may be required. Alternatively, if you are purchasing your equipment off the shelf, it is common to request each vendor being considered to supply a unit for evaluation. In either case, you should choose one or more elements of each vendor's system that appear to be "representative" of his equipment. Since you will examine these submissions from both an engineering and a clinical viewpoint, choose items that illustrate quality of construction, maintainability, reliability, patient safety, user safety, human factors engineering, and ease of in-service training. With vendor proposals and/or equipment in-house, the engineering and clinical evaluations can be carried out in detail.

Engineering evaluation

The engineering evaluation will include a quantitative comparison of each vendor's proposal and/or equipment with regard to the following factors:

1. System performance.
2. Physical construction and its quality.
3. Reliability.
4. Maintainability.
5. Safety.
6. Human factors.

System performance. System performance studies can usefully employ a "comparison matrix." Here, those significant quantitative electrical and environmental parameters supplied by each vendor in his proposal text or data sheets are brought together on a single page for direct comparison. You must be the judge as to what parameters are significant, based on your knowledge of user needs and the specification that reflects them. If there are missing data in your matrix, request that vendor to supply it in writing so that his reply can later become part of the contract or purchase document. With this matrix in hand, you should be able to detect significant differences or even fatal flaws in the performance of the equipment with which the vendors are proposing to meet your specifications.

Physical construction. Physical construction and its quality can be judged by reference to both the vendor's proposal and his submitted equipment. Each vendor should be responsive to any restrictions your specification placed on the equipment. If your specification stressed modularity, evaluate the degree to which this concept is met. Module, subassembly, and system size and weight should be evaluated. Ease of mechanical disassembly and repair may also be of importance, and equipment ruggedness almost certainly will be. Be sure that any unusual mechanical environment requirements (e.g., vibration, water resistance) are met. Another comparison matrix may be useful in this evaluation.

Reliability. Reliability of an instrument is a concept that can be discussed quantitatively in terms of that instrument's "mean time between failure," or MTBF. It is possible to mathematically estimate the MTBF of equipment, or even to develop statistically significant tests to determine it. In most patient-care devices, this information is not available. One can, however, develop some qualitative feel for the probable engineering reliability by examining component quality, evidence of sound manufacturing processes, and evidence of mechanical resistance to shock, impact, and vibration. Judicious sealing of circuitry against dust, dirt, and moisture may be desirable depending on your specification, and at any rate is evidence of thought given to reliability by a manufacturer. Another valuable resource in reliability estimates is consultation with other clinical engineers in other hospitals who are users of this vendor's equipment. Information is also available in *Health Devices*, a publication that reports comparative data on medical devices (Sec. 15.6).

Maintainability. Maintainability is an important consideration, for all patient-care equipment degrades or fails with time. Here again, statisticians refer to the "mean time to repair" (MTTR) of equipment. Minimization of MTTR requires attention to the availability of quality maintenance documentation, the spacing and ease of removal of components when they do fail, the rapid availability of any vendor-specific components in the equipment, the availability of a local vendor-supported maintenance facility, and the availability of specialized training for your technicians at the vendor's plant. All these items should be covered in the vendor's proposal. A maintainability evaluation should consider scheduled service as well as corrective maintenance. Does the manufacturer recommend a calibration or performance verification interval? Do manuals detail methods for performance testing and calibration? If not, remember that you will have to develop these schedules and procedures.

Safety. Safety of patient and staff are covered in detail in Chap. 7. Be sure each vendor adheres to some accepted specification in this regard, insofar as both leakage to ac power system ground and patient lead leakage are concerned. Also, check the provisions each vendor has made for (mechanical, thermal, chemical, radiation, etc.) safety.

Human factors. Human factors in the context of an engineering evaluation are mainly concerned with whether the maintenance of the equipment can be carried out in a straightforward manner. Items such as ease of module replacement, availability and accessibility of electrical test points, and the availability of internal module and system test features should be considered. The presence of, or lack of, adjustments on the front panel impinges on both the engineering and the clinical staff, and vendor philosophies vary considerably. These variations should at least be noted, and they may become a considerable factor in the ultimate choice of vendor. Ease of in-service training is of concern to you, and the presence or absence of controls, and their spacing, will impact on this area also.

Clinical evaluation

Clinical evaluation involves hands-on use of the submitted equipment by the medical staff. This evaluation is carried out with respect to the following factors:

- Quality of in-service training.
- Clinical performance
- Human factors evaluation.

In-service training. Having selected vendor equipment for the engineering and clinical evaluation, ask the vendor to provide his normal in-service training on that particular equipment. You and the clinical staff will be able to perform a valuable evaluation of the quality of his in-service training if he were to supply the whole system. Evaluate his curriculum, his approach and demeanor, and any training aids he provides. Verify that the medical staff feels comfortable using the equipment after the first training class is completed (Chap. 12).

Clinical performance. The clinical performance itself should address several areas. Factors such as accuracy and repeatability should be verified, on a variety of patients and/or in a realistic environment. Mechanical design features (size, weight, ruggedness) should be clinically acceptable.

Human factors. Human factors are an inseparable part of the clinical evaluation, but special attention should be paid to them, for the clinical staff must feel comfortable with the equipment or they will not use it. Necessary controls should be easily accessible and properly spaced. Any appearance of complexity should be avoided, and engineering controls should be hidden or located internally. If the equipment is attached or applied to the patient, it should be easily done, with minimum patient discomfort. The clinical evaluation should be of such a duration that any human factors shortcomings become fully evident.

The clinical evaluation is probably more important than the technical one. Ultimately, the people who must be happy with the final decision are clinicians–nurses and doctors. Don't be surprised to find that engineering enthusiasm for the technical evaluation exceeds the enthusiasm for the clinical evaluation. Particularly in the busy environment of a special-care unit, patient-care personnel may regard an equipment evaluation as a chore, and a great deal of handholding may be required on your part to achieve useful results.

8.11 Vendor Selection

With the technical and clinical evaluations complete, and with final proposals and cost estimates in hand, the task of selecting the vendor can begin. Since these recommendations of the team will be subject to scrutiny at higher levels, it is best to prepare a memorandum of justification putting forth the reasons for your recommendation. This memorandum will normally be divided into five sections: (1) engineering evaluation and ranking, (2) clinical evaluation and ranking, (3) cost evaluation and ranking, (4) additional information, and (5) recommended vendor.

Engineering evaluation and ranking. The engineering evaluation and ranking section should contain a brief description of the technical approach offered by each vendor. Details that make any proposal definitely superior, or definitely inferior, to the others, should be explicitly stated. The section should end with a listing of vendors in order of acceptability, based on the engineering evaluation alone. If any of the vendors are ranked completely unacceptable as a result of this evaluation, the reasons for this should be explicitly stated.

Clinical evaluation and ranking. The clinical evaluation and ranking is similar in approach, and should encompass inputs from all components of the clinical staff. Here again, a ranking based on only the clinical evaluation should be included.

Cost evaluation. The cost evaluation should be relatively straightforward, since the effect of the preliminary proposal review of Sec. 8.9 should have been to place all vendors on an equal technical footing. If acceptable technical alternates were proposed by any of the vendors, the cost of these alternatives should be ranked appropriately, in this section.

Additional information. The section on additional information should include such items as interviews with other users, any pertinent information available from device-rating organizations, and the like. If any of this information modifies the team's technical or clinical ranking of vendors, this should be explicitly stated.

Vendor selection. Vendor selection, with this information in hand, is based on a balanced evaluation of the three rank-ordered lists. Unless the three lists arrive at a common first choice, which is unusual, the process can quickly become qualitative and subjective. A realistic method of proceeding in such a case is to ask the same question that your administration will ask. "Are there overriding engineering or clinical reasons why we should not choose the lowest bidder?" Generally, you must relate such reasons to the "life-cycle-cost" of the system in order to have any impact on the hospital administration. If your team feels that selection of any vendor other than the lowest bidder is justified, say because of equipment maintainability or reliability differences, or because of differences in the in-service training requirements, you should attempt an estimate of the total cost of each proposed system to the hospital over its useful lifetime. Be realistic in these cost estimates, and consider all the factors that may impact life-cycle cost. Continued discussion by team members during this process is mandatory, for all members must agree on a vendor recommendation before it is forwarded to the administration.

Once the vendor is determined, this vendor should be asked to make a formal presentation to the team, covering the technical and cost details of his proposed system. Any final technical clarification that is required should be obtained in writing, and a final cost estimate should also be submitted at that time. The technical proposal, with addenda, and the cost estimate, should be forwarded to hospital administration, along with your memorandum of justification. This information will become the basis for their concurrence with your recommendations.

8.12 Contracts

The next step in the equipment acquisition process is the preparation of a contract with the chosen vendor. Such a contract normally consists of a cover/signature sheet, a schedule of work, and a set of general provisions. The general provisions cover the legal rights of both parties to the contract, and this material, often called "boilerplate" by the layman, is the province of your organization's legal department. The schedule of work, on the other hand, is of vital interest to the clinical staff and to the clinical engineering group. You should be asked to contribute information to the schedule, but you should forward this material to the contracts department even if you are not asked. Generally, the schedule of work is a precise document, containing, in as succinct a fashion as possible, the following sections: (1) a list of applicable documents, (2) a statement of work, (3) terms of delivery and acceptance, (4) schedule of the work to be performed, (5) cost and terms of payment, and (6) warranty and service.

Applicable documents. The section on applicable documents should reference your original specification, the vendor's original proposal, and any proposal addenda or other clarifying documents, by title, document number, and date. The priority of the specification in case of inconsistencies or questions should be stated.

Statement of work. The work statement should briefly outline the tasks to be accomplished during the contract. If this is a development contract, delivery of some intermediate hardware (e.g., breadboards or engineering models) may be required for your evaluation, and a list of these should be included. A complete list of the final deliverable system components, and the number of each, should also appear here. Any documentation required, either during the contract or at hardware delivery, should be listed, along with the quantities required.

Terms of delivery and acceptance. The section of the work schedule pertaining to delivery and acceptance should start with a delivery schedule for every hardware and documentation item listed in the work statement. This is generally stated in terms of so many weeks or months after the effective date of the contract (e.g., "5 months AED"). This section should then repeat the terms of acceptance of these deliverable items that were a part of the original specifications, but including any modifications that may have resulted in the interim negotiations. Generally, the locale and level of testing is defined, as are the criteria to be employed in acceptance.

Work schedule. The schedule of delivery has been called out in detail in the preceding section. Usually, however, an overall statement is also included to indicate the expected contract closeout date.

Cost and terms of payment. The section on cost is a reiteration of the final cost estimate submitted by the chosen vendor. If progress payments are required, this information is included also.

Warranty and service. The warranty terms submitted by the vendor are generally also repeated in the statement of work with an aim toward making the document a self-standing one. This section may clarify the vendor's standard warranty if your contracts group feels it is desirable to do so. The warranty should start at system acceptance, not system delivery, and this point should be so stated. Also, the response time for vendor service during the warranty period should be spelled out in this section.

8.13 Purchase Requisitions and Purchase Orders

In those instances where the equipment survey of Sec. 8.6 has led to a decision to buy "off the shelf," the clinical engineer follows a much easier route than the contractual one outlined above. He simply forwards a standard purchase requisition to the hospital's purchasing department, via a standard signature route. The purchasing department, in turn, issues a standard purchase order to the selected vendor for the desired equipment.

Although the implication in this process is that the clinical engineer makes use of a well-established paper flow system in acquiring the desired instrumentation, he may find himself charged with setting up or modifying the system and its forms from time to time. A brief review of the process is therefore included here.

Purchase requisitions. The purchase requisition is normally a multicopy form containing the information your purchasing department will need to prepare a purchase order. An example of a well-constructed four-copy purchase requisition is shown in Fig. 8.2. The top of the form identifies the originator of the requisition, the date, the required delivery date, and the consignee (i.e., where and to whom the material is to be delivered). The body of the requisition is self-explanatory. However, be sure the description includes any vendor model and option numbers that must be used to completely describe the equipment you desire. If you are willing to accept the equipment from one of several suppliers, the description is usually more general, and contains the words "or equivalent." The form has a "reminder area," asking among other things if you have checked the stockroom, and notifying the safety committee if you are ordering controlled substances. Space for required signatures completes the form.

For the purchase of major capital equipment items, most organizations exercise additional control. Generally, this is done by requiring that for individual items costing more than $X, the items have been approved for purchase on the current capital equipment budget. For these items, the requisition must usually be forwarded with a letter of justification, stating that the item has been approved for purchase, and reiterating the reasons for its need. Most patient-care equipment falls into the capital equipment category, and so you will probably become expert at preparing such letters.

REQUISITION
FOR
EQUIPMENT, MATERIALS AND SERVICES

H 84024

ORIGINATOR ______________________ DATE ____________

REQUIRED DELIVERY DATE ______________________

CONSIGNEE ______________________

ADDRESS ______________________

ESTIMATED TOTAL COST $
TECHNICAL ASSIGNMENT NUMBER
DATE ENTERED PROBLEM BUDGET
ORIGINAL FILED WITH CONTRACT

ITEM	QUANTITY	DESCRIPTION SUPPLY COMPLETE INFORMATION OF PRELIMINARY NEGOTIATIONS, IF ANY, WITH VENDOR(S) (ORAL AND/OR WRITTEN)	ESTIMATED COST		PURCHASE CONTRACT NO.
			UNIT PRICE	EXTENSION	

SUGGESTED SUPPLIER: ______________________

Have You Checked the Stockroom?——*Yes* ☐ *No* ☐

WILL this Order involve Vendor Research or Development?——*Yes* ☐ No ☐

Check if *Explosive* ☐ *Flammable* ☐ *Toxic* ☐ *Radioactive* ☐

APPROVED BY:	DATE	APPROVED BY:	DATE
SECTION		SECURITY	
GROUP		PURCHASING OFFICE	
BRANCH		BUDGET OFFICE	
DIVISION		DIRECTOR	

ORIG.—PURCHASING
1—PURCHASING
2—PROBLEM BUDGET
3—ORIGINATOR
ORIGINAL

URGENCY APPROVED BY ______________________
AUTHORIZED SIGNATURE—$1000.00 OR MORE

ACP FORM NO. 4
20M—8/74

Figure 8.2 Typical purchase requisition.

Purchase order. The purchase order form falls within the purview of your purchasing department. However, you should receive a copy when it is issued, to be compared with, and then filed with, your copy of the original requisition. The purchase order normally contains information transferred directly from the purchase requisition. In addition, it will notify you of the seller's promised delivery date, his terms of payment, and the order number that has been assigned. A typical purchase order form is shown in Fig. 8.3. It is to your advantage to be sure that your requisition has been correctly translated to a purchase order, and this comparison should be carried out as soon as your copy arrives.

8.14 Equipment Acceptance

During the time between the ordering of equipment and its delivery, the details of subsystem and system tests should be worked out. The purpose of such tests is to assure that all system components, and the overall system, perform in a manner that complies with the specification.

There are three sources of information that will be useful in preparing the required test procedures.

1. The specification.
2. Vendor data sheets on off-the-shelf components.
3. Vendor-supplied test procedures, a deliverable contract item.

In instances where off-the-shelf instruments are purchased directly, of course, only item 2 above pertains. The comparison matrix of Sec. 8.10 should also reflect the quantitative characteristics of the chosen vendor's system, and may be quite useful to you.

At this point, a review of these information sources can now be carried out for the purpose of extracting a list of quantitative parameters that determine the performance of the system as a whole, as well as each of the important subsystem components. It will be these parameters that will be subjected to evaluation when the equipment arrives.

You are now in a position to prepare the actual test procedures you will require. It is generally advisable to test at the subsystem level first, and then at the system level, in order to assure that the overall system to be tested is made up of properly performing subassemblies. Taking the illustrative specification of this chapter as an example, consider the testing of the acute-care patient monitoring cart (Table 8.5). This cart, as proposed by the vendor, will probably be made up of five basic components: (1) a waveform display module, (2) an ECG/pulse rate module, (3) a respiration module, (4) a cabinet

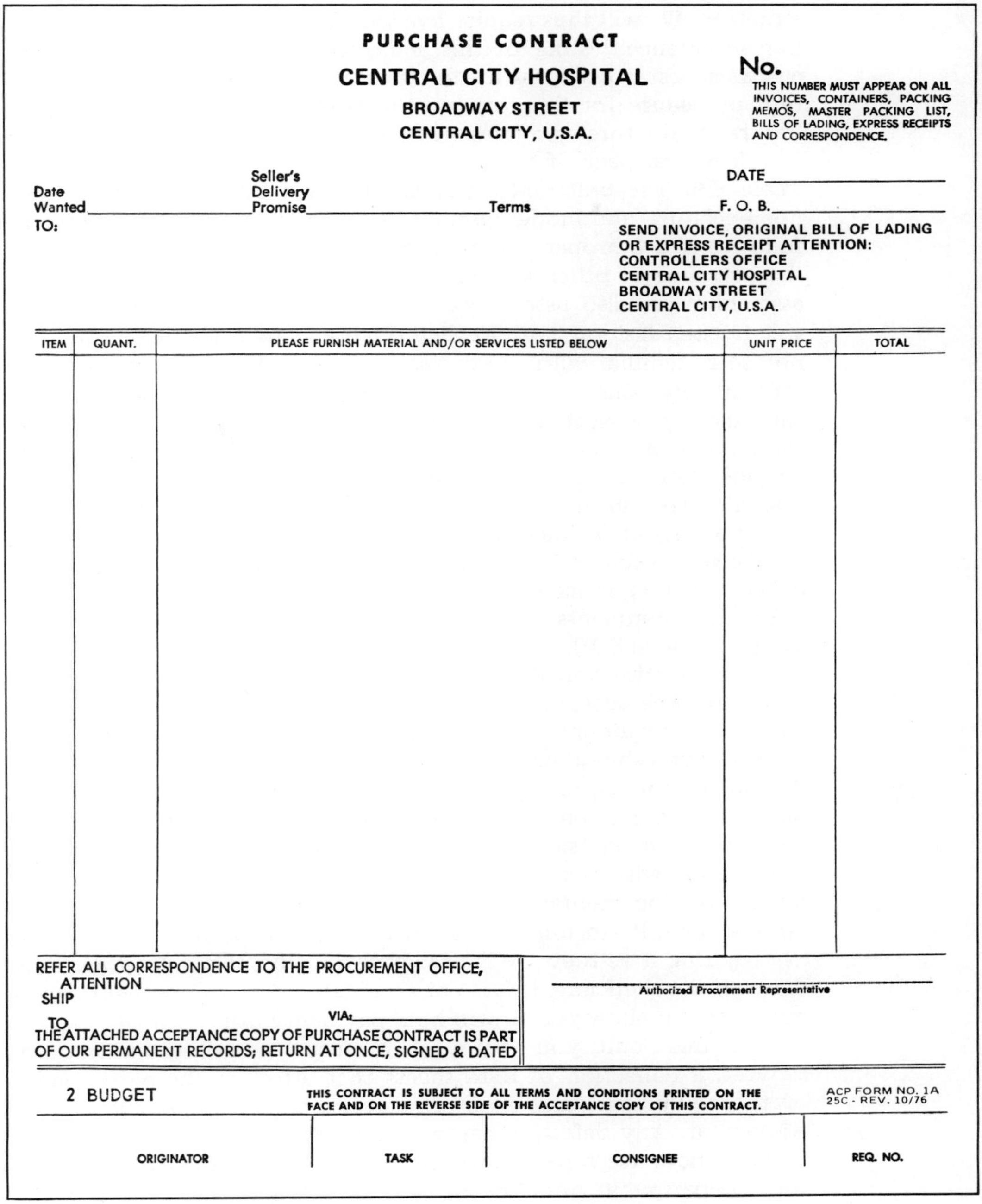

PURCHASE CONTRACT

CENTRAL CITY HOSPITAL

BROADWAY STREET

CENTRAL CITY, U.S.A.

No.

THIS NUMBER MUST APPEAR ON ALL INVOICES, CONTAINERS, PACKING MEMOS, MASTER PACKING LIST, BILLS OF LADING, EXPRESS RECEIPTS AND CORRESPONDENCE.

DATE__________

Date Wanted__________ Seller's Delivery Promise__________ Terms__________ F. O. B.__________

TO:

SEND INVOICE, ORIGINAL BILL OF LADING OR EXPRESS RECEIPT ATTENTION:
CONTROLLERS OFFICE
CENTRAL CITY HOSPITAL
BROADWAY STREET
CENTRAL CITY, U.S.A.

ITEM	QUANT.	PLEASE FURNISH MATERIAL AND/OR SERVICES LISTED BELOW	UNIT PRICE	TOTAL

REFER ALL CORRESPONDENCE TO THE PROCUREMENT OFFICE,
ATTENTION__________

SHIP TO__________ VIA:__________

THE ATTACHED ACCEPTANCE COPY OF PURCHASE CONTRACT IS PART OF OUR PERMANENT RECORDS; RETURN AT ONCE, SIGNED & DATED

____
Authorized Procurement Representative

2 BUDGET

THIS CONTRACT IS SUBJECT TO ALL TERMS AND CONDITIONS PRINTED ON THE FACE AND ON THE REVERSE SIDE OF THE ACCEPTANCE COPY OF THIS CONTRACT.

ACP FORM NO. 1A
25C - REV. 10/76

ORIGINATOR	TASK	CONSIGNEE	REQ. NO.

Figure 8.3 Typical purchase order form.

to house items (1) through (3), containing a wiring harness and perhaps a power supply, and (5) a mechanical support and transport structure. We will thus require five individual procedures to carry out the acceptance testing of this acute-care cart. There will be three electrical test procedures for the three modules, a mechanical inspection procedure for the structure, and an overall electrical test procedure for the three modules mounted in the housing.

The first page of a typical electrical test procedure is shown in Table 8.9. The unit under test is identified by the manufacturer's nomenclature and model number. A list of required test equipment follows. If the proper electrical testing of this subassembly requires the presence of other subassemblies of the system, these other subassemblies are also listed. The test procedure itself is prepared in a step-by-step fashion, with explicit directions to guide those who may not be as familiar with the system as you are. Prepare this procedure thoughtfully, since it can also be employed in troubleshooting this subassembly when it fails, and will enjoy the same useful lifetime as the equipment itself. Each step in the test procedure is followed by an indication of what the proper response for that step should be. A separate data sheet is provided, keyed to the test procedure. One such data sheet is filled out for each subassembly of this type that is received. Table 8.10 shows part of the data sheet for the signal delay unit test procedure of Table 8.9. The data sheet contains a space for comments and an acceptance signature which are not shown in Table 8.10.

The vendor will deliver a set of documentation along with his hardware. The acceptability of this "software" must also be checked against the requirements of your specificaiton. If you are buying off the shelf, you should have examined the acceptability of the vendor's documentation before purchase. Any vendor who supplies incomplete documentation, regardless of his motive, should have been removed from consideration at that point. If you are buying in accordance with a specification, be particularly observant of the quality of documentation on the overall system. Generally, this is "one-of-a-kind" documentation, and some vendors are not as careful in preparing it as they are in preparing documentation for their standard off-the-shelf line. If it is not acceptable, the contractual requirements should allow you to withhold payment until it is made so.

At this point, you have a set of subsystem and system test procedures, a collection of data sheets that attest to the performance of the equipment, and a set of documentation pertaining to the equipment. Any defective hardware or software should be sent back to the vendor for repair. Upon its return and acceptance, your contracts department should be notified in writing that all aspects of the system are in compliance with the specification, so that final payment may be made. This will also initiate the warranty period.

TEST PROCEDURE

Signal Delay Unit, Model 881734-1

Test Equipment Required

1. Hewlett Packard 203A Function Generator, or equivalent.
2. Tektronix 561A Oscilloscope, with 2A63 and 2B67 plug-in units.
3. Triplett Model 601 VOM.
4. Special test cable.

ICU Equipment Required

1. Signal Delay Unit.
2. Brush 4-channel recorder.

Test Procedure

1. Connect a test cable to the 24 pin Blue Ribbon connector at the rear of the delay unit. The connector should be wired as follows:

Pin #	Connection
1 & 24	115V ac
12	Chassis & signal ground
2, 11, 13, 23	No connection
3 thru 10	Signal output, Ch 1-8, respectively
14 thru 21	Signal inputs, Ch 1-8, respectively
22	Connect to pin 12

Proper Response ✓

2. Connect ac power, and turn front switch on.

Proper Response ✓

3. Connect pin 14 of the test cable to ground with a clip lead, set the VOM to the 0.1 volt dc scale, and measure the dc voltage on pin 3. Repeat for the other seven inputs and their respective outputs.

Proper Response $\leqslant$ 0.02V on all outputs.

4. Set the HP function generator at 10 Hz, and adjust its output to 4.0 volts p-p using the Tektronix oscilloscope. Connect the function generator to pins 14 through 21, in succession, and record the output on pins 3 through 10 respectively.

Proper Response 4.0V ± 0.1V on all outputs

Table 8.9 Typical electrical test procedure.

DATA SHEET

Signal Delay Unit
Model 881734-1

Serial Number ______________

		Date		By	Date		By	Date		By
1. Connect test cable										
2. Apply ac power										
3. Dc offset with	Pin 3									
inputs grounded	4									
$\leqslant \pm$ 20 mV.	5									
	6									
	7									
	8									
	9									
	10									
4. Unity Gain Check	Pin 3									
In = 4.0 V p-p	4									
Out = 4.0 ± .1 V p-p	5									
	6									
	7									
	8									
	9									
	10									
5. Frequency and Transient	Pin 3	BW	Rise Time	Over-shoot	BW	Rise Time	Over-shoot	BW	Rise Time	Over-shoot
Response	4									
$>$ 50 Hz B.W.	5									
$\leqslant$ 6 ms delay	6									
$\leqslant$ 15% overshoot	7									
	8									
	9									
	10									

Table 8.10 Typical test procedure data sheet.

The information contained in the test procedures and data sheets will also form the basis for the equipment control program described in detail in the next chapter. Some, or all, of the steps contained in the acceptance test procedures will undoubtedly be employed in your preventive maintenance program, and in the repair of defective units.

8.15 User Training

In parallel with the preparation of acceptance test procedures, the procedures to be employed for in-service training should be finalized. For contractually procured equipment, the vendor will be providing initial in-service training, and you should review and approve his syllabus. Most vendors of off-the-shelf equipment also supply such training, and its adequacy should be ascertained. Chapter 12 treats this subject in detail.

8.16 Control of the Acquisition Process

Up to this point we have dealt in detail with all aspects of equipment acquisition from clinical problem definition through acceptance testing and user education. A few words are now in order concerning overall responsibility for these processes, and finally a word about payment.

Coordination responsibility and payment

Whenever either a single piece of equipment or an entire system is purchased by a hospital, someone should be responsible for assuring that all appropriate phases of the acquisition process are carried out before the vendor is paid. While one would expect this to be a matter of policy built into standard operating procedures, the principle can and does break down in practice where communications are poor. The coordination responsibility is one for which the clinical engineer is uniquely qualified; he is able to effectively communicate with clinical users, the purchasing department, contract office, accounting department, and the vendor. He is able to assure the hospital, through an equipment quality assurance program, that equipment received meets its technical specifications, and that vendor payment is therefore in order.

One mechanism for providing the control required for an effective equipment acquisition system is to establish policy agreements between clinical engineering and both the purchasing and accounting departments. The purchasing department should agree to require clinical engineering approval on all requisitions from the clinical services seeking patient-related capital equipment. On generating an approved purchase order, a note similar to "Pay only on acceptance by Clinical Engineering Services" should be a part of the order. Another stipulation added to the order should be "Warranty begins on acceptance by Clinical Engineering Services." The second note alerts the vendor to the hospital's terms of warranty initiation; the first note alerts your accounting department to hold payment pending your approval. This arrangement is usually a very workable one for an accounting department, since they usually match incoming invoices with purchase orders and receiving tickets. Thus, the purchasing department's note to accounting is very visible. Your copy of the purchase order allows verification that the suggested stipulations are included.

Appendix 8.1

DRAFT

(Organization)

Attention: (Local Rep)

Subject: Request for Proposal—Cancer Center Patient Monitoring Instrumentation

Enclosures: (1) Specification for Cancer Center Patient Monitoring Instrumentation, dated July 30, 1978

(2) Construction Drawing E-22, "Physiological Monitoring System," dated May 13, 1976

(3) Sketches CB155 and 155A, "Nurse's Station Casework," dated April 5, 1978

Gentlemen:

A Regional Cancer Center is presently under construction at the Central City Hospital. We desire to purchase a system of instrumentation for acquiring physiological variables from selected Cancer Center inpatients. The system to be purchased is described in detail in the three enclosures. You are invited to submit a technical and cost proposal for the design, fabrication, test and delivery of this system, in accordance with the information contained herein.

Guidelines for Proposal Preparation

1. It is believed that the requirements of the system to be procured are general enough that they can be met, in the main, with a variety of commercially available assemblies and subassemblies, packaged and interfaced in a way appropriate to the particular requirements of the Cancer Center.

2. The vendor is encouraged to propose a system that is directly responsive to the specification, or to propose alternative solutions that appear to meet the intent of the specification.

3. The proposal should be submitted in two separable sections, (1) technical, and (2) cost.

4. The technical proposal should contain the following information:

(a) A complete and detailed description of the electronic modules/subassemblies being proposed to meet the requirements of the specification. Commercial data sheets should be included, if applicable.

(b) A complete and detailed description of the switching system that is proposed to support the use of these modules/subassemblies in the manner described in the specifications.

(c) A description of the mechanical configuration proposed for the nurse's station monitoring console and the two monitoring cart types, complete enough so that we may evaluate them from the human factors viewpoint.

(d) An indication of the size and weight of the nurse's station monitoring console and each of the carts.

(e) Details of the precautions that have been taken to assure the electrical safety of patients and staff, including the details of such protection when modules are not in place in the two types of carts.

(f) A description of your module, subassembly and assembly test philosophy.

(g) A sample of the type of documentation you plan to supply with the system.

(h) A description of the in-service training that will be provided with the system.

(i) Terms of the warranty that will apply to the system.

(j) A description of the maintenance facilities that you can provide for the system and its components, and a copy of the service contract that is available.

(k) A list of local health-care facilities that employ your equipment, with names and phone numbers of contacts.

5. The cost proposal submitted should be itemized in the following manner:

(a) The cost of four (4) nurse's station monitoring consoles, identical and complete.

(b) The cost of acute-care monitoring carts, in quantities of one (1) through thirty (30).

(c) The cost of intensive-care monitoring carts, each equipped with a waveform display module, in quantities of one (1) through ten (10).

(d) The cost of: Arterial pressure modules
ECG modules
Venous pressure modules
Airway/pleural pressure modules
EEG modules
Blood oxygenation modules
Respiration function modules

in quantities of one (1) through ten (10) each.

(e) The delivery schedule proposed for the equipment.

Proposal Evaluation

Proposals will be evaluated for technical responsiveness to the specification (or for the responsiveness of proposed alternatives to the intent of the specification) and for cost factors. Following this evaluation, you may be asked to make available to CCH certain of the equipment you propose to employ in the monitoring system so that cancer center personnel can evaluate its technical and operational characteristics. This clinical evaluation is expected to require no less than two weeks time.

CCH reserves the right to reject any or all proposals received as a result of the above evaluation.

Proposal Submittal

Four copies of your poposal should be submitted to the attention of:

Ann Brown
Central City Hospital
100 Administration
Broadway Street
Central City, USA

and should be received no later than August 31, 1978.

Any technical questions concerning specific items of the specification should be referred to:

J. B. Forest
Clinical Engineering Division
Regional Cancer Center
Central City Hospital
Broadway Street
Central City, USA

Very truly yours,

8.17 Study Questions

8.1 List the steps involved in the equipment acquisition process.

8.2 Describe the team approach to the definition of clinical requirements for medical equipment.

8.3 What equipment considerations are most important to the physician? To the nurse? To the clinical engineer? To the administrator?

8.4 List the steps to be carried out in the specification of a piece of medical equipment.

8.5 List the major steps carried out in an environmental survey regarding medical equipment procurement.

8.6 Describe the major steps to be carried out in a literature search regarding purchase of a piece of medical equipment.

8.7 How can vendor contacts be utilized in procuring medical equipment?

8.8 What is a "running cost estimate" and how is it used in equipment acquisition?

8.9 Describe the method of synthesizing information from a literature search.

8.10 Describe the four directions that equipment acquisition may take.

8.11 List the five components of an equipment specification.

8.12 Describe the major items to be included in the "requirements" section of the equipment specification.

8.13 Should a specification define what is to be done or how it is to be done? Why?

8.14 What is the purpose of the system testing and acceptance section of the system specification?

8.15 List the major components to be included in a request for proposal (RFP).

8.16 What steps should be followed in submitting an RFP to vendors?

8.17 Describe the three steps involved in a preliminary review of vendor proposals.

8.18 Summarize the activities accomplished in each of the six steps of equipment evaluation.

8.19 List the five sections of a vendor selection memorandum.

8.20 Describe the processes of vendor selection.

8.21 List the three major sections of a contract.

8.22 What are the six sections contained in the schedule of work?

8.23 What is the difference between a purchase order and a purchase requisition?

8.24 Describe the process of equipment acceptance.

8.25 What roles should the clinical engineer play in overall coordination and control of the equipment acquisition process?

8.18 References

Alliance for Engineering in Medicine and Biology (1977). Summary guidelines for technology procurement in non-federal health care facilities. A study (NSF Grant No. APR74-19126 A03). See especially Chaps. 4-7.

Cooper, R. (1974). Limitations of instruments and pitfalls in use. In C. A. Ray (ed.), Medical engineering. Chicago: Yearbook.

Greatbatch, W. (1974). Designing for high reliability in medical electronic equipment. In C. A. Ray (ed.), Medical engineering. Chicago: Yearbook.

Simmons, D. A. (1972). Instrument evaluation. Chapter 8 in D. A. Simmons, Medical and hospital control systems. Boston: Little, Brown.

9
The Equipment Control Program

EMANUEL FURST, WILLIAM F. BETTS,
and ALAN K. REETER

9.1 Learning Objectives

Upon completing this chapter you should be able to:

1. Explain the benefits of an equipment control program in a hospital.
2. Describe the major goals of an equipment control program.
3. List the major elements of an equipment control program.
4. Plan and implement an equipment control program for a hospital.

9.2 Introduction

The equipment control program permits the clinical engineer and the hospital to organize conceptual and managerial techniques for control of the ever-increasing number of increasingly complex instruments which contribute significantly to health care. The clinical engineer provides the technical expertise to free nurses, physicians, technicians, and others from concern over such things as specifications and repair, thus permitting them to concentrate on patient care. It allows centralization of hazard, quality, and cost control functions and provides for regulatory compliance efforts. Most of all, it provides a structure for assigning responsibility and authority to each department for appropriate elements of the hospital's efforts to manage the application of technology; and it allows the recruitment of all employees in this effort.

9.3 Overview of Equipment Control Program

The control program provides a structure for the clinical utilization of equipment in the hospital, and directs the effort by the entire institution to apply technical competence, management techniques, and organizational skills to the control and application of technology. Many of today's advances in medical care are based upon technological successes as much as they are based upon advances in medical knowledge. However, medical personnel, though highly trained and experienced in their own fields, frequently have little technical knowledge. Therefore, they usually do not have the breadth of experience or education to appreciate all the technical efforts required to assure the selection, acquisition, maintenance, and use of equipment in a safe and effective manner. Nor do they appreciate its interaction with other hospital equipment or utility systems. The clinical engineer, by bringing a "systems" concept to the management of technology, helps make hospital equipment more effective and more readily available for patient care.

The concept of an equipment control program received major emphasis with the publication of a comprehensive plan in *Health Devices* (Emergency Care Research Institute, 1971). Many of the programs in place today utilize elements of this effort.

Needs and applicable code requirements

There is a need for an overall institutional effort to manage equipment in the modern hospital facility. The hospital's budget contains substantial funding for the acquisition of instrumentation, ranging from small, portable, electronic thermometers to massive, fixed, radiological systems. The annual preventive maintenance and repair costs may average between 5 and 15 percent of the equipment acquisition costs. With advancing technology, there is enormous pressure to buy new devices in order to maintain a high level of medical care. The very substantial hidden costs of managing and maintaining these instruments is often overlooked.

Technical support in the management of technology can minimize costs and help allocate resources to real needs, as evidenced in the description of equipment acquisition in Chap. 8. Financial, quality, and hazard control help assure that the hospital is getting everything it pays for, and that the patient is receiving the best possible care. Unfortunately, external recognition of the need for technical and managerial controls regarding equipment and facilities has led to the development of cumbersome and expensive regulations, codes, and standards. These attempts to mandate application of good engineering practice, as described in Chap. 4, are partly the result of the slow progress of individual hospitals to develop effective internal controls.

The requirements of the Food and Drug Administration (FDA) promulgated under authority of the Medical Device Amendments (Federal Register, 1976) are also of interest to the clinical engineer. These will impact upon medical equipment from the conceptual and research stage through manufacturing and sales into utilization (Chap. 4).

At the present time, however, for the majority of hospitals, the major requirements are defined in the accreditation standards of the Joint Commission on Accreditation of Hospitals (JCAH), elements of codes and standards issued by the National Fire Protection Association (particularly the *National Electric Code* and the standard on *Inhalation Anesthetics*) and appropriate regulations issued by state and local authorities (such as the State Health Department) (Chaps. 4 and 7).

Accreditation requirements

Section 4.7 points out that the voluntary accreditation program of JCAH requires the hospital to document periodic testing of equipment. This should include tests for safety, calibration, and function. While the JCAH requirements may be independently carried out by many separate departments, placing these responsibilities in a central clinical engineering group results in more efficient management of the program.

The "optimal achievable" standards of the JCAH are intended, among other things, to move hospitals along a planned program to develop and implement methods for providing safe and reliable instrumentation and electrical systems.

Of particular significance in the standards are requirements of the Functional Safety and Sanitation section as well as others distributed throughout the document that relate to electrical safety and medical instrumentation. These require an active safety program with monthly meetings of a safety committee; training of staff in the safe use of electricity and devices; formal scheduled maintenance and calibration programs; technical assistance available to all employees; and substantial documentation.

The result is either a large number of distinct requirements with associated uncoordinated demands upon each hospital department, or the development of a structured equipment control program. It is this cohesive framework, with responsibilities delegated to each hospital department, with a central focus in the clinical engineer, that helps promote efficient management of the complex, instrumented hospital.

Definition of the program

The equipment control program is a system employed by the hospital to insure efficient and effective procurement and utilization of medical devices and instrumentation. Significant program components are policies, procedures, interdepartmental relationships, integration with other clinical engineering functions, and communications. It is also the framework for defining the clinical engineer's roles, responsibilities, and authority. We will describe these aspects of equipment management, which must be understood by all personnel who depend upon instrumentation to provide health care.

An equipment control program, to be effective, must be consistent with the overall mission and integrated within the organizational structure of the hospital (Sec. 2.5). This program must be designed from the start for effective interaction with, and support of, other hospital departments; and closely integrated with all other clinical engineering responsibilities.

The equipment control program, by virtue of its function as the framework for managing equipment, must incorporate a significant data base for managerial decisions and formulation of institutional policies and procedures. Figure 9.1 provides a schematic representation of many of the components that must be established in order to build a cohesive structure.

Policies and procedures. A policy may be considered as a formal recognition of an institution-wide set of desired behavioral objectives. We can perform admirably by means of informal authority and the rapport established between individuals, but in a large organi-

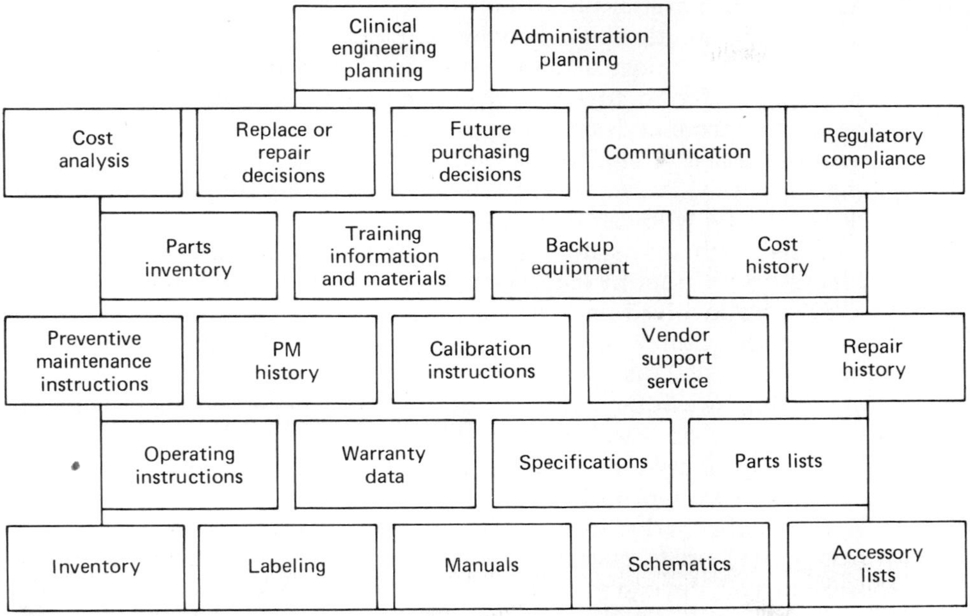

Figure 9.1 The multiple elements of an equipment control program provide the framework for daily operations, decision making, and regulatory compliance at many levels.

zation, we must spell out mutual responsibilities and authorities. Further, we must formalize the expectations of management in order to legally protect the institution and its employees; to provide the patient assurance of safety and quality of care; and to provide a framework for common communication and, if necessary, disciplinary action. The result is usually a series of policies and procedures that publicly identify the important elements and organizational entities responsible for bringing about expected performance. Currently, as examples, the JCAH requires policies controlling patient-owned appliances, extension cords, the handling and reporting of incidents (any unusual events that affect patient care—specifically of interest are those incidents involving malfunctioning or dangerous equipment), and training in the safe use of equipment (Sec. 4.7). Sample policies that may be useful in developing an equipment control program are discussed in Sec. 9.5.

We can measure the effectiveness of the policies and procedures in terms of the number of occasions on which they are correctly implemented, as compared to the number of occasions on which they are not. We can obtain an earlier measure in terms of the staff's awareness of the policies and procedures, particularly those that are not "intuitively obvious." The staff of any organization would prefer not to read material unless it is directly relevant to the performance of a given function. The policy and procedure manual falls in this category. It is most often considered a reference to be kept on the shelf and referred to only when something unusual happens, when protection from undesirable requests is needed, or when a disciplinary action is taken. Therefore, the effectiveness of a policy depends upon the support that can be mustered in developing the policy, the number of people who can be effectively involved during its development, and the mechanisms for disseminating the policy and educating people as to its significance. This is to say that as many key people should be involved as is practical, without spending too much time getting the policy written or adopted. The Hospital Safety Committee (Sec. 4.7) is a very useful and, in general, a required element in developing a policy. This committee will also help in winning acceptance and visibility for the policies. Policies that are found to be ineffective should be rewritten or eliminated. A policy should be considered as appropriate and accurate when it is written. However, it should be periodically reviewed and revised as necessary. If the policy needs revision more often than once a year, it may be poorly written and conceived, while if it does not need revision after five years, it is either an unusually well-written policy or the environment has become static.

Interdepartmental relationships. The equipment control program has two broad objectives: (1) control of cost, quality, and hazards; and (2) establishment of a focus for the management of technology in the hospital. The clinical engineer must be involved in each phase of the instrument's life, from acquisition to retirement, in order to achieve these objectives. Figure 9.2 shows that nearly every phase of the equipment's lifetime of use in the hospital will require some technical support and possibly documentation. Each block in Fig. 9.2 represents a function normally carried out by the clinical engineer or biomedical equipment technician (BMET). Appropriate engineering expertise can enhance the utility of a piece of equipment. Further, equipment exists in virtually every department, is often shared by departments, and is also often mobile. Therefore the clinical engineer must know how equipment functions, how and where it is used, and its typical modes of failure. This overall view is provided by the equipment control program which defines each department's role and responsibility.

Defining the equipment control program is a large undertaking which will require months. The program must provide a clear structure adaptable to a changing program. The equipment control program, like the technology it manages, is an evolutionary activity; if it changes too slowly, it will be out of step with medical practice, while if it changes too rapidly, the ensuing confusion will impede lasting relationships with other departments.

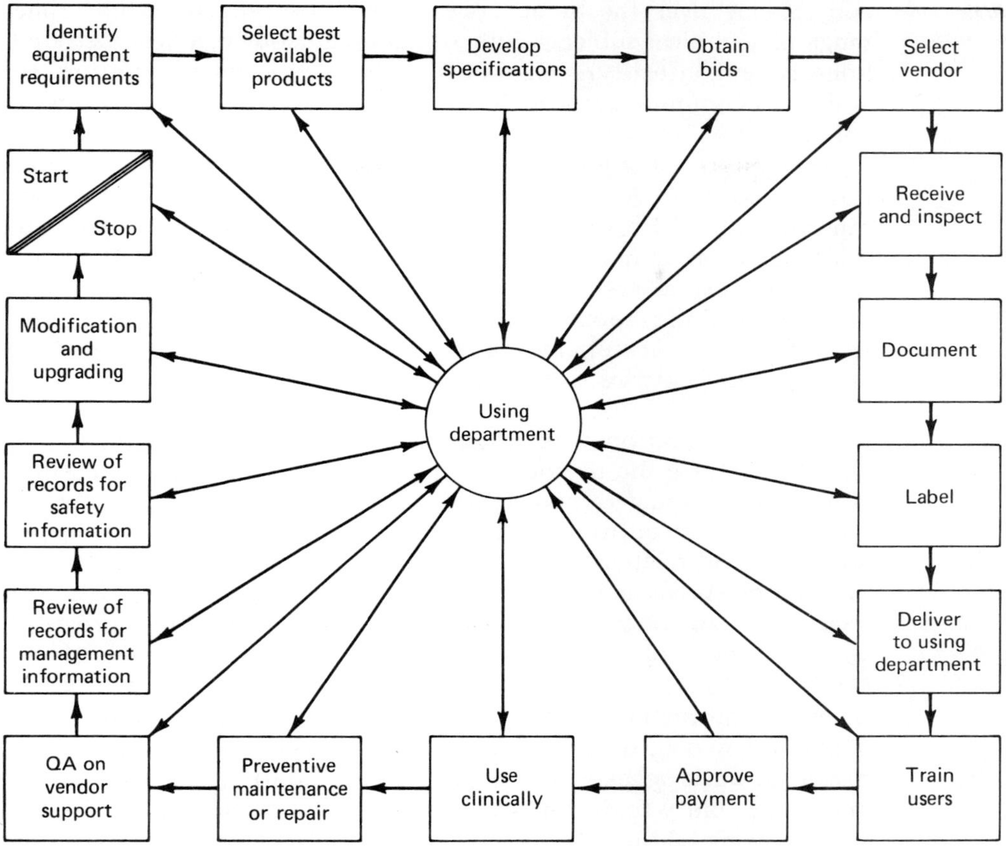

Figure 9.2 The normal flow of new equipment showing opportunities and requirements for communication with using department. Each block represents an area of sole or shared responsibility of the clinical engineer. For simplicity, we do not show (1) the interactions internal to the clinical engineer's office which "close the loop," e.g., experience in maintenance will affect future vendor selection and user training, or (2) the essential interactions with others who may share responsibilities in these areas with the engineer.

The clinical engineer will often find that he fills a unique position within the hospital. As described earlier (Sec. 2.5), this presents formidable problems regarding reallocation of responsibilities and other changes in the hospital structure. The clinical engineer will find the hospital employees' dedication to the care of the patient and to the hospital most impressive. If the engineer develops and shares this same dedication, he will become an accepted member of the health-care team.

There are several elements that assist the engineer in gaining acceptance from the hospital staff. These enable him to communicate informally and achieve desired results. Key items include technical competence over the broad areas of expertise and over the broad range of technical subjects with which he must deal, a demonstrated honesty and integrity such that his motives are not in question, an ability to communicate and to respect others' points of view, and a willingness to be of help and to assume whatever responsibilities are necessary to get the job done. However, the clinical engineer's visible position as a broadly technically competent individual also makes him vulnerable. He is often newly graduated, therefore lacking applied technical and, most importantly, clinical experience. There are also pressures inherent in caring for patients, in being thrust into decision-making responsibilities early in his career, and in entering an institution that initially may contain persons or departments who are opposed to him or to the new program.

Integration with other clinical engineering functions. The ability to utilize the personnel, functions, and data relating to other clinical engineering responsibilities are fundamental to a successful equipment control program (Sec. 6.3). For example, investigation of an instrument-related "incident" will often require access to records of the preventive maintenance and repair performed on a particular instrument, to the schematics and maintenance instructions provided by the manufacturer, to the user and his supervisor and possibly to the training programs conducted for the user. Decisions on procurement of additional instrumentation from a particular vendor (Sec. 8.11) will depend in part upon the maintenance frequency and the historical costs revealed in the equipment control file. The clinical engineer's ability to document his efforts plays a role in accreditation and licensing of the hospital and in obtaining his budgetary requirements. Thus, the materials generated in one aspect of the clinical engineering program will be of great importance to other aspects of the program (Chap. 6).

Communication. In achieving the objectives described in the preceding section, the need for communication stands out. The engineer must gather information. He must then communicate his data and recommendations in a clear and forthright manner. There are a number of communication channels available, although the importance of each varies from situation to situation. These include written memoranda and reports, scheduled meetings with administrative personnel, committee meetings and frequent informal discussions with department heads, unit managers, and special-care nursing staffs. For many people, an open office providing easy access to engineering support or frequent rounds of nursing units is a major part of the communication process.

It may be useful to illustrate each of the communication channels: written reports may be used to provide inspection results and recommendations; memoranda may provide an outline of future plans for instrumenting a new clinical unit; informal discussions can give an opportunity to raise problems facing a department manager; and rounds present many opportunities for staff education. Each line of Fig. 9.2 indicates a need to communicate data, plans, results, or recommendations between the clinical engineer and the user (client) department. These channels of communication are the mechanisms for defining priorities, communicating results, and obtaining the necessary resources with the help of client departments and the administration.

The engineer also must be careful to maintain a professional attitude and to be accurate and complete in developing communications. Credibility is hard to establish and takes months to secure. A lack of professionalism or a posture of knowing more than he actually does will prevent or destroy the essential credibility required. The rapport established will define the success of the program and its integration into the health-care team. It will also provide the documented communication required by JCAH and others (Sec. 4.7).

The clinical engineer must develop two other communication paths. The first of these is a communication and data analysis system that provides reports of work performed, recommendations, cost and quality control information, and in turn solicits the client's perception of quality and cost, and his analysis of program results.

An additional feedback mechanism can be obtained through the committee structure. The clinical engineer is generally a member of the Hospital Safety Committee and may find that he is a member of additional committees. He can be aided by these committees to the extent to which he can educate and communicate, so that committee members understand his role and responsibilities and become aware of regulatory influences. In turn, the engineer must understand the needs of the various departments and administrators represented on the committee.

9.4 Inventory Control

Existing equipment

The first step in the effort to manage the variety and number of instruments utilized within the hospital is to identify what they are, where they are located, and who is responsible for the ownership and use of each device. An orderly method for recording this and all other information related to each instrument must be established so that existing and future data can be retrieved, analyzed, and applied to future decisions and solutions of problems. The form shown in Fig. 9.3 illustrates the types of information and record-

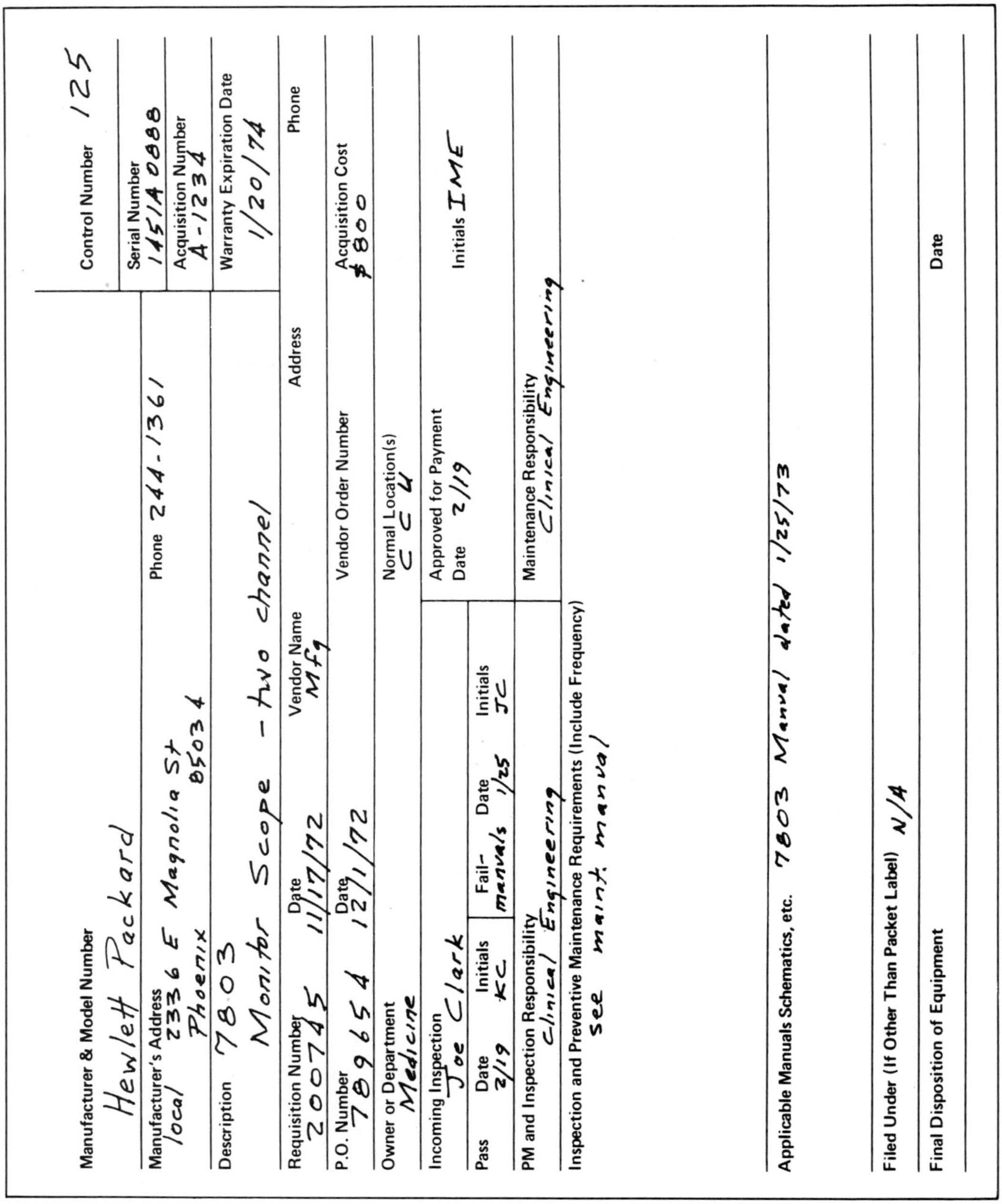

Manufacturer & Model Number
Hewlett Packard

Manufacturer's Address
local 2336 E Magnolia St
Phoenix 85034
Phone 244-1361

Description 7803
Monitor Scope - two channel

Control Number 125

Serial Number
1451A 0888

Acquisition Number
A-1234

Warranty Expiration Date
1/20/74

Requisition Number	Date	Vendor Name	Address	Phone
200745	11/17/72	Mfg		

P.O. Number	Date	Vendor Order Number	Acquisition Cost
789654	12/1/72		$800

Owner or Department
Medicine

Normal Location(s)
CCU

Incoming Inspection
Joe Clark

Pass	Date	Initials	Fail–	Date	Initials
	2/19	KC	manuals	1/25	JC

Approved for Payment
Date 2/19 Initials IME

PM and Inspection Responsibility
Clinical Engineering

Maintenance Responsibility
Clinical Engineering

Inspection and Preventive Maintenance Requirements (Include Frequency)
see maint. manual

Applicable Manuals Schematics, etc. 7803 Manual dated 1/25/73

Filed Under (If Other Than Packet Label) N/A

Final Disposition of Equipment

Date

Figure 9.3 The equipment control card is a summary of major data for managing the instrument. The card is part of an individual file for each instrument or module. Filing is by control number. Adapted from *Health Devices,* Vol. 1, No. 4, p. 91.

keeping procedures that may be employed in building the basic information base required.

The clinical engineer should fill out a form such as Fig. 9.3 for each piece of equipment that is located as part of the inventory. During the initial inspection of the device he can obtain the manufacturer, model number, serial number, and description of the equipment. He should carefully describe the equipment and recognize that there are many pieces of medical equipment that are uncommon. When he is unsure as to the function of a piece of equipment, he should ask someone (a nurse, for example). The description should be included when filling out the index card, shown in Fig. 9.4, and it should be consistent for all pieces of similar equipment. The clinical engineer should add the control number and place a tag bearing this number on the equipment. The acquisition number, if any, is generally an institutional property control tag. The clinical engineer often adds a new control number instead of relying on this acquisition number. Reasons for this apparent duplication include differing criteria between equipment acquisition and control number assignment (acquisition numbers may be assigned only to expensive items), modularity of equipment (which may mean modules are moved, but the mainframe remains fixed), and the need for a consistent numbering system for both existing and new equipment. The acquisition number may help locate other information filed in purchasing or property control departments.

From the department having possession the clinical engineer can obtain items such as the warranty expiration date, purchase order (PO) number, acquisition cost, owner, manufacturer's address, and equipment location. The acquisition cost may be important in depreciating the equipment, in third-party billing, or in decisions to repair, modify, or buy new equipment.

For example, Medicare's reimbursement to the hospital may be based upon the actual cost of providing patient care. This includes the annual depreciation, preventive maintenance, and repair for equipment used in patient care. Thus, it is useful to determine the acquisition cost and date for all equipment so that depreciation costs are known. The manufacturer's address is important in procuring manuals, schematics, and technical support. However, before obtaining these items from the manufacturer, the clinical engineer should check with maintenance or plant engineering to see if they have appropriate manuals. If so, he can enter this information in the appropriate space on the form.

The card illustrated in Fig. 9.4 contains similar information, but is completed more easily. The clinical engineer should normally file the cards by manufacturer. He may file second copies by equipment type. He should also transfer the information contained in the control and index cards to an inventory list such as that shown in Fig. 9.5. The three forms shown in Fig. 9.3 to 9.5 have sample entries for illustration.

Mfr. H/P **C/N** 125

Model # 7803

Description 2 channel monitor scope

Location CCU

S/N 1451A0888

Req. # 200745

P.O. # 789654

Mfr. *H/P* C/N *125*

Model # *7803*

Description *2 channel monitor scope*

Location *CCU*

S/N *1451A0888*

Req. # *200745*

P.O. # *789654*

Figure 9.4 Index cards are completed for each instrument. One is filed by manufacturer (and model number under this category). A second copy could be filed by equipment type.

DEPARTMENTAL INVENTORY LIST

Date: Jan 1973 | Department: Medicine

Equipment Description	Manufacturer	Model	C/N	S/N	Location	Purchase Date	Connected to
ECG Monitor	HP	7830	0125	1451A088	ICU	12/1/72	central station
			0126	1451A0889		12/1/72	central station
			0127	1451A0912		12/1/72	central station
			0128	1451A0913		12/1/72	central station
			0511	1451A1234		12/1/72	central station
			0716	1451A1962		7/6/73	central station
Arterial Pressure Monitor	HP	7809	0205	1114A0352		12/1/72	float
			0206	1114A0355		12/1/72	"
Temperature/ Pressure Monitor	HP	7819	0207	1114A0123		12/1/72	"
			0208	1114A0124		12/1/72	"
Monitor Scope	HP	7803	0209	1104A6221		12/1/72	"
			0210	1104A6222		12/1/72	"
Defibrillator	HP	7802	0211	1116A5113		12/1/72	"
Electric Bed	Simmons	H570	0273	60023		11/17/72	N/A
			0274	60024		11/17/72	N/A
			0275	60025		11/17/72	N/A
			0276	60032		11/17/72	N/A

Figure 9.5 Cross indexing of the inventory file by department is useful in budgeting and scheduling of clinical engineering workload.

Perhaps the most time-consuming aspect of inventorying existing equipment is obtaining as complete a list as is possible. Locating and identifying the equipment as to manufacturer, model and responsible (owner) department is but part of this task. The clinical engineer must come to know the use to which this equipment is put by each of the individuals who utilize it. For example, an ECG machine may be used during the day by the ECG technician and left at night on a particular nursing unit. This same ECG may or may not be suitable for use with catheterized patients. It may be safe when used by ECG technicians, while the same unit may not be safe or reliable in the hands of the various persons who will make use of it on the evening and night shifts if they have been inadequately trained.

As the clinical engineer attempts to gather existing documentation on the instrument, he will find that much of this information doesn't exist. For example, some devices may have been the property of particular physicians rather than the hospital. It may no longer be possible to obtain documentation on older equipment. Another difficulty is determination of whether or not maintenance contracts exist, who provides the maintenance coverage, and at what cost. Some of this information may be readily available from the purchasing department. This search for information that is difficult to locate should not be tackled in a rush or at a single point in time.

Obtaining the information required to initiate this file on existing equipment is not only time-consuming but frustrating, because the typical hospital usually has not centralized this information prior to the development of the clinical engineering program. The vendor is often not the manufacturer, but a local sales representative whose name bears no resemblance to that of the manufacturer. In short, the records that are of most use to the purchasing agent may be of little use to the engineer. Another possibility is that each clinical department may be responsible for the purchase and maintenance of its own equipment. Therefore, the engineer may have to deal with each department in obtaining the requisite information and will often find that data regarding current maintenance contracts, preventive maintenance, repairs, and calibration are scattered, if they exist at all. His ability to locate or to obtain these data will depend upon his communication of the importance of this information to each department as well as to the administration. Once again, the clinical engineer's ability to find the "silver lining" in the cloud of required regulation and documentation will determine his effectiveness in obtaining the required equipment data base. One "silver lining" may well be an identification of hidden costs, or of methods by which to improve service or reduce costs. He should be alert, however, to the existence of established friendships or satisfactory, long-standing business relationships which can be hazardous when threatened. He must not assume that external vendors are incompetent, dishonest, or uncaring.

There is room for external vendors even within large hospitals with well-established CE programs.

The clinical engineer should be alert to the nuances of user satisfaction and dissatisfaction with the instrumentation as he performs his inventory, identifying such causes as instrument condition, vendor support and training. It will take experience to be able to sift through the feelings expressed and ask appropriate questions to correctly identify the real source of any dissatisfaction encountered so as to be able to eliminate the problem. The clinical engineer should also identify sources of user satisfaction so that in-house effort can replace external vendor services without loss of confidence or credibility. As rapport is established with individual clinical departments, as clinical engineering staff are acquired and trained, and as proficiency and organization develop, the clinical engineer can examine the existing maintenance programs and analyze these efforts for quality and cost effectiveness.

Initial inspection. An initial inspection can supplement the data base obtained during the inventory. High on the priority list is a general safety inspection to identify malfunctioning, damaged, or worn instruments, or those whose leakage current or other safety aspects need attention. Following this, the clinical engineer can acquire or develop tests utilized for incoming inspection (Sec. 8.14) or preventive maintenance and apply these to obtain the baseline data for future performance testing. He can compare these test results with the manufacturer's specifications and test criteria and then set priorities for effort and corrective action. We suggest that the criteria established for existing equipment be similar to those utilized in incoming inspection. This requires the clinical engineer, in collaboration with the administration and the appropriate clinical departments, to evaluate the risks and costs of corrective action when older equipment does not measure up to new standards.

Following the initial inventory and initial inspection, the clinical engineer, biomedical equipment technician, and supporting clerical staff can set about labeling the equipment and filling the gaps in the documentation. The desire for complete documentation should be tempered by the benefit of the information; where little or no benefit exists, little or no effort should be expended to acquire the data.

Departmental interaction

The clinical engineer should anticipate concerns on the part of the client departments when he first expresses an interest in their equipment. There will be curiosity as to why there is a new, high level of interest in the instrumentation. There may be concern for the use of the information obtained. For example, a department may see the clinical engineer as taking away the department equipment. We have earlier described the need for the administration and the clinical engineer to communicate to the hospital staff the significance of the clinical engineer's role and the method by which the program will develop. It is possible for the staff to become personally and perhaps emotionally concerned when the clinical engineer begins to implement his program. The first sign of this may well be the inventory stage. We suggest that the clinical engineer keep in mind the fact that hospital departments have traditionally owned and been solely responsible for the instrumentation. Therefore, he should utilize tact and patience in acquiring the information. It is not necessary to learn all there is to know about a particular piece of functioning equipment at this first stage. Repeated visits should elicit all the information necessary during the first six months of the program's life.

In a given hospital, the clinical engineer may find much of his documentation available from the purchasing agent and the financial officer. The approach should be open and tactful, as these people are of utmost importance to the success of the clinical engineer. In most institutions, they will welcome sharing the responsibilities of acquiring instrumentation or of negotiating maintenance services, while in other institutions the initiation of a clinical engineering activity will be seen as a major threat to the power that traditionally resided in one or both of these offices. We suggest that early and frequent communication be established.

With the data base established, the clinical engineer is in a position to work with the nursing personnel and the in-service training offices to provide some personnel training and to help arrange for vendor support of their training efforts (Chap. 12). He will also, by this time, having identified priorities, be able to provide assistance in those areas where maximum return can be obtained for a given investment of time or money. This effort will assist greatly in his ability to generate rapport with client departments, particularly if he can show his clients the benefits they have received.

New equipment

Chapter 8 points out that much of the information required for the data base described above is obtained as a portion of the purchase specification or of the internal purchasing procedures. Information regarding the manufacturer's support in the areas of warranties, guarantees, maintenance availability, and preventive maintenance and repair requirements helps provide the basis for the preventive maintenance repairs, and consulting services offered by the clinical engineer. The information required for the forms shown in Figs. 9.3, 9.4, and 9.5 should then be available from the equipment acquisition process described in Chap. 8.

Once the clinical engineer has obtained a data base on existing equipment, knows the hospital organization and personnel and has become known to them, he will find it advantageous to initiate procedures for purchase of new equipment. Implementation of the specification and incoming inspection programs, utilization of warranty and other vendor services, and related elements allows the clinical engineer to "put his finger in the dike" and stem the flow of additional equipment into the hospital unless it meets the hospital's safety and performance requirements. This will make his long-term effort somewhat more manageable. Those elements of Chap. 8 regarding establishment of a high level of quality for newly purchased equipment will be of particular importance.

9.5 Hazard Control

One of the prime reasons for establishing a clinical engineering service relates to fulfilling the hospital's obligation to provide safe equipment for patient care (Chap. 7). Control begins with identifying the source of the equipment and continues with participation in all significant aspects of equipment utilization.

Sources of hazard information

When acquiring new equipment, the engineer can turn to a variety of sources for information regarding the safety and effectiveness of the equipment. Continuing surveillance is aided by three publications. *Health Devices* and *Health Devices Alerts*, publications of the Emergency Care Research Institute, contain evaluations and hazard notices (Sec. 15.6). The Food and Drug Administration recall notices are published in the *FDA Enforcement Report* (Sec. 15.9). Manufacturers voluntarily or under compulsion also issue recall, warning, and other notices. The clinical engineer is the logical focal point for this information, which can also be used in the selection of new equipment. The clinical engineer may maintain a file of all hazard reports by equipment type. He can use an index card to list the equipment type, manufacturer (if appropriate), and source of the hazard warning (with date, source, and summary of the warning). If a device under consideration for purchase historically has problems associated with it, it becomes the responsibility of the clinical engineer to evaluate the significance of the published data as well as any experiential information he can get from engineers and physicians in other institutions. He must then make this information, along with his technical recommendations, available to the purchasing agent, clinical department, and perhaps to the hospital administration. Some hospitals have a Capital Expenditure Review Committee which makes the final decision in the purchase of equipment, and recommendations to this committee will be of importance in reaching decisions.

Sources of acquired equipment

There are various means by which equipment enters a hospital. In addition to purchase or lease of equipment through formal administrative channels (Chap. 8), there is equipment owned by the various health professionals who practice within the institution, new equipment loaned for presale evaluation, purchases made by affiliated organizations such as the Hospital Auxiliary, donations made directly to physicians or to hospital departments or to the administration, and equipment that was designed, manufactured or modified by engineers, technicians or others in the institution. Equipment from each of these sources comes within the clinical engineer's responsibility for assuring the acquisition and safe use of instrumentation. Once the various sources and techniques are identified, he can establish hazard control policies.

Hazard control policies

The hazard control policies must provide for mandatory application of clinical engineering skills in the evaluation and testing of each instrument before its use on patients. The policies must also provide for documentation of this effort and for identification (tagging) of each approved instrument before its use on patients. They must also provide for prohibition and removal from service of disapproved equipment. Table 9.1 illustrates the kind of controls that may be part of one such policy. The enforcement of this policy obviously requires the cooperation of the purchasing agent.

It is not possible for the clinical engineer to initiate, promulgate, and enforce hazard control policies solely by virtue of his position within the institution. The JCAH requires written policies. In addition, the engineer is carrying an unnecessary burden, which is subject to rapid change in acceptability, unless the hospital administration and the safety committee have adopted written policies and procedures that provide the clinical engineer with the authority and the responsibility required. With this support, he will interact with hospital departments and with individual nurses and physicians as one professional to another, rather than as a policeman.

Table 9.1. An Example of a Basic Policy in Hazard Control

Central City Hospital
Sample Policy

General Policies and Procedures

Effective Date: Policy Number:

Subject: Technical Review of All Electrical Medical Equipment.

Policy: Clinical Engineering (CE) support will be provided to assure technical review of electromedical equipment which is purchased, leased, rented, borrowed, fabricated in-house, demonstrated, staff owned or modified for use at the Central City Hospital. Where applicable the technical review will consist of reviewing requisitions, provision of written specifications and quality assurance testing for safety and performance.

Technical review is provided to assist in (a) assuring the safety of patients, research subjects, and staff, (b) ensuring the best equipment for the intended applications, (c) defining interface requirements with existing equipment, (d) suggesting alternative cost-saving measures and (e) assuring compatibility with utility and other systems.

Definitions: For the purpose of this policy: *Electromedical equipment* is defined as electrically powered equipment used in the care or treatment of medical subjects or electrical equipment that is likely to come in contact with hospital patients. This includes instrumentation used to process signals or samples that will result in diagnostic or therapeutic action (examples include ECG displays and blood gas instruments).

Medical subjects is defined as hospital patients or human research subjects.

Procedure:

1.0 Medical Purchasing will provide a copy of requisitions for electromedical equipment to Clinical Engineering.

1.1 Requisitioning departments may transmit a copy of the requisition to CE, simultaneously with its submission through existing channels, to eliminate delays in the writing of specifications.

1.2 All departments are responsible for indicating on their requisition whenever the electrical equipment is *not* intended for use as medical equipment by typing the following statement above the first item to be purchased: "Material listed hereon will not be used in care, treatment, or research involving medical subjects." All other equipment requisitions will be subject to the procedures for medical equipment.

2.0 Clinical Engineering purchase specifications will accompany the request for quotation for all electrically (including battery) powered medical equipment.

2.1 Any delays, increased costs or changes required due to the vendor's response to the requisition or to the specifications will be communicated to the requisitioning department.

2.2 In the event that the requisitioning department, Clinical Engineering, or Medical Purchasing do not agree on the specifications, determination of the appropriate specifications will be discussed in a meeting of the appropriate parties and the Safety Committee. The committee will then provide a recommendation to the appropriate administrative authority.

3.0 Acceptance testing will be performed by Clinical Engineering for *all* electromedical equipment. This includes all such items purchased under CE specifications, and all other electromedical equipment that may be leased, rented, borrowed, demonstrated, staff owned, and equipment that is modified or fabricated by hospital personnel.

3.1 Medical Equipment owned by the hospital and cleared for use shall be tagged with a control number.

Table 9.1. (Continued)

3.2 All electromedical equipment shall be labeled when cleared for use to identify the application for which it is approved. These labels will have an expiration date as described in the policy on labeling of medical instrumentation.

3.3 The cost of equipment testing and inspection shall be budgeted by the hospital.

4.0 Upon receipt of electromedical equipment for acceptance testing, Clinical Engineering shall establish a file for each piece of equipment receiving a control number.

4.1 This file will be turned over to the appropriate department(s) in order to develop a permanent record of all maintenance performed over the life of the instrument.

4.2 The user department responsible for the equipment shall notify Property Control and CE when instrumentation having a control number is transferred or disposed of. This notification by memorandum, addressed to Property Control, should include acquisition, control and serial numbers, brand name, model, and details of the change being made. A copy of the memorandum shall be sent to CE. If equipment is being traded in, the same information as above should be noted on the purchase requisition for the new instrument. A separate memorandum is not required.

Labeling of equipment. A device useful for controlling equipment and for communicating approvals and constraints to the user is a label affixed to the equipment. Labels provide the means by which the user can identify acceptable equipment for the purpose at hand, and can be assured of its safety. They should be simple and their purpose obvious. They should assist in getting everyone involved in implementing appropriate aspects of the equipment control program. In turn, they permit the hospital to place the responsibility for assuring that the policies are enforced where they logically belong—on the user.

These labels include "control numbers" which provide a unique number for each device, module, or major component of a system so that a historical record is available from the date of purchase to the date of retirement (Sec. 9.4). A second label, showing the date of scheduled inspection (and preferably also the date when the next scheduled inspection is required), should be affixed so that the user has confidence that incoming inspection and/or PM has been performed. A third label is required to inform the user as to where a piece of instrumentation may safely be used. The last two labels may be combined.

There are two approaches to equipment classification. In the first approach, equipment meeting AAMI's safe current limits classification for isolated patient leads, which can safely be attached to catheters routed to the patient's heart, is labeled to indicate that this is acceptable [Fig. 9.6(a)]. Equipment not intended for this purpose is labeled differently. An intensive-care unit, therefore, may contain an ECG with a green label and a second ECG with a yellow label.

PATIENT CARE EQUIPMENT
SUITABLE FOR USE ON ANY PATIENT
DO NOT REMOVE LABEL

GREEN

PATIENT CARE EQUIPMENT
NOT FOR USE IN VICINITY OF PATIENTS
WITH CATHETERS CONNECTED TO THE HEART

YELLOW

(a)

MEDICAL EQUIPMENT
SUITABLE FOR HOSPITAL-WIDE USE
CHECKED: DATE ______ INITIALS ______ EXP. DATE ______

GREEN

MEDICAL EQUIPMENT
NOT FOR USE IN CRITICAL CARE AREAS
CHECKED: DATE ______ INITIALS ______ EXP. DATE ______

YELLOW

(b)

Figure 9.6(a) Illustration of wording for labels of Method I where the criteria is acceptability for use with a particular patient. These are taken from labels developed by Barry Altman, available from the Brady Co.

Figure 9.6(b) Illustration of labels for Method II where the criteria is hospital area designation.

The former unit would have an isolated input stage while the latter would not.

A second approach [Fig. 9.6 (b)] is to label all equipment that could safely be used in the intensive-care environment with a label indicating its approval for this environment rather than for use with particular patients. Since patients with catheters to the heart are restricted to a limited number of nursing units in many hospitals, it is an easy matter to identify them. These areas can be "restricted" in the sense that *all* equipment in these areas is safe for use with such patients. Equipment that will not conceivably be connected to conductive pathways to the heart, such as a ventilator, is also labeled as acceptable in this environment. In this second approach, equipment that could conceivably be connected to the catheter or the pacemaker lead of the susceptible patient, but that is not approved for this purpose, is prohibited from the entire nursing unit.

Table 9.2 compares these techniques. We assume, of course, that all instrumentation meets contemporary safety standards for its category.

Both approaches to categorizing the equipment are acceptable. The first relates more directly to equipment characteristics. The second, however, appears simpler and easier to communicate to the nursing staff with its susceptibility to turnover and to temporary employees. The second method is more expensive in an institution where the amount of older equipment may not permit the luxury of restricting older equipment (such as ECG monitors with nonisolated input) from the ICU.

Delivery of the equipment. The delivery of the equipment presents a focal spot for the efforts that the clinical engineer has provided up to this point. The documentation file for the history of the device has been established, and it may be transferred to the client department or retained by the clinical engineer, depending upon who has responsibility for the continued maintenance of the device or of its documentation. This file would include forms such as those shown in Figs. 9.3, 9.5, 9.7, 9.8, and 9.9. Figure 9.9 shows a copy of the inspection report, which should be delivered with the equipment and a copy sent to the purchasing agent. The report should include certification of safety and possibly certification

Table 9.2. System for Risk Categorization by Patient Classification or by Hospital Area ("Critical care" as used here would include cardiac catheterization, emergency, and operating rooms.)

	Method I	*Method II*
Technique	Categorize each instrument as safe (or unsafe) for direct connection to cardiac catheter.	Categorize each instrument as appropriate (or inappropriate) for use in hospital areas which may contain a patient with a cardiac catheter.
Illustration		
Label for an ECG having isolated input:	Safe for connection to cardiac catheters. Or, safe for use with patients with cardiac catheters.	Safe for use in all hospital areas.
Label for an ECG not having isolated input:	Not for use with patients having cardiac catheters.	Not to be used in critical care areas.
Label for a ventilator:	Safe for use with patients having cardiac catheters.	Safe for use in all hospital areas.

INCOMING INSPECTION REPORT

Date 1/24/73 — C/N 125

Inspector J. Clark — Req. No. 200745

P. O. No. 789654

Item Monitor Scope — Model 7803

S/N 1451A0888

1. Does equipment appear to be undamaged? yes
2. Have all accessories been included in the shipment? yes
3. Have the operating instructions been included? Qty? yes / 1 only
4. Have the maintenance instructions been included? Qty? yes / 1 only
5. Check power cord. Correct plug (Hubbellock, Hospital Grade, etc.), quality wire, strain reliefs? yes
6. Is the primary power line fused? yes
7. Ground pin to chassis resistance. 0.1 ohm or less yes
8. Electrical leakage from chassis to ground (in rms microamperes as listed by U. L. Standard 544 (76). Limits: Patient care – 100 or less, Nonpatient care – 500 or less.

	Normal	Open Gnd.	Rev. Polarity	Open Gnd. Rev. Polarity
OFF	1	9.5	1	24
ON	1	12	1	20

9. Electrical leakage from patient leads to ground (in rms microamperes as listed by U. L. Standard 544 (76). Limits: Isolated terminals – 10 or less (with cable – 20 or less), Non-isolated – 50 or less.

OFF	N/A			
ON	N/A			

120-Volt, 60-Hz signal applied to Isolated Patient terminals – 10 or less (with cable – 20 or less).

OFF	N/A	
ON	N/A	

10. Check for noise induced onto power line (if indicated). Results ok

Figure 9.7 A standard form for use with the majority of equipment received. It can be supplemented with specific tests for the particular device type as appropriate.

Item ______________________ Control Number__________

Date__________

Equipment Support Items

List all: Cables
Power cords and jumpers
Electrodes
Batteries
Manuals
Paper supplies
Special tools needed for operator performed adjustments
Frequently replaced components
Etc.

needed for operation and maintenance support of this equipment.

Item	Part Number	In Stock Now?

Figure 9.8 The accessory equipment list makes maintenance and engineering support more efficient.

Date Jan. 25, 1973

C/N 125

MEMORANDUM

TO: Administrative Assistant, Medicine

FROM: I. M. Engineer
Director, Clinical Engineering

SUBJECT: INCOMING EQUIPMENT INSPECTION

Requisition # 200745 P.O. # 789654

Delivery Ticket # 73-357

Vendor Hewlett Packard

The following equipment:

1 - model 7803 Monitor Scope

XX Has been inspected for compliance with safety requirements.

XX Has been inspected for proper operation and calibration accuracy.

____ Equipment has met requirements.

____ Equipment is defective/does not meet requirements and is being held by Clinical Engineering pending action by vendor.

XX Equipment has been released for use; however, the following items are missing/do not meet requirements:

1 each maintenance and operation manuals

____ If you are satisfied with the equipment, we have no objections to payment of the invoice.

XX We recommend that payment be withheld until the equipment passes inspection/missing items are delivered.

Copies to: Purchasing Agent, User Dept. Original to Clinical Engineering.

Figure 9.9 The inspection report keeps all the appropriate individuals informed of the equipment's status.

of the functional testing performed on a device. In the evolutionary stages of the clinical engineering program, functional testing may not be performed on all equipment purchased. In fact, although more equipment is tested for functional performance each year, it will be several years before the clinical engineering program functionally tests all equipment. This report to the department, administration, and/or whoever is financially responsible should include a recommendation to pay or withhold payment to the vendor.

We recommend that payment be withheld for any lack of compliance with specifications, such as failure to include the instruction and/or repair manuals. There will, of course, be occasions when exceptions are made to this policy, or when partial payment is made pending completion of the requirements. However, the general policy should be to withhold payment until all terms and conditions are met.

Written policies. The clinical departments must understand the importance of hazard control requirements in relation to the safety of the patients and to the regulations that are described in preceding chapters. If this communication and training effort has been successful, it should be possible to adopt hospital policies in compliance with JCAH recommendations (Sec. 4.7), making it possible for each user to be confident that the instrument has been tested and certified within the recent past as being safe for use in appropriate applications. Tables 9.1, 9.3, and 9.4 provide useful examples of policies required by the JCAH.

Reduction of hazards through training. A vitally important factor in the hazard control area is the need for trained user and maintenance personnel. It is up to the hospital to assure that the vendor and the supervisory staff provide adequate in-service education of user personnel. The tendency is, however, to ship the majority of equipment to the hospital by an appropriate freight agency, and for the hospital to unpack and install most portable equipment without additional effort on the part of the manufacturer. In contrast to radiologic technologists, who have years of training in the use of X-ray equipment, the critical care nurse is placed in charge of comparably complex instrumentation without similar education. It therefore falls to the clinical engineer or the vendor's representative, in cooperation with the clinical and nursing departments, to assure training of hospital employees. The importance of competent user and maintenance personnel is treated elsewhere in more detail (Chap. 12).

Hazard investigations. Every hospital will experience "incidents" in which a piece of equipment fails or is misused in such a way that a patient is injured, or his care is compromised. It is neces-

Table 9.3. Policies on Labeling of Medical Instrumentation (This policy describes implementation of labeling in accordance with Method II, described in Table 9.2)

Central City Hospital
Sample Policy

General Policies and Procedures

Effective Date: Policy Number:

Subject: Labeling of Medical Instrumentation

Policy: All patient related electrical devices shall be limited to the uses approved by the Safety Committee and appropriately labeled by Clinical Engineering.

Definitions: For the purpose of this policy, ***patient-related equipment*** is defined as equipment that is likely to come in electrical contact with patients. All patient-related electrical devices will be approved for one of the three following equipment categories.

Risk III equipment is defined as electrical equipment approved for uses in which it may come in electrically conductive contact with patients having a foreign conductive pathway to the heart or to the great vessels, such as pacing leads or catheters. This includes equipment intended for connection to those conductive cardiac pathways as well as all other electrical equipment not intended for such connection, but for use with such patients.

Risk II equipment is defined as electrical equipment approved for uses in which it will be intentionally placed in electrically conductive contact with patients *not* having foreign conductive pathways to the heart or to the great vessels.

Risk I equipment is defined as electrical equipment approved for uses not requiring Risk III or Risk II equipment, in which it is likely to come into only incidental contact with patients.

Procedure:

1.0 Clinical Engineering is responsible for labeling devices approved for patient related application.

- 1.1 Electrical equipment approved for patient related applications will have a colored label affixed to the device by Clinical Engineering indicating its use category as either Risk I, Risk II, Risk III.
- 1.2 Labels and the corresponding approval of equipment is void if:
  - (a) The equipment is removed from Central City Hospital (CCH) and passes out of the control of CCH employees.
  - (b) The equipment is modified.
  - (c) It is determined that the equipment may no longer meet the standards required for such approval.

  Requests for relabeling of equipment with voided labels should be submitted to Clinical Engineering.
- 1.3 Any requests for a change in risk category should be submitted to Clinical Engineering.
- 1.4 Any appeals of equipment categorization should be directed to the Safety Committee.

2.0 The label indicating the equipment category shall have a provision for an expiration date.

- 2.1 The initial dating of labels shall be the responsibility of Clinical Engineering during acceptance testing.
- 2.2 The expiration date will not be placed on hospital-owned equipment that is not on the preventive maintenance program.

Table 9.3. (Continued)

2.3 Expired labels on equipment maintained by the hospital may be renewed by the group responsible for the maintenance of that equipment.

2.4 Expired labels on equipment not maintained by the hospital must be renewed by Clinical Engineering.

3.0 Using Staff shall be responsible for insuring that all patient related electrical equipment used has the proper label for its intended use and that the label has not expired.

3.1 Unless patient care will be jeopardized or unless otherwise approved by Clinical Engineering, equipment not having a current label appropriate for its application should be removed from service. The Administrative Assistant should then be notified.

3.2 Labeled equipment may be used in applications described in the definition of its category or in the definitions of lower numbered equipment categories. For example, Risk III equipment is approved for use in all other categories. Labeled equipment shall not be used in applications requiring equipment with a higher numbered category. For example, Risk I equipment may not be used in applications described in the definitions of Risk II or Risk III.

3.3 No other electrical devices, except those labeled as Risk III shall be allowed in the vicinity of patients with foreign conductive pathways to the heart or to the great vessels. Equipment is considered to be in the vicinity of a patient if it is within reach of the patient or of an individual who can also reach the patient.

3.4 Equipment found to be hazardous by hospital staff, shall be tagged with a completed red DANGER label.

3.5 All staff are responsible for notifying the Administrative Assistant or Clinical Engineering if they become aware of any reason why a label should be voided. Refer to section 1.2 of this policy.

Table 9.4. Policies for Electrical Patient Appliances

Central City Hospital
Sample Policy

General Policies and Procedures

Effective Date: Policy Number:

Subject: Electrical Patient Appliances

Policy: The use of electrical appliances brought to the Central City Hospital by patients for their own use of for use by visitors in patient rooms shall be limited to only those items approved as electrically safe by the Clinical Engineering Division and described on a properly executed *Electrical Appliance Use Permit* (sample attached).

Procedure:

1.0 The use of personal electrical equipment by patients shall be limited to those items which have a completed *Electrical Appliance Use Permit* and have been appropriately approved and labeled by Clinical Engineering (CE).

1.1 The *Electrical Appliance Use Permit* (Form 30119, attached) may be obtained in the patient's care unit. This form must be properly signed, witnessed, and added to the patient's medical record.

1.2 The labeling of electrical devices will be accomplished following approval by Clinical Engineering and shall be in accord with the policy on *Labeling of Medical Instrumentation.*

Table 9.4. (Continued)

2.0 The nurse in charge is responsible for assuring that no electrical equipment is used by patients unless it is properly approved as described in this policy.

3.0 Clinical Engineering will be responsible for safety inspection and approving the use of any electrical equipment used by patients.

- 3.1 In general-care areas, the only line powered devices that may be approved are electric razors and hair-grooming appliances.
- 3.2 In intensive-care areas, the use of personal, line-powered devices is prohibited.
- 3.3 Battery-powered items will be approved for use in both intensive- and general-care areas on an individual basis. (Battery chargers and battery-powered devices with built-in line-powered chargers are considered as line-powered devices.)
- 3.4 Battery chargers may be approved by CE provided that they are kept at the nursing station and are not handled by patients.
- 3.5 Certain devices that would otherwise be prohibited under this policy may be approved by CE, provided that (a) the patient's physician orders the use of the device in the patient's medical record as necessary for the patient, (b) a hospital-owned medical device cannot be found to meet the patient's needs, and (c) the device meets the safety standards required for its application.

CENTRAL CITY HOSPITAL

ELECTRICAL APPLIANCE USE PERMIT

I hereby agree that in using ______________________________
an electrical appliance(s) brought by me or others for use in my room while a patient at Central City Hospital, . I do so at my own risk and hereby absolve the said hospital from any and all responsibility from burns, injuries, or property damage which may result from or because of said appliances(s).

Date ______________ Patient's Signature ______________

Time ______________ Witness (Employee's Signature) ______________

Approved by Clinical Engineering: Date ____ Signature __________

FORM # 30119

sary to have a policy that ensures that the clinical engineer is promptly informed when such an incident occurs, and that he is authorized to investigate all aspects of the situation. His investigation should result in determination of probable cause(s) and recommendations for action to prevent a recurrence of the incident. His actions should be closely coordinated with appropriate departmental and administrative personnel, and his reports should contain sufficient detail to be useful in any resulting legal action (Sec. 14.3). Outside consultants may be helpful in protecting the hospital's legal position. Where litigation may be anticipated, the clinical engineer and the administration may prefer to "seal" the untouched equipment for independent analysis. The probable cause, the cost of such investigation, and effects of the loss of use of the equipment must be weighed against the likelihood that the plaintiff's attorney will question the credibility of the hospital investigating its own incident.

9.6 Quality Assurance

The clinical engineer should adopt a philosophy, "Is there anything that can possibly go wrong with this equipment or can it be misused?"

Quality assurance begins with procurement

In its broadest sense, we can consider prepurchase evaluation of the manufacturer's literature or of the instrument itself (Sec. 8.10), as the first step in quality assurance (QA). However, for most equipment, the first step beigns with the incoming inspection to the requirements of the specifications, and to the clinical engineer's requirements. Equipment acceptance testing was described in Sec. 8.14.

Once a clinical engineering program has become fully established, incoming inspection is an appropriate stage to define the preventive maintenance routines and schedule (Sec. 10.3). The technicians who are responsible for incoming inspection can perform these new test procedures and evaluate them for appropriateness, clarity, and other practical considerations. The data obtained can then be used in future inspections as a basis for comparison (Sec. 10.3).

Quality assurance of maintenance services

A ticklish aspect of the QA program is that concerned with services provided by outside vendors, such as warranty work, preventive maintenance, and repair, whether contractual or noncontractual. The selection of outside vendors involves, to some extent, the same kinds of criteria employed in the selection of instrumentation (Sec. 8.11). Table 9.5 (especially items 1.2 and 5.0) provides some useful criteria.

Table 9.5. This form is useful for comparison or evaluation of contractual maintenance services. Adapted from a form provided by Hospital Shared Services of Arizona, Inc.

Furst, Betts, and Reeter

Hospital: ______________

Dept.: ______________

Service Contract Evaluation Form

Vendor: ______________

____

Date of Evaluation: ______________

Evaluated by: ______________

Phone: ______________

1.0 Content: What service does the contract offer?

	Yes	No	Not Specified
1.1 Detailed equipment list provided? (Type, Vendor, Model No., Serial No., Property Control No.)	____	____	____
1.2 Scheduled Maintenance Specifics			
○ Number of service visits ____			
○ Interval between service visits ____			
○ Scheduled maintenance procedures for each equipment type provided	____	____	____
○ Do maintenance procedures include calibration?	____	____	____
○ Time (avg.) for each scheduled maintenance specified	____	____	____
○ Samples of service records provided, tags	____	____	____
1.3 Demand Repair			
○ Response time (specified ________)	____	____	____
○ Guarantee on response time	____	____	____
○ 24 hr. service specified	____	____	____
○ Weekend coverage specified	____	____	____
○ List of available loaners provided	____	____	____
○ Safety and calibration checks after repair specified	____	____	____
2.0 Pricing: How much will service cost?			
○ Basic hourly labor rate (normal hours)	____	____	____
○ Overtime – Weekdays	____	____	____
○ Overtime – Weekends	____	____	____
○ Minimum charge per visit	____	____	____
○ Travel charge (base to hospital)	____	____	____
○ Extra travel charge if serviceman is in field (portal to portal)	____	____	____

Due to JCAH standards (Sec. 4.7) and legal liability (Sec. 14.3), the documentation of work performed by outside vendors is no longer an optional matter. Table 9.6 summarizes these require-

Table 9.5 Continued.

Service Contract Evaluation Form

	Yes	No	Not Specified
2.0 Pricing (Continued)			
○ Period over which price is quoted			
○ Escalation clauses			
○ Parts (extra, included)			
○ "Usual and customary" clauses			
○ Parts markup specified			
3.0 Billing Procedures			
3.1 Service Order Contents			
○ Date work performed			
○ Labor hours			
○ Parts (include part numbers)			
○ Equipment serial no. & description			
○ Sign off by requestor			
○ Separate sales tax			
4.0 Contract Terms			
○ Advanced payments			
○ Automatic renewal			
○ Termination option			
○ Bundled services			
○ Blanket purchase order			
○ Liability provisions specified			
○ Parts warranty ______ (1 year typical)			
○ Automatic new equipment inclusions			
○ Parts repair authorization			
5.0 Qualifications			
○ Reference provided			
○ Number of technicians available for type of work under contract			
○ Qualification or certification specified			
○ Test equipment specified			

6.0 Summary Remarks

ments and is useful for informing the vendor of the documentation required (Sec. 8.7). With this documentation, historical analysis will provide evidence of the quality of service, of capabilities of repair personnel, of the need for replacement of equipment, or of the need for training of the user—in short, the same kind of information we would expect to get from a well-organized in-house maintenance program. The clinical engineer's supervision of outside vendors will

Table 9.6. Equipment Documentation Requirements

This form is attached to each purchase order and made a part of each service contract and rental or lease agreement.

Central City Hospital

Outside Service Contracts — Temporary Lease or Rental

The Joint Commission on Accreditation of Hospitals issued new standards December 1975 which became the accreditation criteria in October 1976. These "optimal achievable" standards include a requirement for periodic "testing and verification of performance and use specifications, for all electrical and electronic patient care equipment, based upon established safety requirements, performance criteria, and manufacturers' claims . . . All new equipment shall be so evaluated prior to its use." The JCAH intends that calibration be included when required, with much the same criteria. Therefore, the following specifications are intended to assist the Central City Hospital (CCH) in its implementation of these requirements.

I. All patient related electrical equipment must be included in a documented preventive maintenance program which ensures its safety, reliable performance, cleanliness, calibration, and proper adjustment.

II. Up to date *documentation* on *each* piece of equipment *must* be provided to the department purchasing this service, which includes, for each inspection or repair:

1. The manufacturer's name and model, and the serial and CCH control numbers.
2. Reports of the results from each test.
3. Descriptions of any repair or corrective action that may have been required by the device.
4. Reports of the results of the retesting following repair or adjustment.

Each instrument requires an individual form.

III. In addition, for each type and model of equipment, at least ten days prior to the first scheduled service the vendor shall furnish:

1. A detailed description of all routine (preventive maintenance) tests and inspections that will be performed.
2. The specific limits that must be met during each test in order for the instrument to pass.
3. The preventive maintenance interval.

IV. The vendor's service personnel will check in with the Clinical Engineering Department in room 123 before and after each visit during normal working hours. At other times he will contact the on-call Clinical Engineer through the telephone operator. This is the only mechanism available to assure payment for services.

also serve as significant notice to the vendors that the hospital expects to get its full dollar value and a high level of quality from these services.

The clinical engineer's role will vary, depending upon the authority vested in him by the hospital administration. The administration may centralize the responsibility for all maintenance, including the use of outside vendors, in the clinical engineering office. They may leave the selection of inside versus outside, as well as the selection of particular vendors, to the individual departments. Regardless, an effective equipment control program requires a centralized responsibility for QA.

With the QA responsibility centralized in the clinical engineer, the contract should describe the procedures to be used by the vendor. These should (but rarely do) include his checking in with the clinical engineer prior to performing any PM and repair work, and checking out with him afterward. He should leave whatever documentation is required and discuss the PM and repair performed and turn over any parts replaced; payment can then be initiated. There must be an understanding that, should the vendors not communicate in this fashion, payment will not be authorized. This assures the clinical engineer of the opportunity to observe, study, and influence the vendor's performance.

In accepting the responsibility for PM and repair, the clinical engineer must determine how he will internalize the QA function of the services he and his employees will provide. One mechanism is to organize PM and repair services in such a fashion that the work performed in each area acts as a check and balance on the other. The central equipment file allows the technicians to flag those events where PM following repair reveals malfunction. Conversely, repairs initiated as a result of scheduled PM, or following it, reveal gaps in the PM procedures. Note that supervisors must utilize this information in a way that does not discourage the technicians from continued generation of internal QA activity. In passing, we note that regardless of who performs PM and repairs, there will be a small percentage of equipment that fails shortly after either maintenance effort.

A more difficult situation is presented when the clinical engineering or equipment maintenance functions are distributed throughout the hospital. Quality assurance of one engineering center by another within the same institution is fraught with difficulty. Although the check-and-balance system would appear to be optimized by separation of QA and PM responsibilities, in practice this will not generally hold, due to the intense competition for limited dollars, space, and visibility. Clinical engineering is established within the hospital to increase the efficiency of use of equipment and simplify its management. Fragmentation of effort has the opposite effect.

The clinical engineer is required by good management practice and by JCAH standards to communicate periodically with the client department regarding significant aspects of the internal maintenance services or of the external vendor services he oversees. If the engineer enjoys good relationships with the "front line" hospital personnel and departmental supervisors and encourages them to communicate rapidly and completely any time they have concerns over the service being provided, the engineer will obtain excellent feedback on the quality of these services.

Another aspect of QA arises from observation, conversation,

and PM experience: discovery of operator errors that result in less than optimum use of equipment, and often are initially identified by the user as malfunctioning equipment. Since incorrectly used equipment can result in erroneous diagnosis or treatment, subsequent communication with supervisors, informal training and other clinical engineering actions can influence patient care for the better (Chap. 6, 12, 13).

It is important to establish the QA function, and therefore the clinical engineering function, in a direct line from the top or second level of the hospital administration. This parallels similar requirements in industry, where there are inherent conflicts at lower levels between production, on one hand, and quality or safety on the other. The visibility and support of the administration must be omnipresent to assure a successful effort.

Regardless of the mechanism established, the QA function cannot be filled by other personnel within the institution who are not engineers. In addition, it is rarely possible for an outside consultant to fulfill this function in all but the smallest institution. These restrictions are related: QA can be effectively performed only if the function has the support of top management and is accomplished by competent personnel. A competent outside consultant may not have continued support for the purpose of providing bad news. He loses access to meaningful information and cannot quietly affect operations from within the system.

Quality assurance through other personnel

The clinical engineer establishing his program should consider the role of other hospital personnel in providing QA. The engineer is only one individual, or perhaps a member of a small department, whose impact must be felt at the front line. His real effectiveness in providing QA must extend to the time and location where the equipment is being used. It has historically been the responsibility of technologists in a clinical laboratory and, more recently, of radiologic technologists to provide a level of QA on a routine basis. Standardized samples are processed daily in the clinical laboratory and the results analyzed to determine if the instrumentation and procedures are functioning as expected. X-ray technologists take test films using special devices that provide information as to focus, X-ray dose, and quality of film processing. Similarly, ICU nurses, OR techs, and others may play a role in routinely determining the quality of performance of the instrumentation they use. This has not, however, been a traditional nursing role. The majority of nurses today feel that this is not a part of their responsibilities, and they certainly do not feel comfortable performing QA tests. Nonetheless, our experience in formal nursing courses indicates that many nurses are becoming motivated and interested in developing a greater understanding of the equipment they use. With this increased knowledge they are more willing to assume additional responsibility for the equipment. Certainly we should be able to expect the user of any equipment to know when it is functioning properly and to distinguish its routine operation from circumstances in which the instrument is malfunctioning, even if in a minor fashion. Any implementation of QA in nursing units should minimize the efforts that the nurses need to make, and build upon existing rapport and adequate training.

Many nurses when hired do not know how to use equipment. Some equipment manufacturers provide a checklist to assist nurses in equipment operation. But many manufacturers do not, so the clinical engineer may need to first provide instruction in equipment operation. However, even though nurses may not understand the technical problems, they can provide valuable QA functions.

For example, if an ECG monitor does not reliably trigger the cardiotachometer, inspection may show that a lead may have open-circuited. The most obvious corrective action may be to ensure that the nurses are educated to treat the leads carefully, to localize and identify the defect, and to substitute a new set of leads. On another level, a routine QA effort by the nurses could prevent many of these failures. Using one of the many hand-held testers now available, the nurse can check the monitor and leads before the patient is first monitored and perhaps when changing electrodes. Should a problem develop at other times, the nurse can use the same device to quickly verify the monitor's operation. Nursing QA is assisted and encouraged if the PM program is sensitive to the common failure modes of the equipment, which in this example might mean periodic replacement of patient cables.

The discussion above was oriented toward the nursing staff who are generally the first priority for clinical engineering support. The philosophy of first-line QA obviously applies throughout the hospital. The engineer should let the staff know that he is eager to assist them in establishing QA procedures for the equipment that they use.

9.7 Cost Control

Cost control represents one of the most important aspects of the clinical engineering program, particularly because dollar costs are meaningful to the client department and the administration. Everyone professes to believe in control of hazards and of quality of the work—at least on an intellectual level. But everyone believes in cost control on an emotional as well as an intellectual level. Every dollar saved may be equivalent to ten or more in additional revenue, since the cost of hospital operations must be subtracted from revenue before there is additional money to spend.

Cost control in procurement

Savings effected in the purchase of instrumentation can be obtained in several ways. Direct savings result when the engineer is able to: (1) identify existing equipment that can be utilized instead of purchasing new equipment, (2) identify lower cost alternatives by determining what the minimum requirements are for the task at hand, (3) assure that the equipment to be purchased fills actual needs rather than perceived needs as identified by the vendor's representative, or (4) reduce the impact of obsolescence by delaying a purchase until a later model is available (Chap. 8). Less obvious, but still significant cost savings occur when the engineer considers the ease of maintenance and reliability of the device, thus providing continued savings. Other savings result from placing a dollar value on the documentation provided with the instrumentation and other services by the manufacturer. Good manuals with PM and repair procedures will save many man-hours in repair and in developing the PM program. Good training services and documentation can be used to save many hours of engineering and nursing time, and undoubtedly will save many false calls for repair that result from operator error.

The incoming inspection program is an important vehicle for assuring that the vendor provides contracted services and that the appropriate file is established to be utilized throughout the entire life of the instrument.

Cost control in maintenance

In developing the clinical engineering program, good financial management demands that continued studies be made to determine which instrumentation should be maintained with in-house personnel as opposed to vendor services. These decisions are not solely economic, but also involve patient-care requirements, the relationships between personnel, and the political situations existing between departments, administration and the clinical engineer. No single prescription can be written for all hospitals.

It is important for the clinical engineer to understand the role a shared service or outside vendor can play in the hospital, with both financial and nonfinancial aspects. Downtime can translate into lost revenue, so availability of personnel and parts is a factor, as is the level of expertise provided. If the hospital is large enough, external vendors should not be relied upon exclusively since their cost is usually significantly higher and travel time reimbursement leads to additional costs.

An important cost-control measure can be implemented by supervising external vendor contracts. First, an engineer can learn a great deal by monitoring the vendor services provided, not only as to quality, but as to effectiveness and cost. He can observe the level of skill and training required of the maintenance technician, the extent to which repair parts are required, their costs and markup by the vendor, and whether they are available through other channels. He should monitor the procedures used in the documentation the vendor utilizes. By requiring that test procedures and test results be spelled out, he is in a position to assist each department and the administration in negotiation of subsequent vendor contracts. In the process, he will develop an opinion as to whether these services can or should be done in-house. Specific attention to certain elements of the contract will provide additional clues, for example:

1. Are a given number of visits required of the vendor each year?
2. Are the repairs done at additional cost or at an additional hourly rate if the instrument is found to be malfunctioning during preventive maintenance?
3. Is the cost excessive because the vendor travels a long distance?
4. Is it so difficult to get the vendor to respond promptly that repairs are performed in-house, even though a contract exists?

Table 9.5 shows additional items. Following analysis of the contract, negotiations for more favorable terms, and observations of vendor performance, the engineer can analyze the hospital's possible actions. Questions to be asked include:

1. Does the dollar cost warrant transfer to in-house support?
2. Are trained personnel on-board and is their time available?
3. What is the cost of training required?
4. Will future acquisitions make training a better investment?
5. Does employee turnover present a significant obstacle to reliable maintenance or escalate the cost of training?
6. Is adequate space and equipment available?

Cost control in administration

The cost of documentation and other aspects of the clinical engineering program should be in line with internal needs, economic results, and regulatory requirements. Paper shuffling can be expensive. Administrative costs should permit planning, responsiveness, and growth, but not grandiosity. Efficient collection and dissemination of data internal to the hospital can save a great deal of time.

Cost control through training

A vital part of the cost-control effort is training. Haslam

and Bruner (1973) established that 58 percent of repair calls to an intensive-care unit were directly related to operator training. Further, approximately 10 percent more of the equipment failures they described could have been solved by the nursing staff had they been provided with knowledge of elementary troubleshooting techniques, spare power cords, and patient cables. Thus, training of nursing staff and nursing support personnel can also be translated into meaningful dollar savings (Sec. 12.6). Damaged pressure transducers are a national hospital problem (Smith et al., 1978). For example, four-hundred-dollar transducers are frequently damaged by dropping, overpressurizing or being cut with scissors and scalpels. It is impossible to reduce the level of damage to zero, but many hospitals suffer from an above average failure rate because training has been less than totally effective.

Section 12.6 describes techniques for effectively implementing training programs. Clinical engineering participation in the nursing in-service program can effectively meld both the clinical and technical requirements for transfer of necessary information to the nursing staff. The clinical engineer's role in the purchase and testing of the equipment can be used to assure that the vendor is not paid until he fulfills his training responsibilities in addition to his technical responsibilities (Chap. 8). Initiation of in-service training at the time of purchase, utilizing whatever services and facilities are available from the vendor as well as the clinical engineer, can be an effective cost-containment measure.

Cost control through analysis

The data generated through the equipment control program, including that developed in the prepurchase evaluation, incoming inspection, PM and repair functions, should be analyzed on a periodic basis to determine when a piece of instrumentation should be replaced because the cost of PM and repair has become excessive or because repair is becoming a too frequent requirement. Some instrumentation will have a large annual PM and repair expenditure, even when it is performing at its best. However, the average instrument will require a five to nine percent annual repair expenditure. When repair costs become excessive, it may be worth considering replacement. The same data base will help identify reliable instruments, and therefore the brands that should be given preference in selection of additional instrumentation.

The duration and nature of the warranty agreements are important factors in the decision to purchase equipment. The clinical engineer should evaluate the additional costs required by warranties that force the hospital to pay transportation to remote locations and lose use of the equipment for extended periods of time, which may mean lost revenue or may require renting, borrowing, or purchasing substitute and backup equipment. Instrumentation with nonstandard parts, parts identification resulting in availability only through the manufacturer, and poor documentation must be considered a liability. Incorporation of repair specifications and response times in the hospital purchase specifications is not always a solution, as it is difficult to enforce these requirements.

With experience and advice from other institutions in the same geographic area, using the same types of equipment, the clinical engineer can maximize the likelihood of making the correct decisions and over a period of time develop a vendor-rating system. This procedure allows for the incorporation of historical evidence in the selection of vendors and the elimination of those who do not perform as they have promised.

9.8 Management

In developing the equipment control program there is a need to generate a variety of documents. Some of these, such as repair reports, are used repetitively for each piece of equipment, while others, such as file cards, are generated only once in the life of the equipment. At any point in this history, as discussed above, analyses can be accomplished that can identify a device whose repair costs are increasing to the point where replacement should be considered, or identify a device under contract PM and repair where the contract requires payment of premiums not justified by the failure history.

The number of devices of a given type is an element in determining whether technicians should be trained to assume in-house PM and repair responsibilities or if it is more economical to retain a contract with the vendor. A shared service program or ambitious engineer may help reduce vendor cost by arranging for several hospitals to share the expense of a specially trained technician or by coordinating the travel of an out-of-state vendor so that his representative may service several hospitals in an area and transportation costs can be apportioned among them. The repair record will identify for replacement the equipment that loses calibration rapidly, is unusually difficult to use and prone to operator error, that repeatedly fails at critical times, or that is out of compliance with codes and standards. The clinical engineer should review files as new codes or standards address performance or safety features. The cost of modification or replacement along with the device's cost and reliability history will aid in making decisions.

There is a desire to distill from the equipment control program a simple output that provides a feedback mechanism to control the safety, the clinical cost effectiveness of the instrument, and the cost effectiveness of the clinical engineering services. This magical formula does not exist, and if it did, it would require alteration or adaptation for each hospital. But the search for this mechanism is implicitly a goal of every clinical engineering operation, for a stable and accurate feedback system can exist only in the closed loop state.

The use of computers

Computers have been used for scheduling preventive maintenance activities and storing the records generated in maintenance. Shepherd et al. (1976) have developed a software effort to identify failure modes throughout a shared service program in northern California. This type of analysis can identify weak points for particular attention in preventive maintenance and can more easily enable the engineer to consider the maintainability and reliability of an instrument as an element in purchase decisions. Sutter Memorial Hospital, Sacramento, CA 95819 has made use of the computer to analyze personnel time allocations and total cost, as well as failure modalities. Periodic reports to the departments and informational messages indicating when equipment needs to be available for preventive maintenance can also be generated. Trends in cost and failure modalities for particular instruments may be useful results. It would appear that a computer is required to economically sift through the data to identify critical values and to provide output in the numerous forms required by the many different people who need access to these data in a large hospital.

9.9 Getting Started

The major requirements in establishing an effective equipment control program relate to organization, management, personnel, costs, and equipment. The equipment control program requires very little equipment, but is usually built into a clinical engineering program that includes PM and repair responsibilities. The documentation exists not only for these PM and repair services but for the cost-effective utilization of the data generated as the basis of the equipment control program. The cost is amalgamated with the cost of establishing the PM and repair activities, since the instruction and repair manuals, inventory and cost of establishing the filing system is also essential to an efficient PM and repair operation and to one that meets regulatory and accreditation requirements. Nonetheless, if considered separately, there is a substantial cost associated with the inventory and initiation of the documentation system. Yardstick estimates may assist in the early stages and can be adjusted on the basis of experience. If the hospital is a 300-bed university-type hospital or a 600- to 900-bed community hospital, we can expect to initially issue 1500 to 2000 control numbers for the medical devices and instruments. The engineer can expect each device to require approximately 30 minutes for establishing the equipment control file and affixing the control number. An elementary safety inspection conducted at the same time could be expected to take another 30 minutes. Obtaining original purchase information from files in the purchasing or financial office can take a considerable amount of time, but should only be necessary for a fraction of the items inventoried.

Phasing it in

An essential ingredient in establishing the equipment control program and the clinical engineering effort in general is the enthusiasm of the staff. Without this, the temptation to quit in frustration over the lack of support, lack of resources, or the presence of hostile elements will be overwhelming. On the other hand, too much enthusiasm can be a problem as well. The temptation to do everything at once is almost irresistible, but must be resisted or the engineer will be "spread too thin." In that case, responsibilities are not met satisfactorily, and credibility is lost.

If at all possible, the clinical engineer should discuss program priorities in some detail during the employment negotiations, so that the administrator will understand the extent of the task, and the engineer can get a handle on the support and commitment available. This does not imply that the program will be spelled out in detail; this is impossible without the experience obtained from months of employment in the particular institution. Also, the experience of engineers nationally has shown that neither the administrators nor the engineers appreciated the magnitude of the job when they began.

One suggestion is to place as high priority as possible on those efforts that improve patient care, those that impact favorably in developing close relationships between the clinical engineer and other health-care professionals, and those that have the potential of saving significant dollars in the near or long term.

Planning for the future

The equipment control program reflects the rapid changes in medical practice, technology, and regulations. The hospital will find it cost-effective to anticipate as many of these changes as possible. This facilitates planning to stay abreast of technology and regulations, minimizing either the expense of hasty catch-up efforts which are difficult to plan or implement effectively, or the expense of overreaction with expenditures that are not functionally justifiable. Additionally, hospital budget cycles are usually lengthy; two years may pass from date of identification of needs to the time when funds become available. In addition to the techniques and resources described in Chaps. 12, 13, and 15, certain salesmen and manufacturer's employees will also be helpful sources of information regarding current and future developments. The engineer must devote sufficient time to long-term planning to ensure that he has the equipment and the support necessary to cope with the increasing amount of complex equipment he is responsible for supporting.

9.10 Summary

A well-conceived equipment control program provides the framework for virtually all activities performed or supervised by the clinical engineer. The data base acquired will assist equipment users, in-service educators, maintenance personnel, and financial planners. Moreover, it is one of the reasons for hiring a clinical engineer as opposed to trained maintenance personnel or technicians. In time, the program should produce a data base and a rapport with the clinical staff that permits meaningful consultation, cost containment, and effective planning for equipment and facilities.

9.11 Study Questions

9.1 Describe the fundamental objective of an equipment control program.

9.2 List three reasons for the existence of an equipment control program.

9.3 Justify the use of a central clinical engineering service in meeting JCAH requirements.

9.4 List the four components of an equipment control program.

9.5 What are the two major objectives of the equipment control program?

9.6 Explain how the equipment control program may be integrated with other clinical engineering activities.

9.7 List three modes of communication available to the clinical engineer in implementing the equipment control program.

9.8 What are likely sources of information for an existing equipment inventory?

9.9 Explain how an equipment control card can be used in a control program.

9.10 What is the importance of the equipment acquisition cost and date?

9.11 List three possible sources of documentation (manuals and schematics) for a piece of equipment.

9.12 Why might a clinical engineer meet resistance from other departments when gathering information for an equipment control file?

9.13 Explain how the equipment acquisition process provides the basis for control of newly acquired devices.

9.14 List three sources of hazard control information.

9.15 List three ways in which equipment may enter a hospital.

9.16 Describe the two basic methods for indicating safety of equipment through proper labeling.

9.17 What policy statements are necessary to insure compliance with JCAH policies?

9.18 Describe the role of training in the hazard control program.

9.19 Explain how cost control can be implemented in procurement of equipment.

9.20 How can costs of maintenance be controlled?

9.21 What is meant by "cost control in administration"?

9.22 Explain how user training can help clinical engineering.

9.23 How can a careful analysis of equipment costs lead to an improvement in cost control?

9.12 References

AAMI (1976). Safe Current Limits for Electromedical Apparatus. Arlington, VA: Association for the Advancement of Medical Instrumentation.

Emergency Care Research Institute (1971). *Health Devices.* 1:75–93.

FDA Enforcement Report. Rockville, MD: Food and Drug Administration.

Federal Register (1976). Medical Devices, Reorganization, Republication and Recodification. February 13.

Haslam, K. R., and J. M. R. Bruner, (1973). The epidemiology of failure in cardiac monitoring systems. *Med. Instrum.* 1:293–296.

National Fire Protection Association (1973). Standard for the use of inhalation anesthetics (flammable and nonflammable). NFPA 56A. Boston.

National Fire Protection Association (1977). National Electric Code, 1978. NFPA No. 70-1978. Boston.

Smith, R. G., L. Lampe, and E. Furst, (1978). Transducer quality control. *Proc. AAMI Annu. Meeting*:250.

10
Preventive Maintenance and Repair

W. SANFORD TOPHAM

10.1 Learning Objectives

Upon completing this chapter you should be able to:

1. Explain the meaning and benefits of PM.
2. Describe the major aspects of a PM and repair program.
3. Set up a PM and repair program.
4. Justify the existence of a PM and repair program on an economical basis.

10.2 Introduction

Repair and preventive maintenance are important to the success of any clinical engineering program. Since the goal of the clinical engineering department is to provide safe and effective equipment for delivery of the best health care possible, preventive maintenance (PM) and repair are the first-line activities that guarantee the medical equipment is calibrated and operational.

Fast and reliable repair is necessary since equipment will malfunction regardless of how well designed; but repair is only brought into play after failure has occurred and inconvenient or ineffective use of the equipment has resulted. A PM program will reduce the number of repairs required and more importantly, will reduce the inconvenience and frustration caused by malfunctioning equipment. PM, while being practiced in many industries, including hospitals for such equipment as heating and air conditioning, has only recently been available for patient-oriented equipment.

While repair and PM are the first-line activities for the proper maintenance of equipment, other factors discussed in Chaps. 8 and 9 will help determine the effectiveness of these programs. If proper specifications are written and the most reliable and effective piece of equipment is purchased initially, then the time spent in PM and repair can be greatly reduced. Also, an effective incoming inspection will provide a baseline of data on which an effective program can be established.

10.3 Preventive Maintenance

A good PM program requires the cooperation and support of the hospital personnel. Proper understanding of the need for preventive maintenance is essential and hospital personnel should realize that no amount of PM can reduce the number of equipment malfunctions that are caused by abuse. It is necessary to conduct the maintenance procedures when it is most convenient to the hospital personnel and least interferes with normal hospital operation.

PM will help eliminate hazards before they develop or become serious. There are many problems that occur suddenly and therefore cannot be detected or prevented by a PM procedure. However, many problems occur because of the deterioration of equipment caused by normal use and can be detected prior to their causing a malfunction of the equipment. There would be few people who would be willing to fly the passenger airlines today if it were not for their very extensive and thorough PM programs.

PM prolongs the useful life of equipment by providing for proper cleaning, lubrication, and repairing of defective parts prior to development of a serious fault. Equipment kept in good operational condition will certainly not deteriorate as rapidly. As the PM program becomes effective and the reliability of the equipment increases, there is also an increase in the confidence of the users in the equipment. They have less difficulty operating the equipment, there are fewer times that it is nonfunctional, and they are sure of the calibration. The equipment is thus used with greater frequency and confidence so that better medical care usually results.

If the PM procedures are done by the same technicians that repair the equipment, they become familiar with the equipment and its operation. Because of this familiarity, when a malfunction does occur, the downtime for repair is frequently less because the technician can more rapidly determine the cause.

While there are many advantages to conducting an active PM program, perhaps the greatest advantage is to ensure that the equipment is always in the best condition possible and its calibration is verified.

There are some problems connected with operating an effective PM program. One of these is the ability to obtain the equipment so that a PM procedure can be carried out. This is particularly true of critical-care equipment that is used daily for monitoring or treatment of critically ill patients. To work out an effective schedule with the medical staff so that this equipment can be maintained properly is difficult. In fact, for equipment that is used almost constantly, it is desirable to have an extra unit so that substitution can be made while the maintenance procedure is being performed.

It is often difficult to find equipment when it is scheduled for PM, particularly in a large hospital. There are many items of equipment that are used throughout the hospital and are transferred from department to department as the need dictates. The location of the equipment is always changing and it frequently cannot be located for scheduled inspection. The equipment control program described in Chap. 9 may help to control this situation.

These two problems may not seem serious but they are mentioned because they are often given as the reason for not completing a PM procedure on equipment at the scheduled time.

Inspection Schedule

The inspection schedule is an important part of a PM program. To set up a program, best-guess recommendations can be made concerning the frequency of inspection for different types of equipment. These can be obtained from other institutions using the same type of equipment and/or from manufacturers' instruction manuals. An initial schedule is provided but most equipment schedules will have to be modified as experience is gained. If PM is performed too often, then valuable time and money are being wasted and if the inspection interval is too long, then the PM is not effective and again, the effort is wasted. In determining the inspection frequency, signs of effective PM should be evaluated. Table 10.1 shows some of the criteria that may be used for determining effective PM. These criteria are not absolute, but do provide guidelines.

On some items of equipment, manufacturers' PM schedules must be followed in order to keep the warranty valid. When this is the case, the schedule is usually based on a significant amount of data obtained by the manufacturer or is a requirement of components within the instrument and should be followed explicitly.

Once a PM schedule has been set up, it should be reviewed at least yearly to determine if the equipment is functioning properly. After two or three years, the schedule will become firm and will then only be modified because of increased use of the equipment or because a marked reduction in the use of equipment permits a reduction in the inspection frequency. Because of the time and cost involved with preventive maintenance a continued effort should be made to optimize the schedule.

Procedures

The details of the PM procedure are the basis of a good maintenance program. If the procedure is inadequate then ineffec-

Table 10.1. Criteria for Determining an Effective PM Schedule

Frequency of PM Procedures		
Too High	*Effective*	*Too Low*
Equipment always in calibration at inspection. No changes needed.	Calibration slightly off but does not affect machine operation.	Always out of calibration. Erroneous results obtained for machine.
Equipment requires no cleaning; hardware and knobs tight, require no lubrication, etc.	Some cleaning necessary; lubrication will keep equipment operating properly; some slight tightening of hardware.	Fan filters dirty and impede good flow of air; wear has begun because of inadequate lubrication; lost nut or bolt; knobs loose causing settings to be off.
	No complaints received about instruments' operation.	Frequent complaints on instruments' operation.
	Reduced repair frequency.	Repair frequency remains same or increases.

tive preventive maintenance will result and an effective inspection schedule cannot be implemented. To arrive at an effective procedure is difficult, and without the availability of recommendations from the manufacturer or procedures based on experience of others, significant effort will be required to develop an adequate procedure. Again, if the procedure is too detailed and requires too much time, resources are being wasted, and if the procedure is too superficial, then adequate PM is not being done, resources are wasted, and an ineffective program results.

Generally there are at least two levels of PM: major and minor procedures. The major inspection is done less frequently than the minor and depending on the type of equipment is usually performed annually or semiannually. Minor inspections are usually performed quarterly, but in some cases may be done monthly or even weekly.

There are four steps in a PM procedure—visual inspection, cleaning, function testing, and safety testing. All steps need not be done during every inspection.

Visual inspection. Visual inspection is done to determine if there are any problems that can be detected visually. In most cases this requires a thorough visual examination of the equipment and includes inspection of the external and internal parts of the device. You should watch for such things as loose hardware, dented or damaged cases, flaking paint, decals that cannot be read, loose components, any signs of electrical shorts or damage, signs of unusual wear or abuse, fluid leaks, frayed cords, broken plugs, deteriorating insulation, dirty fan filters, and any other problems that might cause malfunction of the equipment.

Cleaning. It is generally necessary after the visual inspection to clean at least some parts of the equipment. There should be a standard procedure outlined for cleaning. The procedure should state those solvents or cleaning agents that can be used on the equipment as well as methods that provide the most effective cleaning. In general, the outside of the equipment should be wiped down and all dirt and oil that has accumulated because of normal use should be removed. Lint and dirt should be removed from the internal parts, particularly the electronic components, by using compressed air or by vacuuming. Fan filters should be cleaned or replaced and any blood, salts, or other residue removed from the equipment.

Function testing. The results of this series of tests guarantee the effectiveness and calibration of the instrument. In some cases, function testing should be done weekly and might be just a limited function test to make sure the equipment is operating properly, or it might include a more involved calibration procedure.

Safety testing. Safety testing of equipment is frequently directed toward electrical safety and is done to determine the level of leakage currents and the quality of grounding (Sec. 7.6). However, the testing of devices that are incorporated into the instrument to provide safety to the operator or the patient should also be done. Such items would include pressure relief valves, low-pressure cutoff valves, protective covering or shields to eliminate radiation or shock, limit switches, emergency mechanical releases, and emergency electrical or gas shutoffs. All safety devices and alarms on a piece of equipment should be understood and tested regularly.

In writing PM procedures, it is important they be as complete as possible to guarantee that the technician performs the inspection the same way every time and that an adequate level of inspection is maintained.

The procedures should not consist of a laundry list with such statements as "Clean unit and inspect"; "Check amplification"; "Measure ground and leakage current." On the other hand, the procedures should not be so detailed as to require extensive effort without increasing the effectiveness of the PM.

Manufacturer recommendations must be considered when the procedures are written, as well as the codes and requirements that regulate the equipment in the hospital. In addition to the codes described in Chaps. 4 and 7 of this text, there may be local codes and requirements which also must be met. Remember that JCAH presently requires that PM be done on every piece of hospital equipment at least semiannually.

Break the procedures into major and minor categories and fit them into the inspection schedule as appropriate. Usually the minor procedures are a subset of major inspection procedures and are not a separate list. Examples of some PM procedures for common types of hospital equipment follow.

PM procedure examples. These procedures were developed at the Down State Medical Center in Brooklyn, New York (Ben-Zvi, 1975).

Preventive Maintenance Procedure for Mercury Sphygmomanometer

Performed by ________ Date ________ Approved ________

1.0 *Introduction*

This unit is used to develop and monitor the pressure in an inflatable cuff wrapped about a patient's limb in order to determine his arterial blood pressure.

The spaces in this procedure must be filled in with the exact type of information called for. Spaces are provided: (a) to record measured or observed values, (b) for check marks where appropriate, or (c) to answer "yes/no" questions.

The purpose of this procedure is to determine if the unit is operating properly. Failure of the unit to meet any of the requirements of this procedure must be corrected before proceeding to the next step. Describe all repair work on the "Record of Services" form.

All steps in this procedure must be performed at each inspection (i.e., there is no minor inspection for this unit).

2.0 *Required Test Equipment*

2.1 Rigid cylindrical form (4.5 inches in diameter — for arm cuff)
2.2 General lubricating oil (SAE 20)
2.3 Stopwatch

3.0 *Visual Inspection and Cleaning*

3.1 The following items should be checked to insure that they are clean and free of damage or deterioration which might interfere with the unit's proper operation.
3.1.1 Rubber tubing, bulb, and cuff ________
3.1.2 Unit body, pedestal, and base ________
3.1.3 Casters (oil lightly) ________
3.1.4 Glass mercury tube ________
3.1.5 Tubing "quick disconnects" ________
3.2 Is mercury silvery? ________
Replace Hg if it is oxidized. If necessary, clean the tube and/or reservoir (refer to the unit's service manual for guidance in changing the mercury).

4.0 *Preparation for Test*

4.1 Wrap the cuff around the rigid form. ________

5.0 *Operational Tests*

5.1 *Zero.* At atmospheric conditions (i.e., no pressure applied at the bulb), insure that the mercury column rests at a reading of 0 ± 2 mm. ________ mm Hg
5.2 *Leak Test.* With a stopwatch ready, inflate the system to 230 mm Hg and then slowly reduce it to 200 mm Hg. When the pressure reaches 200 mm Hg, close the airflow valve and start the stopwatch. After 30 s note and record any drop in the pressure reading. The drop shall be less than 10 mm. ________ mm
5.3 Insure that the pressure can be "dumped" at the airflow valve. ________
5.4 Check for excessive Hg bounce or sluggishness. The kid skin diaphragm at top of tube may require replacement if above are excessive.

6.0 *Final Inspection and Certification*

6.1 Reinspect the unit for general cleanliness and clean if required. ________
6.2 Insure that the unit has its proper complement of components. ________
6.3 Affix inspection tag to unit. ________

Preventive Maintenance Procedure for Cardiac Monitor Oscilloscope

Major ________ Minor ________

Performed by ________ Date ________ Approved ________

1.0 *Introduction*

The Model 7803 Monitor Scope is a general-purpose, two-channel dual sweep speed, long-persistence oscilloscope especially suited for displaying physiological phenomena. Many units have been equipped with a heart-rate graticule; some have three sweep speeds.

The spaces in this procedure must be filled in with the exact type of information called for. Spaces are provided: (a) to record measured or observed values, (b) for check marks where appropriate, or (c) to answer "yes/no" questions.

The purpose of this procedure is to determine if the unit is operating properly. Failure of the unit to meet any of the requirments of the procedure must be corrected before proceeding to the next step. Describe all repair work on the "Record of Services" form.

For major inspection, perform all steps in this procedure. For minor inspection do not perform those steps indicated by an asterisk (*) in the left-hand margin.

2.0 *Required Test Equipment*

2.1 Function generator.
2.2 Electrical safety tester.

3.0 *Accessories*

Check that the following accessory is with the unit:
3.1 A.C. line cord. ________

4.0 *Visual Inspection and Cleaning*

4.1 Clean exterior of unit. Remove all dust, unnecessary stickers, rust, corrosion, and other deposits. Clean face of graticule as required.
4.2 Remove top cover plate. Clean and vacuum the interior as required. Inspect the interior for damage or deterioration: components, wiring, etc. ________
4.2.1 Spray contact cleaner on all switch contacts, rotating each switch as it is being sprayed. *Caution:* Remove excess cleaner; do not allow contact cleaner to get on to the circuitry.
4.3 Inspect the following to insure that they are clean and free from damage or deterioration which might interfere with the unit's proper operation or safety:
4.3.1 Line cord and plug. ________
4.3.2 Connectors. ________
4.3.3 Controls. ________

5.0 *Operational Tests*

5.1 *Position controls*
5.1.1 Depress the POWER button. The indicator lamp shall light. ________
5.1.2 Set the SWEEP slide switch to 25. With CHANNEL 1 POSITION and CHANNEL 2 POSITION both at OFF, there shall be no trace on the screen. ________
5.1.3 Turn CHANNEL 1 POSITION on and obtain a baseline. This control shall be capable of positioning the baseline at least 3 cm above and below the center of the screen. ± cm; - cm
5.1.4 Turn CHANNEL 2 POSITION on and obtain a second baseline. This control shall be capable of positioning the second baseline at least 3 cm above and below the center of the screen. ± cm; - cm
5.2 *Intensity.* Rotate the screw-driver, adjust INTENSITY control on the rear panel. At its extreme CCW position, both baselines shall disappear. At its extreme CW position, the baselines shall be very intense and out of focus. By adjusting the control, a fine trace shall be obtainable. ________
5.2.1 Cover the light sensor. The trace shall become less intense. ________

5.2.2 Uncover the sensor. The trace shall return to the previous intensity. ________

5.3 *Gain*

5.3.1 Apply a 1-V p-p 10-Hz sine wave between the tip and ring conductors of the upper CHANNEL 1 INPUT jack on the rear panel. ________

5.3.2 Rotate the screwdriver, adjust CHANNEL 1 GAIN control on the rear panel to its extreme CW position. Record the amplitude of the trace seen on the screen; it shall be at least 4 cm p-p. ________ cm p-p

5.3.3 Rotate the CHANNEL 1 GAIN control to its extreme CCW position. Record the trace amplitude; it shall not exceed 0.5 cm p-p. ________ cm p-p

5.3.4 Rotate the CHANNEL 1 GAIN control so that the trace amplitude is exactly 1 cm, peak-to-peak. Leave in this position for the remainder of the test. ________

5.3.5 Repeat Steps 5.3.1 through 5.3.4 for Channel 2. Record the maximum amplitude ________ cm and the minimum amplitude ________ cm . Leave the gain at 1 cm/V.

*5.4 *Frequency Response*

5.4.1 With the same test setup as in section 5.3, apply a 10-Hz sine wave to CHANNEL 1 INPUT and adjust the function generator amplitude control for a 6 cm p-p deflection.

5.4.2 Increase the input frequency until the deflection decreases to 4.2 cm p-p.

5.4.3 Record the generator frequency (upper 3 dB frequency) at which this occurs. It shall be 100 Hz or greater. ________ Hz

5.4.4 Repeat steps 5.4.1 through 5.4.3 for CHANNEL 2 INPUT. Record the upper 3 dB frequency. ________ Hz

5.5 *Sweep Speed Accuracy*

Note: The remainder of the procedure can be performed on either channel 1 or channel 2.

5.5.1 Apply a 1.0-Hz, 1-V p-p, square wave to the input.

5.5.2 On the scope trace note the distance in mm between corresponding points on two adjacent square wave cycles.

5.5.3 With the SWEEP slide switch at 25 mm/s this distance shall be 25 ± 2 mm. Record the distance. ________ mm

5.5.4 With the switch at 50 mm/s, the distance shall be 50 ± 4 mm. ________ mm

5.6 *Triggering*

5.6.1 With the same input as in 5.5.1, set the rear panel mode switch to FREE RUN and the SWEEP to 55 mm/s. The scope square wave pattern shall continue to retrace automatically. ________

5.6.2 Set the mode switch to SYNC; in the absence of a synchronizing input, the pattern shall trace across the screen and then be seen to "wait" for about 1.5 s before beginning to retrace.

5.6.3 Apply a 5.5-V square wave in *synchronism with the input square wave* to the SYNC INPUT (make use of the "high" and "low" output terminals if the function generator is so equipped). The sweep shall now be triggered by the SYNC square wave. ________

5.6.3.1 In addition, on those units equipped with a heart rate graticule, the rising edge of the first full square wave on the display shall cross the heart-rate scale at a reading of 60 ± 3 bpm.

6.0 *Electrical Safety Test*

6.1 *Ground Circuit Resistance.* Measure the dc resistance from the ground pin of the unit's line plug to 3 different exposed metal parts of the unit, e.g., chassis screws, control knobs, etc.

In order to check for possible intermittent continuity, when the first measurement is made, flex and bend the unit's line cord at several places while noting the resistance reading. It is particularly important to do this at the point where the line cord enters the plug.

Record the three resistance measurements. Each one shall be (and during flexing shall remain) less than 0.2 Ω.

Run 1	______ Ω
Run 2	______ Ω
Run 3	______ Ω

6.2 *Leakage Current.* Measure and record the leakage current to the unit's chassis (ground open) with power both on and off and with both normal and reverse polarity of the ac line. The leakage current shall not exceed 25 μA.

	Polarity	
	Normal	*Reverse*
Power on	______ μA	______ μA
Power off	______ μA	______ μA

7.0 *Final Inspection and Certification*

7.1 Affix inspection tag to unit. ______

10.4 Repair

The best and most effective PM program cannot eliminate the need for repairs, since it is impossible to detect and prevent all equipment failures before they occur. Servicing and repairing equipment is a significant part of a clinical engineering department's activity, and some standard procedures and routine methods should be established to make equipment repair as efficient as possible.

Figure 10.1 shows a flow diagram of the troubleshooting procedure when faulty equipment is reported to the clinical engineering department. When the report is first made it is best to obtain as much information as possible concerning the equipment failure; a standard form should be available for reporting faulty equipment. This form is generally filled out by the staff member who reports the failure, but should be done as soon as possible after the failure is discovered.

Frequently an item of equipment needs to be repaired immediately and its failure is reported to the clinical engineering staff by phone. If this is the case, then the form should be filled out by the person receiving the report. Regardless of who fills out the form, questions that should be answered on the form are the following:

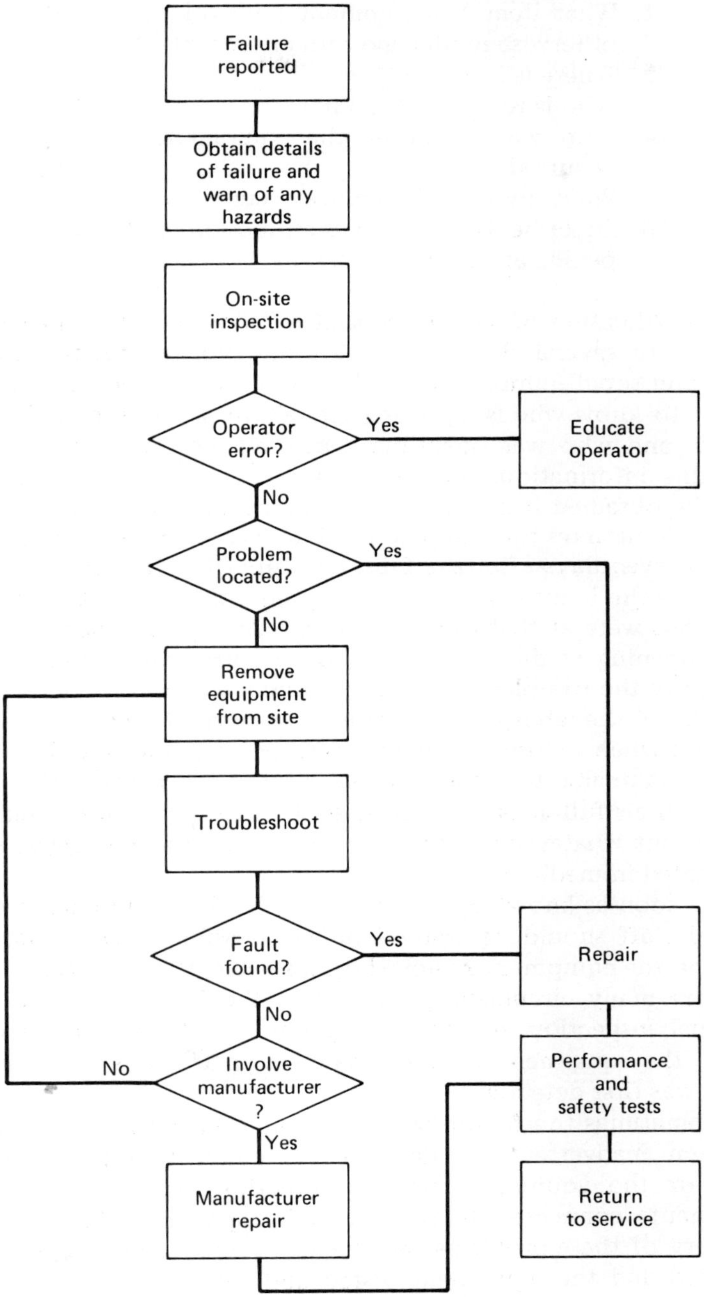

Figure 10.1 Flow diagram for repair procedure after equipment failure has been reported to the clinical engineering department.

1. What item of equipment (control number if possible, otherwise model and serial number) failed?
2. Where is it located?
3. Who is reporting the failure?
4. Who was operating the equipment when the failure occurred?
5. What time did the failure occur?
6. Describe briefly the malfunction and what was happening at the time of failure.

Identification of the device that failed is necessary, particularly if there are several devices of the same type within the hospital. Model and serial numbers make this identification positive. It is important to know who is reporting the failure, where the equipment is located, and who was operating it at the time the failure occurred. With this information, additional facts concerning the failure can easily be obtained if problems arise during the troubleshooting procedure. Sometimes the equipment failure is not reported until several hours or even days after the failure occurs. It is important to know what time the failure did occur to help determine what the prevailing conditions were at that time. And finally, a brief description of what was happening at the time the failure occurred will frequently help to simplify the troubleshooting procedure.

All of the above information is important, but there will be instances when failure occurs in emergency situations and the medical staff is unable to respond to such questions because the patient requires their full attention. In these situations there is limited information, but whatever can be done to alleviate the emergency should be initiated immediately.

As soon as knowledge of the failure has been obtained, then the medical staff should be warned of any possible hazards that might exist and the equipment removed from service, if necessary. However, there are many circumstances in which the failure can be detected by visual inspection and easily corrected. If there is no significant hazard, the equipment should remain set up as it was at the time the failure was first detected.

Sometimes the failure will be nothing more than a control that has been inadvertently maladjusted, a connection that has come loose, or the equipment may malfunction because of the other instruments connected to it. A visual inspection will lead to the discovery of these problems and the malfunction can be immediately corrected and the equipment tested and returned to service. However, once the equipment has been removed from the environment in which the failure occurred, some problems may never be discovered and it will be very difficult to determine whether the malfunction was caused by an intermittent problem or was due to operator error or improper connection.

If the fault is not discovered by visual inspection and limited testing, then the equipment should be removed from service and the troubleshooting procedure carried out. If the fault is found then it should be corrected and a limited performance test run to guarantee it is operating properly. A safety test is usually necessary before the instrument is returned to service.

If a significant amount of time has been spent trying to troubleshoot the piece of equipment and the fault cannot be found, it might be necessary to involve the manufacturer. The manufacturer may offer suggestions in the troubleshooting procedure or he may assume total responsibility for repairing the instrument. If the instrument is repaired by the manufacturer, a performance test and safety test should be carried out prior to returning it to service.

The decision to consult the manufacturer involves many considerations and all should be carefully reviewed. If the apparatus is complicated and the clinical engineering staff has had little experience with that type of equipment, it is probably wise to contact the manufacturer after only preliminary troubleshooting, even though a repair cost will result. If the clinical engineering staff is familiar with a piece of equipment, then additional troubleshooting time should be spent in an effort to determine the malfunction because their time is less expensive than that of the manufacturer. However, unproductive troubleshooting should not continue indefinitely because there will be a point at which more time and expense will result than if the manufacturer had been involved. This assumes that the manufacturer is more efficient in troubleshooting because of his familiarity with the design of the equipment or that the repair is contracted at a fixed cost regardless of the length of time it takes.

In the new, modular equipment, it may be less expensive to purchase a new modular board and replace it than to spend the time troubleshooting the module. For example, if a new board costs $100 and we assume that it costs an average of $10/man-hour for the members of the clinical engineering department (see Sec. 13.5 for method of cost calculation) and it is necessary to spend 10 hours troubleshooting the board, a new board may as well be purchased. In difficult problems, 10 hours is not an unreasonable length of time.

These factors should be weighed by the clinical engineer. His goal is to provide the most rapid repair at the least cost, whether done in-house or with the aid of the manufacturer or other outside service.

Troubleshooting

Troubleshooting can be inefficient and a time waster. Many hours can be wasted using a detailed troubleshooting method when a much simpler method would resolve the problem. Simple methods should be tried first and complicated, detailed procedures only brought into play when the fault cannot be found. For example, first visually inspect the equipment for loose wires, arcing, and burnt resistors.

Figure 10.2 shows a suggested troubleshooting flow diagram. The three top blocks, symptom/function, signal tracing, and voltage/resistance are ways in which the problem can be isolated:

Symptom/function—knowing how the device should operate and observing the characteristics of the failure, determine the location and probable cause of failure.

Signal tracing—apply a known signal to the input and trace it serially through a device until it becomes abnormal. Characteristics of the abnormality will indicate the probable failure.

Voltage/resistance—make voltage and/or resistance measurements at test points where the normal values are known. Abnormal values will help locate the failure.

The diagram in Fig. 10.2 is directed toward troubleshooting electronic devices, but notice that the concepts apply to other devices as well. For instance, signal tracing may be analogous to tracing variations in pressure or flow through a pneumatic device. The voltage measurements might be gas pressure, or the resistance measurements could be resistance to flow in the device.

In almost all devices, it is the active components that usually cause difficulty. They are checked first, usually by substitution, after using one of the three methods just described to determine the location of the fault. Substitution of components is a very effective way of determining the faulty component. If the number of active components are few, then direct substitution even without the preliminary analysis may result in the most efficient troubleshooting technique. However, the substitution of solid state components soldered to the printed circuit board are frequently difficult to remove and they might be destroyed or rewired incorrectly—only adding to the troubleshooting problems. Before making substitutions in this case, it should be determined that the component to be replaced is probably malfunctioning.

When the above methods do not work, write down the troubleshooting procedures that have been completed and review the results to see if there is an error in the logical sequence, some part of the circuit has not been evaluated, or abnormal signal or voltage relationships were not observed. Compare the measured values to those on a schematic or functional diagram. Even better, if there is a working unit available, comparison of values in the malfunctioning unit and the working unit will help to point out where the error occurred.

If the results of this evaluation are still negative, then look for unusual problems: cold-soldered joints, hidden broken connections,

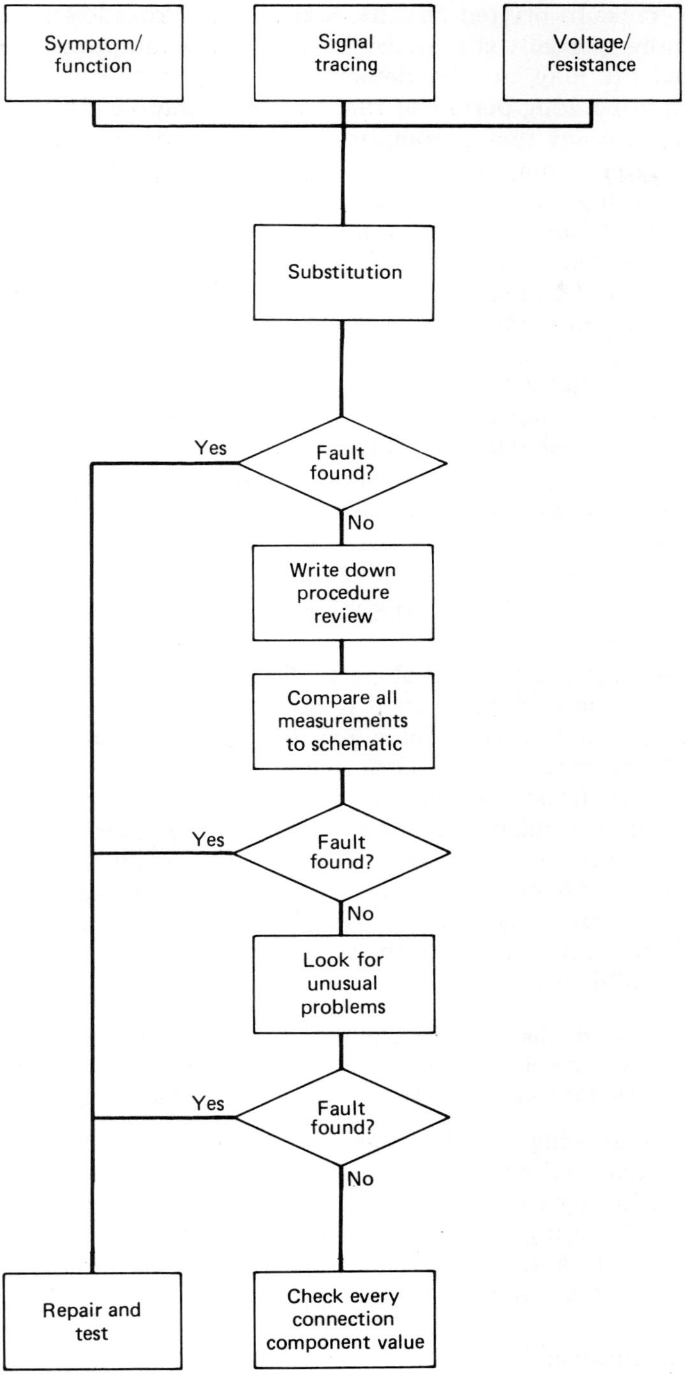

Figure 10.2 Flow diagram for troubleshooting procedure.

hairline cracks in printed circuits, high-voltage breakdown caused by dirt or other partially conductive media. Wrong component value is a possibility. It may not be detected until after the unit has been operating for a long period of time and the components have deteriorated to a point that a component with the wrong value causes a failure. Wrong components could have been put in during any troubleshooting procedures. A component may have been wired incorrectly. All connections should be checked on any component that has been replaced during this or other procedures.

If the fault still cannot be found, then as a last resort every connection and every component value should be checked. This includes a detailed check of all passive and active components and their connections, pin-by-pin and wire-by-wire. During these second and third levels of troubleshooting, it may be desirable to return at any time to the initial troubleshooting procedures, particularly that of substitution. Substitution of whole modules or cards is often preferable to point-by-point or detailed checks, even if added expense is incurred.

10.5 Records

Recordkeeping is an essential part of the PM and repair program. Without adequate records there is no way to determine if PM has been accomplished and valuable assistance is lost in troubleshooting if there is no record of repairs. Records are necessary should legal problems arise because of equipment malfunction; they are necessary to determine the reliability of equipment and to determine the operating costs; they will indicate when a piece of equipment has served its useful life and should be replaced; and they are required by standards and regulations described in Chap. 4. For instance, the Accreditation Manual for Hospitals published by the JCAH states (Chap. 4, 1976 ed.)

> Records shall be maintained to reflect the dates of inspection and maintenance, as well as the status of all equipment, including the need for replacement and the individual notified of this need.

Recordkeeping can be time-consuming and laborious, but if properly organized and if the records are written at the time of evaluation, the burden can be greatly lightened. Since recordkeeping must be done, it is an important part of the activities of the clinical engineering department. Careful consideration should be given to the organization of records.

Required information

There are many ways in which information can be recorded but there are certain items of information that should be part of any equipment record. A summary of this information is shown in Table 10.2. Organization of this required information will be discussed in the next section, but to provide complete records it should be included. *Topham*

The required information may be divided into four categories: equipment identification, dates, cost, and descriptive information. Section 9.4 presents methods of obtaining this information and examples of recordkeeping procedures. Section 13.5 discusses cost considerations.

Forms

There are many ways to record PM and repair data; each institution may have its own set of forms to fit local needs. There are two obvious ways to create forms for recording information. One is to create a general form that can be used on many, if not all, types of equipment, and the other is to generate a specific form for each type or category of equipment within the hospital. A general form tends to be long and cumbersome and because of its general nature consists of many blank sections for any given piece of equipment. The specific form for each type of equipment is sometimes expensive and requires a significant amount of effort to develop. A compromise is to generate a form that has a similar format and sequence of recording information but is specific for a type of equipment in that only those parameters or information necessary are included on the form.

Not all the information concerning a piece of equipment needs to be on the form used routinely for the recording of PM or repair procedures. Much of the identification information can be entered on a form that is filed centrally and is updated with dates and costs

Table 10.2. Required Information for Equipment Records

Equipment Identification	*Dates*	*Costs*	*Descriptive Information*
Description	Date purchase order issued	Initial cost	Description of failures and action taken
Model number	Date of incoming inspection	Cost for repair (parts and labor)	Unusual comments by user — praise or complaints
Serial number	PM schedule	Cost for PM (parts & labor)	Parts needed for repair
Hospital tag number	PM performed	Improvement costs	
Manufacturer	Warranty dates		
Local representative	Failure and repair dates		
List of accessories	Date removed from service		

as the work is done. The detailed information is recorded on a working form. For example, in recording the results of a PM inspection, the actual PM procedure may be listed with the appropriate spaces available to record the parameters as they are measured (see PM procedure example, Sec. 10.3). This is a long form and after experience the procedure material tends to be redundant. However, it does help guarantee that the PM is always done in the same way because the procedures are not being done from memory but are described on the inspection form for reference during the test.

A second and more common type of PM form only provides a place for recording the measurements taken. A word or a brief statement indicates what procedure should be followed. An example of this is shown in Fig. 10.3. Many forms have been developed for the recording of PM data and are available upon request. For example, ECRI provides other standard forms for all equipment tested and reported in *Health Devices* (Sec. 15.6). Manufacturers of electrical safety testing equipment also provide standard forms for recording the data obtained during inspection. Some manufacturers of medical equipment provide forms. Before new forms are developed by a hospital, available ones should be reviewed and if not adopted for use, the best parts of them can be incorporated into the forms developed within the hospital.

Filing

Filing all of the records such that the data can be retrieved rapidly and easily is an important consideration. There are many ways to organize files but it should be done with minimum duplication of record information and there should be a central index file which has key information on each item of equipment and provides reference to the details of inspection, repair, etc.

The main types of information to be filed are the following: equipment identification, PM procedures and results, records of failure and repair, operation manuals, service manuals, incoming inspection data, and cost information. The filing system is often set up in such a way that it aids in scheduling of PM and provides reminders when PM is due on a specific device. A card-sort system with key information on the sorting card can provide this capability. A good example of this method is that used by Down State Medical Center, Brooklyn, New York (Ben-Zvi, 1975).

A system might be set up that uses the file for assistance in scheduling PM as shown in Fig. 10.4. There are three parts to this program: the scheduling file, the PM inspection forms, and the individual equipment files.

The scheduling file consists of 48 folders—two folders for each month for a two-year period. For each month there is an "A" folder and a "B" folder. Folder A has a list of equipment that should be

Defibrillator/Monitor Inspection Form

ACTION
Not Needed ☐ Needed ☐ Taken ☐
Removed from Service ☐
Control No.
Date of Inspection:
Next Inspection Due:

LOCATION____________________ SERIAL NO.____________________

MODEL____________________ MANUFACTURER____________________

	OK	ACTION NEEDED	ACTION TAKEN (Date & Initials)
1. GENERAL CONDITION OF UNIT & CHARGER			
2. CONDITION OF MOUNT			
3. POWER CORD, ATTACHMENT PLUG, STRAIN RELIEFS			
4. FUSE OR CIRCUIT BREAKER			
5. POSITION OF CONTROLS			
6. CONDITION OF CONTROLS, INDICATORS, BATTERY			
7. CONDITION OF CABLES, PADDLES, CONNECTORS			
8. ACCESSORIES (Check box if deficient): Electrodes ☐ Gel ☐ Saline Pads ☐			
9. GROUNDING RESISTANCE: Chassis, ____ ohms; RL lead, ____ ohms			
10. LEAKAGE CURRENT TO GROUND (Circle unacceptable values):			

	Max. Leakage	Wiring/Operating Mode
Chassis	μA	
Electrodes		
Paddles		

	OK	ACTION NEEDED	ACTION TAKEN (Date & Initials)
11. INTER-ELECTRODE LEAKAGE:			
Max. Leakage: ____ μA Wiring/Operating Mode:			
12. INPUT ISOLATION: Current ____ μA			
13. CHARGING TIME TO MAXIMUM ENERGY: (Previous value ____ sec.)			
Battery Operation: ____ sec.; Line Operation: ____ sec.			
14. INTERNAL DISCHARGE OF STORED ENERGY			

Continued on Reverse Side

Health Devices Program
Form HD 318-276

ECRI

THE EMERGENCY CARE RESEARCH INSTITUTE
5200 BUTLER PIKE, PLYMOUTH MEETING, PA 19462 / (215) 825-6000

Figure 10.3 Defibrillator/monitor inspection form. From Emergency Care Research Institute, Plymouth Meeting, PA 19462.

Defibrillator/Monitor Inspection Form, Side 2

	OK	ACTION NEEDED	ACTION TAKEN (Date & Initials)
15. DELIVERED ENERGY (watt-seconds):______			

CONTROL SETTING	INDICATED ENERGY	DELIVERED ENERGY	PREVIOUS VALUE

16. ENERGY DELIVERED AFTER 1 MINUTE, MAXIMUM SETTING:__________W-Sec.______

17. OUTPUT OF TENTH REPEATED DISCHARGE, MAX. SETTING:__________W-Sec ______

18. QUALITY OF DISPLAY (Check box if deficient):______
Focus ☐ Slope ☐ Bow ☐ Baseline position (Auto/Manual) ☐
Intensity (Auto/Manual) ☐ Burn spots ☐ Interference ☐

19. EXTERNAL 1-MILLIVOLT PULSE (Check box if deficient):______

Amplitude (__________mm @ 10 mm/mV) ☐ High-frequency response ☐
Vertical linearity ☐ QRS indicator ☐
Non-fade display ☐ Through-the-paddle monitoring ☐

20. SAG: Time ☐ Distance ☐ to half amplitude________@________mm/sec______

Approximate lower 3-db point,__________Hz ______

21. INTERNAL CALIBRATION PULSE:________mm @ 10 mm/mV______

22. SWEEP SPEED: Control setting,________mm/sec______

Sweep time,________sec for ________mm______

OTHER TESTS (Describe):______

______

______

COMMENTS & DESCRIPTION OF DEFICIENCIES (Refer to item numbers):______

______

______

______

☐ INSPECTION TAG AFFIXED: SUITABLE ☐ NOT SUITABLE ☐ FOR INTRACARDIAC CONNECTION

☐ EQUIPMENT SERVICE REQUEST FORM COMPLETED

INSPECTED BY:______

Figure 10.3 Continued.

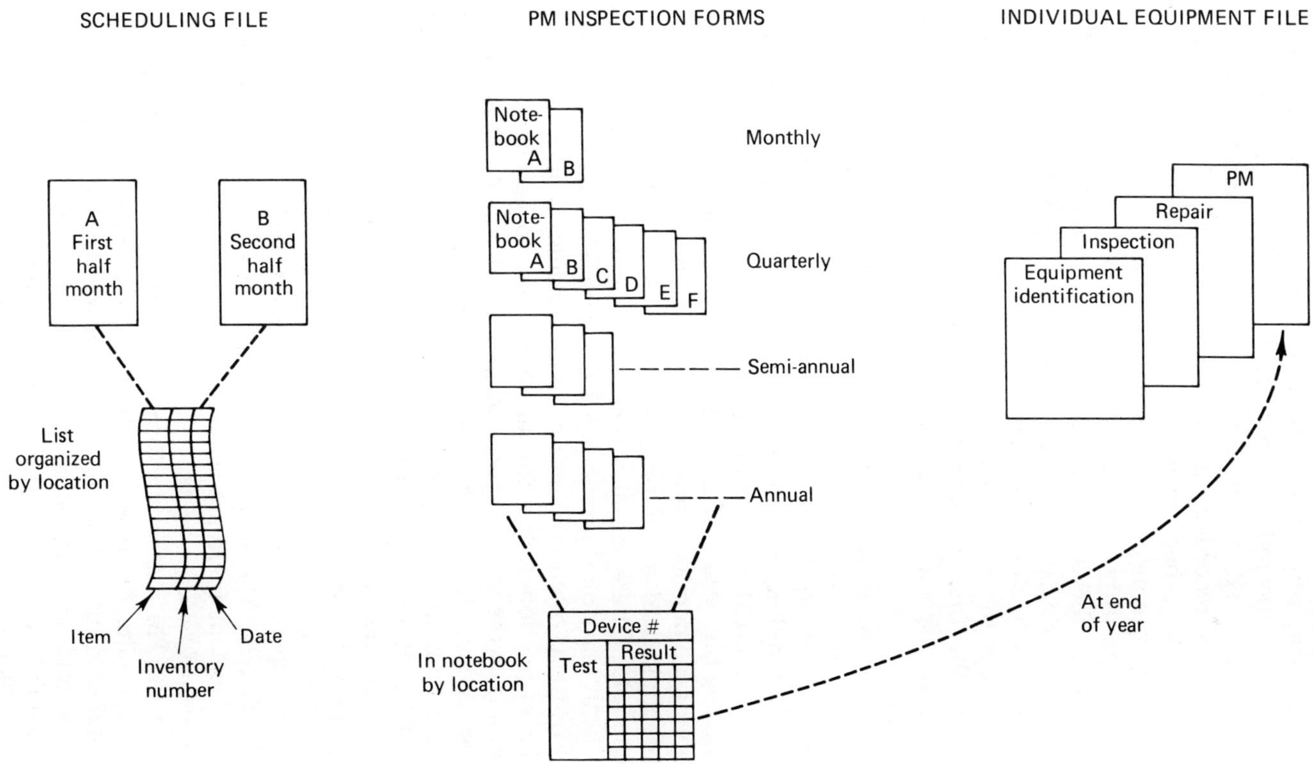

Figure 10.4 Diagram of the filing and PM scheduling system used at William Beaumont Hospital, Royal Oak, Michigan.

inspected during the first half of the month, and folder B lists the equipment that should be inspected during the last half of the month. Each list, which is organized by location in the hospital, lists the item to be inspected along with its inventory control number. As the PM is completed, the date of the inspection and the initials of the inspector are recorded on this list. Once the list is complete and all listed items have been inspected, then it is placed in the folder of the month for the next scheduled inspection. There are four lists—one each for monthly, quarterly, semiannual, and annual inspections. For example, at the first of January, the quarterly list is pulled from folder A and during the next 15 days all the equipment on that list is inspected. When the inspection is complete, this list is then placed in folder A for the month of March. If the monthly inspection list is pulled at the first of January, then after the equipment on the list has been inspected, it is placed in folder A for the month of February.

A file capacity of two year's data is necessary so that the list can be advanced into the next year's folders. Two years is sufficient time because the largest interval between inspections is annual.

The scheduling list indicates what equipment should be inspected and the PM inspection forms provide the details and a place for recording the results of the inspection. These forms are contained in notebooks. There is a monthly set of notebooks which consists of two notebooks and a quarterly set of notebooks which consists of six notebooks. There are also semiannual and annual sets of notebooks, or at least one notebook with separate divisions because the number of items of equipment on these schedules is much less.

The inspection forms for all equipment to be checked during the first half of the month every month are contained in the monthly notebook A and those to be inspected during the last half of the month are contained in monthly notebook B. As the equipment is inspected (specified by the list from the scheduling file), the inspection forms are pulled and the data recorded. They are then returned to the supervisor, who notes that the inspection has been done on the list from the scheduling file, and on the 15th or 30th of the month the inspection forms are all returned to monthly notebook A or B respectively. The same procedure is followed with the quarterly inspections except there are six intervals in which inspection can be done instead of two.

Since the PM inspection forms have sufficient room on them to record the results for one year, they are kept in the PM inspection form notebooks until the end of the year, at which time they are transferred into the individual equipment file and become part of the permanent record for that item of equipment. The individual equipment files contain the standard information such as equipment identification, incoming inspection, repair and failure record, and then once a year the PM record is added.

If the PM record of a device is needed before the end of the year, quick reference to the individual equipment file will provide that device's inspection schedule and the PM inspection form can then be pulled from the proper notebook.

Computerization

As can be seen from the examples given and the type of records kept, if the facilities are available, computerization of these records would be very helpful and could automatically provide PM scheduling as well as the capability for a rapid search of any information needed. The information might include the number of items receiving PM during the month, the cost of the PM, how many items experienced one or more failures, the cost of repairing these failures, an inventory list of all the equipment available in a given hospital location, and a complete inventory list of all equipment in the hospital.

There are various computerized systems and many are presently in place (Barnard, 1974; DesRoches, 1975; Ben-Zvi et al., 1978). Frequently, because of the current system under which the hospital is operating, it is necessary to modify these systems to fit the characteristics of the individual hospital (Sec. 5.9; Sec. 6.9).

Most computer systems have the following features: a list of equipment on which PM is to be done is usually generated automatically on a regular basis; when the PM is complete the results are stored in the computer and the costs calculated. The computer will turn out inventory lists on command along with identification numbers of the equipment. Repair and failure data are stored as are the costs for making repairs.

While computerization may solve many of the manual filing problems of records and probably provides more rapid access to the data within a file, it is frequently much more convenient to reach for a card file in a department office rather than conduct a computer search either by remote terminal or by initiating a formal request to the computer service.

10.6 Facilities

The size of physical facilities for a clinical engineering department is highly dependent on the responsibilities given the department and the number of personnel needed to carry out those responsibilities. For a basic staff comprised of one engineer, five technicians, one secretary, one clerk-typist and one clerk, 1200 sq ft of space is recommended. Each additional technician would require an additional 100 sq ft. This would provide space for work benches, test equipment, places for equipment under inspection or repair, file cabinets, desk and chairs. It is estimated that another 100 sq ft of space would be necessary for equipment in active storage (Ben-Zvi, 1975).

For two people, say a biomedical equipment technician and clinical engineer, at least 300 sq ft of space should be provided. This includes room for desks for each individual, filing cabinets for keeping the records, a workbench area for the repair and PM of equipment, and a limited storage area for parts and for backup equipment and equipment to be repaired. This is a room about 15 × 20 ft. The recommendation for 100 sq ft of additional space for each technician would still be applicable and if a secretary is added to the department, an additional 100 sq ft would be required.

The test or repair laboratory area should contain the following facilities:

1. Forty square feet of bench space for one engineer and one full-time technician—approximately 25 sq ft of bench space for each additional full-time technician.
2. A small desk for each technician in which he can keep his personal items as well as make reports and keep records up-to-date.
3. Each workbench should have a 120-V power strip with at least 20-A capability.
4. Vacuum and compressed air for each work bench.
5. Sink located somewhere in the test area but placed in such a position that no safety hazard would exist because of the wet area.
6. A central supply section containing a parts cabinet, some laboratory chemicals such as Quick-freeze, contact cleaner, degreasers, etc., which might be used in repairing or troubleshooting the equipment. The parts cabinet should include a wide variety of hardware, nuts and bolts, a good selection of resistors and capacitors, and a complete set of all fuses that might be used in the repair of medical equipment.
7. An adequate supply of connectors and adapters and various types of probes and leads which are used for testing and for connecting test instrumentation to the equipment.

The minimum instrumentation needs for the laboratory are listed in Table 10.3. The tool kit should include at least the following: screw drivers (Phillips and slotted blade of about three different sizes), pliers (long nose), diagonal cutters, wire strippers, an adjustable wrench, a set of hex and spline wrenches, a pair of slip-joint pliers, electrician's pocket knife, an alignment tool, a solder

Table 10.3. Minimum Instrumentation Required for PM and Repair

Instrument	*Approximate Cost in 1976*
Oscilloscope (dual trace, 15 MHz)	$2800
Digital multimeter (3½ digit)	275
DC power supply (variable to ±20 V, 1.5 A)	230
Volt–ohm multimeter	95
Function generator (0.1 Hz to 1 MHz, 4 waveforms)	295
Electrical safety analyzer	800
Defibrillator tester	650
ECG simulator	185
Tool kit	65
Total:	$5395

aid fork and reamer, a soldering iron, tweezers, and a nut driver set. This is a basic tool kit but will be adequate for a majority of the repairs and PM procedures. Each individual should have his own tool kit, which can be locked, consisting of at least the basic tools outlined.

It is desirable to have on hand at least one complete tool kit. These instrument repair kits are put together by many tool companies for about $350–$400. They provide all of the tools necessary for repair in a convenient case which can be locked and carried from place to place in the hospital, if needed.

Other items of equipment will be purchased as the PM and repair program expands and could include such items as a counter (up to 10 MHz); oscilloscope camera; small portable oscilloscopes; pulse generators; various standards for frequency, voltage, and current; recorders and calculators. These items should be added as needed and as they become economically feasible.

There may be specialized test stands that will be set up within the laboratory to allow the testing of particular devices of equipment more readily. Frequently these test stands can be made in-house and save significant cost. An example is a pneumatic test stand which can be built for approximately $400 but would cost $1500 if purchased. The development and construction of such test stands are left to the ingenuity of the engineer and his staff.

Unless there is a significant amount of construction of special devices for use in the hospital, the need for machine-shop types of equipment (drill presses, metal lathes, etc.) is frequently unjustified. It is usually cheaper to contract this work out than to provide the equipment necessary. Some hand power tools, such as an electric drill, are beneficial if money permits.

The equipment, as it has been discussed here, is for use in the test and repair laboratory and does not provide for portable equipment which can be carried to the device site for repair. A complete instrument test unit that is constructed in a modular fashion (usually containing a power supply, function generator, digital multimeter, small oscilloscope, etc.), can be purchased from several manufacturers and provides a convenient portable instrumentation system for repair or PM of equipment on site. This type of equipment should be purchased as soon as possible and may even be used to satisfy the initial laboratory requirements. A modular system which allows plug-ins to be used in different pieces of test instrumentation is a good way to purchase equipment. Careful consideration should be given to the capabilities provided by a given manufacturer. Even though only limited capability is initially needed, a coordinated instrumentation system will prove invaluable in the future.

10.7 Costs

Costs for the PM and repair of equipment are an important factor in the overall expense of operating it. Preventive maintenance costs may be reduced by good prepurchase evaluation but it is necessary to provide for repair costs in the budget. In the past, PM costs have usually not been included in the budget, but without an adequate PM program, quality of performance of the equipment will be reduced and may interfere with providing the best medical care possible.

Estimated cost for PM and repair

Estimating the PM and repair cost for equipment is usually done by taking a percentage of the purchased price of the equipment. This percentage figure will range from 5–10 percent of the original cost, with 10 percent being the figure most frequently quoted by hospitals (Frost and Sullivan, 1975). These figures come from equipment manufacturers or from firms that provide PM and repair contracts. Variation depends not only on the firm but also on the type of equipment they service. For instance, a manufacturer of patient monitoring equipment claims that its PM and repair costs should be about 7 percent; a firm that provides PM and repair contracts for X-ray equipment estimates a cost of 4.5–6 percent of the new equipment; a manufacturer of hemodialysis systems requests 8 percent of the total equipment cost; and a manufacturer of ECG machines will only sign PM and repair contracts that amounts to 17 percent or more of the cost of the equipment. As the age of the equipment increases, the percentage may also increase.

Another way to estimate the PM and repair cost is to review the actual costs that have been incurred by the equipment within the hospital during the past years. Since PM has not been emphasized, this estimate will be primarily oriented towards the cost for repairs. *Topham*

A study was done in which the repair costs in 11 hospitals of various size in a large metropolitan city were evaluated (Topham, 1976). The frequency of repair and the cost per repair for 21 different types of commonly used medical equipment were determined. The average repair values are presented in Table 10.4. Notice that with all equipment considered, the average repair frequency was one time per year at a cost of about $56. This repair cost is representative of outside service contracts and might be significantly lower if it had been done with in-house services. From this table, a good estimate of repair costs can be obtained for the common types of hospital equipment by using the number of the different units located in the hospital for which the analysis is being done. Estimation of the cost for PM of this equipment must then be added in order to budget an effective PM and repair program.

Table 10.4. Equipment Repair Frequency and Costs (labor only) for 1976

Equipment Type	*Units*	*Repair Frequency (repairs/unit/year)*	*Repair Cost (avg. cost/repair)*
Anesthesia machine	68	0.77	$62.00
Blood gas analyzer	24	1.50	41.60
Blood pressure instrument	190	1.79	19.00
Defibrillators	130	0.80	67.00
ECG machines	179	1.76	46.60
EEG machines	8	0.80	33.50
External pacemakers	45	0.46	47.50
Electrical surgical unit	104	1.96	45.29
Hypo and/or hyperthermia machine	206	0.58	42.33
Incubators/warmers	143	1.01	74.00
Infusion pumps	143	1.03	41.00
Microwave diathermia	11	0.91	47.00
Nebulizers (ultrasonic)	136	0.85	37.71
Oxygen analyzers	12	0.79	75.00
Patient monitors (fixed)	228	2.08	76.13
Patient monitors (mobile)	165	1.17	80.43
Recorders	26	0.85	53.00
Suction machines	412	0.50	32.00
Pressure transducer	30	0.75	110.50
Ultrasonic therapy	10	1.68	52.00
Ventilators	162	0.50	86.33
Clinical lab	189	1.22	52.40
X-ray equipment	99	6.21	137.00
All equipment	2720	1.30	59.10
All equipment (without X-ray)	2621	1.08	55.56

It is estimated for a given set of equipment that the time it takes to perform the PM will be approximately equal to twice the amount of time it takes to provide the repair on that same equipment. A PM program is expensive but may save as much as 45–50 percent on the repair cost. This however is not enough to offset the cost of the PM program and other cost considerations must be evaluated.

There is a decrease in the amount of downtime for equipment under a PM program, thus reducing the amount of wasted time in the use of the equipment. When equipment is running and reliable, tests do not need to be rerun, and the medical staff's time is not taken up with efforts to recalibrate a piece of equipment or make it operate. These all contribute to increasing the effectiveness of the medical staff and reducing the amount of frustration and lost time. The inconvenience and frustration caused by malfunctioning equipment is very difficult to quantify in terms of cost savings, but certainly can distract from quality medical care. There are definite benefits to be gained from a good PM program, but in order to acquire these benefits, some cost must be incurred.

10.8 Study Questions

10.1 List three advantages of a PM program.

10.2 List three criteria that indicate an effective PM schedule.

10.3 List the four parts of a PM procedure.

10.4 What is the longest PM interval permitted by JCAH?

10.5 Describe one operational test for the mercury sphygmomanometer PM procedure.

10.6 Describe two electrical tests for the cardiac monitor PM procedure.

10.7 List three sites where equipment may be repaired.

10.8 List four items that should appear on an equipment failure form.

10.9 Give a cost advantage and a cost disadvantage of sending equipment to the manufacturer for repair.

10.10 Name and briefly describe three ways of troubleshooting an equipment problem.

10.11 List three unusual problems that should be considered during equipment troubleshooting.

10.12 List three reasons for keeping records of PM and repair.

10.13 Name four categories of information that should appear on equipment records.

10.14 Describe two types of forms for recording PM data.

10.15 List four types of equipment information that should be filed.

10.16 List and describe the three parts to the filing and PM-scheduling system.

10.17 List three characteristics of a computerized equipment-filing system.

10.18 List five facilities that should be found in a clinical engineering test or repair area.

10.19 List five items of instrumentation required for PM and repair.

10.9 References

Barnard, J. Q. (1974). Computerized record of maintenance activities. *Hospitals, JAHA*, 48:102.

Ben-Zvi, S. (1975). Preventive maintenance program, final report. Publication PM-5. Chicago: American Society for Hospital Engineering.

Ben-Zvi, S., S. Gross, and F. Burnett (1978). Computerized control of the PM and repair of medical instrumentation and the maintenance of service records. *Proc. AAMI Annu. Meeting*:31.

Buchsbaum, W. H. (1974). Tested electronics troubleshooting methods. Englewood Cliffs, NJ: Prentice-Hall.

DesRoches, H. B. (1975). Computer monitoring of preventive maintenance. *Hospitals, JAHA*, 49:86–92.

Frost and Sullivan (1975). Safety and maintenance of biomedical equipment in the USA, 1974–1983. Report #259. New York.

Health Devices. A test, evaluation and advisory service. Plymouth Meeting, PA: Emergency Care Research Institute.

Topham, W. S. (1976). The cost of a clinical engineering department. *J. Clin. Eng.* 1:24–26.

Veterans Administration (1973–74). Maintenance procedures, G–29. Department of Medicine and Surgery supplement, MP-3. Washington, DC.

11
Facilities

BURT DODSON, JR.

11.1 Learning Objectives

Upon completing this chapter you should be able to:

1. Explain why a systems approach is necessary for facilities planning in a hospital.
2. Describe the four methods of facilities analysis.
3. Describe the special facilities necessary in operating rooms.
4. List the four most important hospital areas that require facilities planning.

11.2 Introduction

Clinical engineering cannot be effectively and singularly practiced in the sterile atmosphere of operating rooms and intensive-care units. Rather, it must be practiced from the hospital basement to the patient room. This concept requires the clinical engineer to have the knowledge and expertise of a *specialist* and the understanding of a *generalist*.

11.3 A Systems Approach to Clinical Engineering

The clinical engineer's need for a broad understanding of the hospital is demonstrated by the following incident: A clinical engineer was contacted by a nurse epidemiologist (infection control nurse) to discuss a cluster of infections in the ICU which had been traced to a gram-negative microorganism (*Pseudomonas*). Although respiratory therapy equipment was the logical suspect as the source of infection, exhaustive evaluations showed the reservoir of the pathogen to be not in the biomedical equipment, but in the compressed air line. The cause of this problem was a "breeding reservoir" resulting from excessive water in the compressed air, caused by failure to bleed the compressed air tank of accumulated condensation. Once the problem was identified, the plant engineer asked, "How do we decontaminate the compressed air lines?" Thus, the clinical engineer found himself with a problem involving both plant engineering and the infection control department, thereby demonstrating the evolving interfaces that the clinical engineer must prepare for. This chapter provides an overview of clinical engineering as it relates to problems that arise in the facilities area.

The systems view of clinical engineering

Experience has shown that decisions involving devices and biomedical equipment technology have often been made in the absence of considerations that encompass the hospital environment, the personal characteristics of staff and patients, existing codes and regulations, or, for that matter, economics. Countless examples exist of equipment purchased and delivered for installation when in fact the hospital lacked adequate electrical power, sufficient space for the equipment, proper shielding, or other essential physical plant characteristics.

The purchase process is further complicated by the overlap of regulations established by multiple standard-setting organizations such as the NFPA and the JCAH. The multiplicity of codes, standards, and regulations requires that the clinical engineer fully comprehend their overall impact (Chap. 4). The various utility and facility codes and regulations traditionally have been the responsibility of plant or safety engineers. The clinical engineer must also be knowledgeable in these codes and regulations in order to understand the plant engineering function and foster good and effective hospital communications.

Solutions to clinical engineering problems may create other problems that must be addressed. A classic example has been the use of conductive floors in operating rooms. The original intent in using such floors was to minimize the possibility that electrostatic sparks might explode flammable anesthetic gases. However, the installation of conductive floors introduced an electrical hazard for the staff and patients by the combination of liquids, electrical systems, and metal surfaces involved. (With the diminished use of cyclopropane the original problem has all but disappeared.) From this potential electrical hazard stemmed the need for ground-fault detectors and isolated power systems. This is not to say that ground-fault detectors and isolated power systems have no purpose in the operating room, but rather to point out that, in solving one problem, another problem may be created.

These examples emphasize the need for the clinical engineer to more thoroughly understand the total-system aspect of clinical engineering and how it relates to his particular role in health care.

11.4 The Hospital as a Systems Problem

Health care has undergone dramatic changes in the past 25 years. Figure 11.1 dramatically portrays one aspect of the change, the exponential growth in the field of medical equipment. This growth, both in terms of complexity and quantity, has presented critical challenges to the field of clinical engineering. The rapid growth of technology has produced substantial improvements in patient care while at the same time it has produced serious problems relating to facility interface, safety, and personnel.

One of the more recent technological developments in medical instrumentation is the computerized axial tomography scanning unit (CAT Scanner) (Gordon et al., 1975). The scanner generates ionizing X radiation and uses a collimated X-ray beam with an array of detectors to make a series of transverse passes through the patients at various angles over the 180 degrees of axial rotation. The penetrating radiation data that are accumulated are mathematically processed by an interfacing computer and an image is produced of

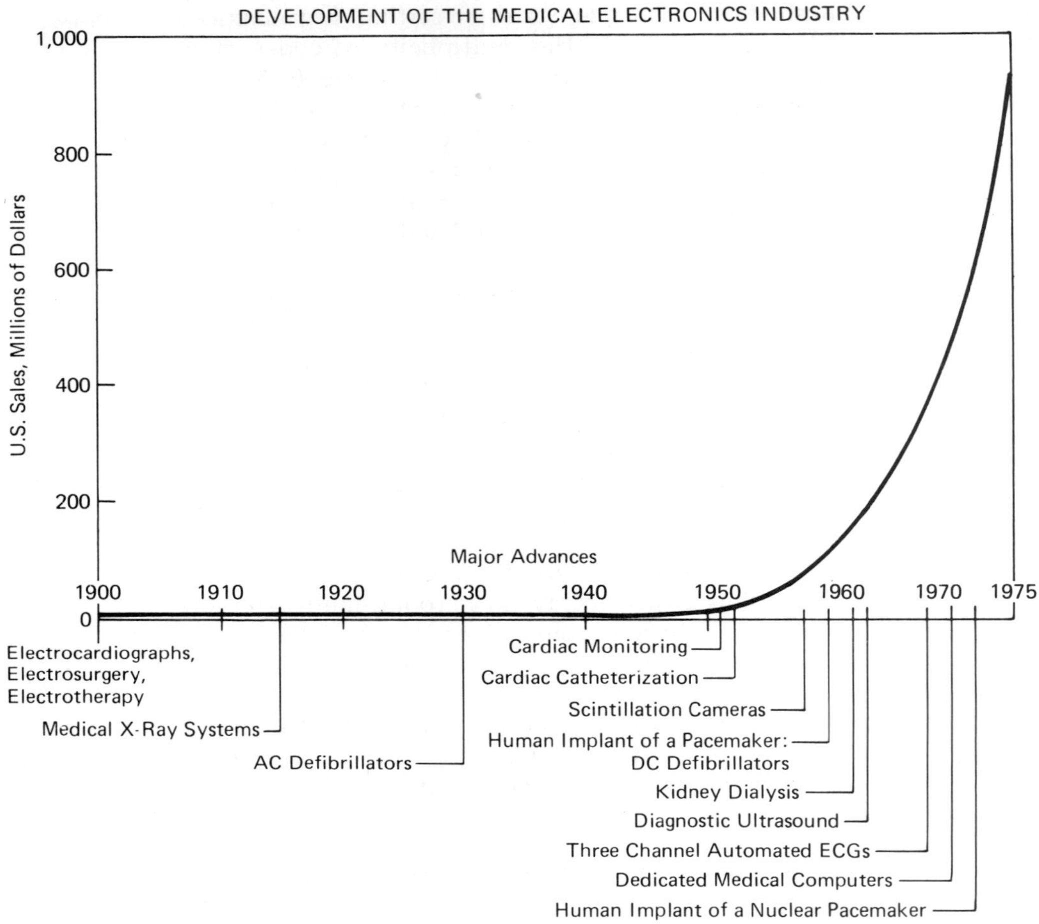

Figure 11.1 The use of medical technology, as measured by sales, has increased exponentially with major advances in equipment and techniques. From *Biomedical Electronics,* Oct. 10, 1975, published by Girard Associates, Mt. Arlington, NJ.

the scanned part which is far more detailed and definitive than previous capabilities in the field of radiology. When we consider the diverse elements of the system, it becomes clear that a broad range of knowledge is necessary to be able to competently discuss the process. The CAT Scanner, depicted in Fig. 11.2, contains high-voltage power supplies, electronic control circuits, heat exchangers, image processors, and display terminals for viewing the images and alphanumeric data.

Over and above the thrust of new technology in the field of clinical engineering and physiology, other problems lying dormant

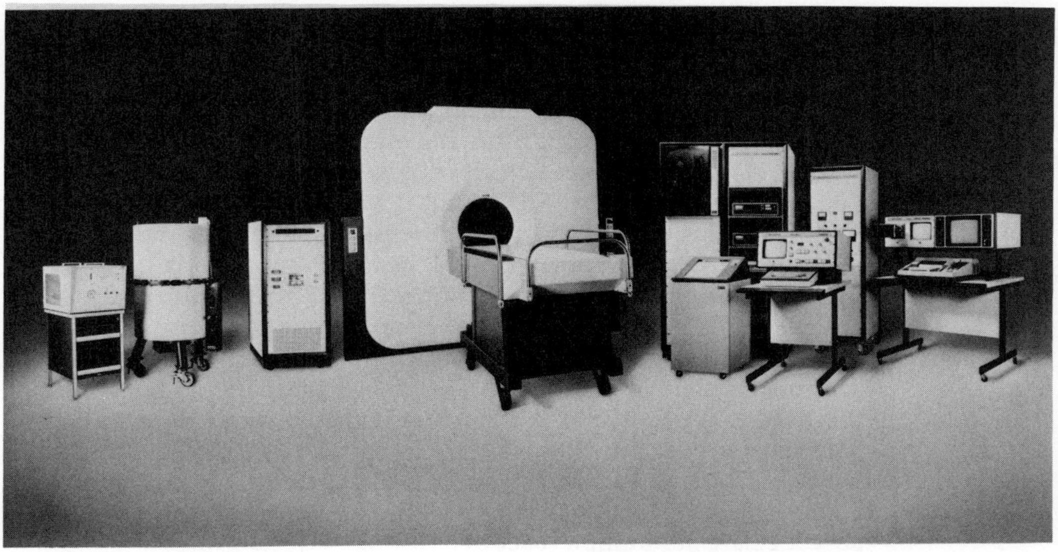

Figure 11.2 A display of the components of a computerized axial tomography scanning unit. Courtesy of Ohio-Nuclear, Solon, OH.

for several years have recently come into clearer focus. For example Case Study No. 1 below demonstrates some of the complexities of clinical engineering from the systems viewpoint.

Case Study No. 1: Residual anesthetic gas problem in anesthetizing locations

Recent publications indicate that a health hazard affecting operating room personnel is probably related to the chronic exposure to trace concentrations of inhalation anesthetics. The anesthetic agents primarily considered are nitrous oxide and halogenated gases, which are the most frequently used and which, although quite different in physical properties, displace similar air distribution patterns in the operating room. Efforts to study more precisely the effects of chronic exposure to inhalation anesthetics were undertaken with a large-scale national survey in 1973 (Martin, 1974).

The solution to this problem is four-fold: (1) proper anesthetic technique to minimize gas leakage; (2) increased levels of equipment maintenance on anesthesia equipment to reduce gas leakage; (3) installation of methods to collect and dispose of residual gases (scavenging); and (4) an air monitoring program in the operating room and assurance that the proper number of air exchanges are occurring as required.

Even when a good scavenging and evacuation system is in use, the air of the operating room is contaminated with anesthetic gas and vapor. The sources of contamination include: release into the air when the gas is turned on before it is attached to the patient; leaks between the face mask and the patient's face; leaks between parts of the anesthetic machine; discharges from some ventilators; and exhalations of the patient after the mask is removed. The scavenging system does not affect these sources of contamination (Jacoby, 1977).

It is possible to test the operating room air for every type of anesthetic gas or vapor. This is generally unnecessary, however, because nitrous oxide is used almost universally, and analyses for nitrous oxide reveal the extent of contamination. Any other agent used along with nitrous oxide will also be present in the air in a proportion similar to that which is given the patient. For example, the patient is given oxygen 32 percent, nitrous oxide 66 percent, and halothane 1 percent. The ratio between nitrous oxide and halothane is 66:1. If nitrous oxide is found in the air of the operating room in a concentration of 66 parts per million, halothane is also present at about 1 part per million (Jacoby, 1977).

A definitive study of operating-room contamination was made by Stanford University School of Medicine for OSHA (1975). This report recommends that the maximum safe level of nitrous oxide be no more than 30 parts per million; the maximum safe level for halogenated agents be no more than 0.5 parts per million. The report also recommends that the air of each operating room be checked at intervals of one week or more often. An infrared gas analyzer is used for detection of ambient nitrous oxide. In addition, there are commercially available services that test for traces of waste anesthetic gases. These services make use of a gas chromatograph to perform the analysis.

Comments. This particular example demonstrates a problem now in focus after lying dormant for the several decades that anesthesia has been in use. It also demonstrates the systems view. Involved here are users who must apply their techniques, clinical engineering personnel who must be involved in some way in equipment maintenance and in measuring residual gas levels, and those who must handle the scavenging problem. Finally, consideration must be given to the facility interface involving the number of air exchanges and the air conditioning and ventilating system, which lie in the plant engineering area.

Other examples demonstrating the requirement for facility interfaces are the use of ultraviolet light in operating rooms for diminishing infections, the problem of isolated power systems in special-care units, the related problems of the electrical environment, the effect of electromagnetic interference on pacemakers, the toxicity of ethylene oxide used for sterilization, and countless other factors.

The clinical engineer will increasingly be involved with a greater array of paramedical specialists, requirements for more intensive clinical engineering assistance, and demands for communication and input into the decision-making process. The clinical engineer, who is a new specialist himself, must communicate with those responsible for infection control, with dialysis technicians, with cardiopulmonary technologists, and a host of others who are users of technology in the health-care setting.

Inevitably, the purchase of any major new piece of medical equipment will produce corresponding system-related problems. If the clinical engineer has primary responsibility for advising in the equipment selection process (Chap. 8), he clearly must be prepared to deal with the problems generated by the domino effect. Figure 11.3 illustrates how equipment acquisitions influence many components in the hospital system—the impact must be dealt with. And, although equipment is a key concern, the matter of materials also can generate questions and considerations for the clinical engineer. Decisions such as the efficacy of carpeting the intensive-care unit and disposal methods for reagents in the laboratory are possible questions facing the clinical engineer.

Finally, the clinical engineer must be aware of two other components of the system problem in the current environment. These are the regulations imposed by both voluntary and federal systems, and the economics of operating a hospital, including the need for cost containment. Clearly, the clinical engineer occasionally will find himself trapped in a cross-fire when quality of care and economy are in conflict. He must be prepared to deal with these issues. In terms of regulations, the clinical engineer in the current environment will find a growing body of codes and regulations primarily aimed at producing a rational use and safe patient environment in the midst of new technology thrust upon hospitals (Chap. 4). The major standards and regulations that concern the clinical engineer are put forth in the NFPA Codes, the standards of the JCAH, federal regulations issued under the OSHA Act of 1970, regulations issued under the Radiation Control and Health Safety Act of 1968, and regulations forthcoming from the Medical Device Amendments of 1976.

In summary, we have posed diverse questions that should demonstrate the clinical engineer's need for a comprehensive view of the hospital as a systems problem. We now turn to certain specific facility considerations with which clinical engineers should be familiar. As the chapter proceeds, other relevant case studies will be presented to demonstrate various points.

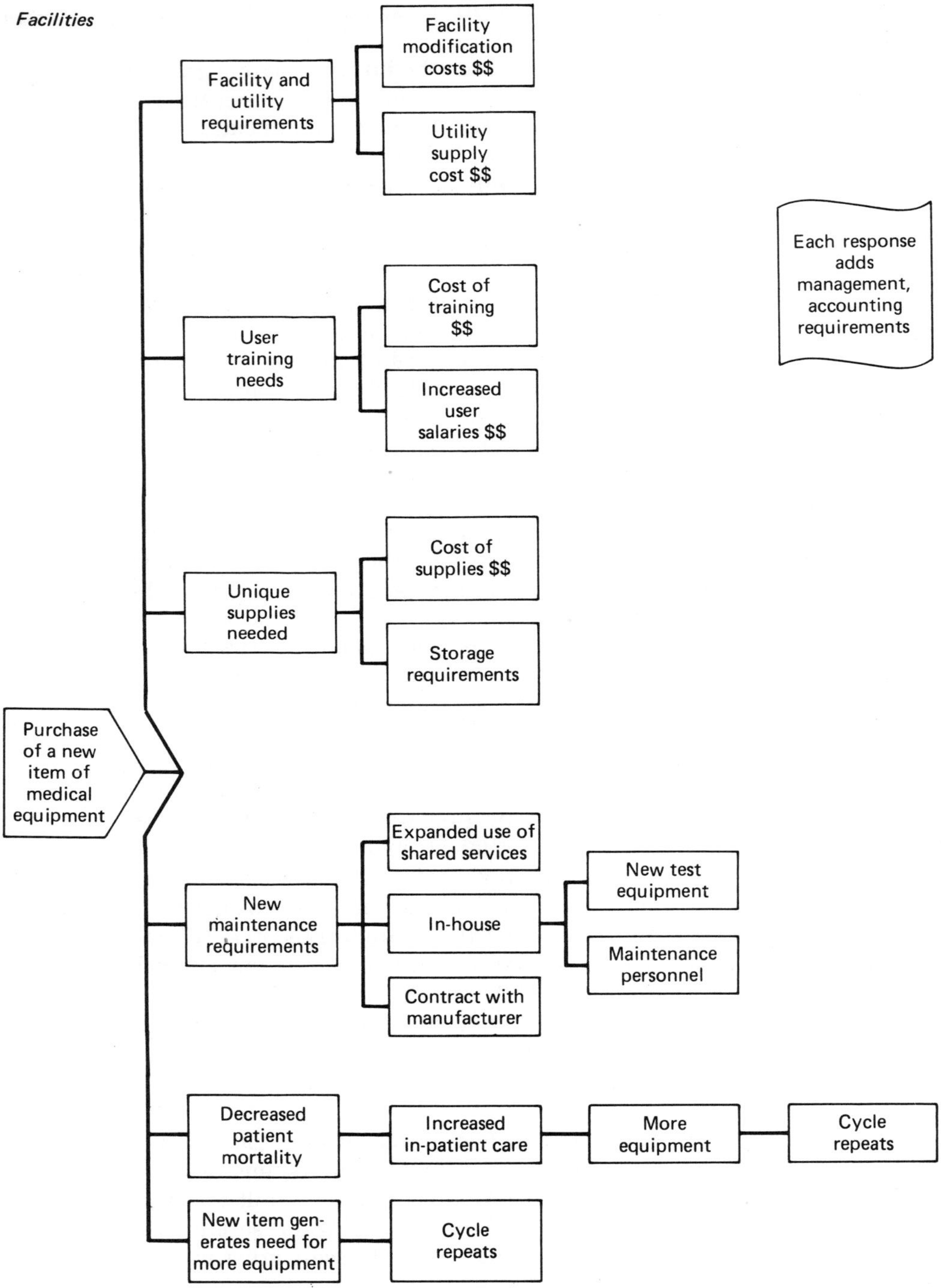

Figure 11.3 The complex considerations that result subsequent to the purchase of an item of biomedical equipment.

11.5 Facilities Analysis

Facilities analysis is directed to the primary function of engineering and maintenance in the hospital and includes the physical plant, utilities system, major nonmovable equipment items, fire protection, safety systems, and electronic communication systems. The clinical engineer should be generally familiar with methods by which facilities analysis is performed and with the important components of this analysis.

Functional analysis

The first method of facilities analysis is *functional analysis*, in which the flow of people and materials is analyzed and evaluated. This component typically falls into the realm of industrial engineering, but frequently requires substantial clinical engineering input to assure valid judgments concerning hospital layout plans, renovations, etc. For example, when construction for ancillary service areas costs \$80 to \$90 a square foot, it is essential that all space requirements in construction projects are justified. A saving of 1000 sq. ft. by good functional analysis with clinical engineering input could save a hospital between \$80,000 and \$90,000 in construction costs alone, not to mention the lifetime maintenance and utility support cost for that amount of space.

The following case study demonstrates this point:

Case Study No. 2: Determining the number of rooms required for fluoroscopy

Hospitals have had considerable difficulties in determining the actual number of diagnostic X-ray rooms needed to meet expected workloads. One method for calculating the number of X-ray rooms needed is contained in a monograph by Lindheim (1971). The following example demonstrates the procedure for one part of the workload.

A 358-bed hospital, which planned to increase its bed complement to 615 and to expand its outpatient services, wanted to plan for a radiology department to accommodate the required load based on a 10-year use projection.

A 45 percent growth in population was projected on the basis of statistics compiled by the local planning body. The hospital administration estimated that there would be roughly 150 admissions annually per 1000 population, not counting mental patients. The outpatient X-ray load was estimated by projecting the recent growth rate in the outpatient department. The chief radiologist made a rough estimate of the growth in the ratio of X-ray examinations to inpatient admissions. The projections for the X-ray work load were broken down into fluoroscopy, chest X ray, general examinations, and special procedures, and time estimates were made for each category of examination. Previous experience showed that 75 percent of the load occurred between 9 A.M. and 4 P.M. An 85 percent utilization figure for each room was desired. Given these parameters, the number of X-ray rooms (N) needed for each category of examination could be calculated by the following formula:

$$N = \frac{\text{(Percent of peak load during workday)} \times \text{(Peak work load)} \times \text{(Average time per procedure)}}{\text{Amount of time each X-ray room or unit is to function daily}}$$

Thus, if the projected peak number of fluoroscopic examinations per day were 40, then 85 percent, or 34 examinations, would take place between 8 A.M. and 4 P.M. The average time for fluoroscopic examinations would be 19.8 hours. If the room were to be regularly available for eight hours, three (2.4) rooms would be required to handle that particular patient load for fluoroscopy.

Comment. The case study demonstrates a facility analysis evaluation based upon the functional needs of the unit. Clinical engineers should be knowledgeable concerning the techniques used in determining space requirements, since improper evaluations ultimately result in problems with which the clinical engineer must deal on a day-to-day basis. For example, it would not be unreasonable to expect the hospital administrator to call the clinical engineer into his office and ask the question, "Do you think we really need 3 rooms for fluoroscopy? I have the radiologist input and now I would like your objective opinion."

Analysis by layout

The second method of facility analysis involves *analysis by layout*, that is, determining affinity relationships among the various locations involved. Figure 11.4 shows proximity values as an example of affinity relationships.

To more specifically analyze the affinity relationship, we can consider the air-pressure relationship of adjacent areas, a highly important consideration in hospitals since it concerns infection control, control of undesirable odors, and fire protection.

Zoning is a typical approach used in hospital general ventilation systems design. This concept provides the isolation of one hospital department from another to typically prevent cross contamination between zones.

Table 11.1 is an extract from an HEW report (1975), which

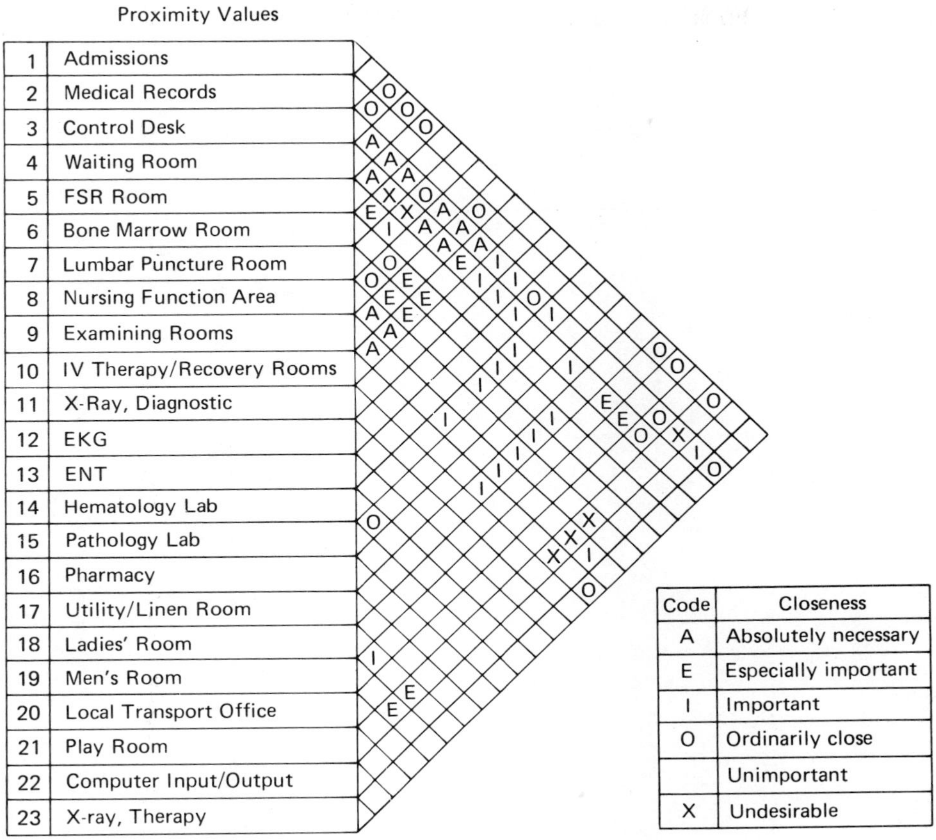

Figure 11.4 Affinity relationships that must be considered in facility layout include the proximity relationships illustrated here. From Examination of case studies in health facilities planning, National Cooperative Services Center for Hospital Management Engineering. Reprinted with permission from the American Hospital Association.

shows the air-pressure relationships of adjacent areas and reflects a type of affinity relationship. Other types of affinity relationships concern the location of departments and movement traffic. For example, the soiled linen room should not be located directly adjacent to the entrance of central sterile supply for obvious reasons.

Support systems analysis

The third area of facility analysis concerns the utility and facility support systems. This includes heating, ventilation, air conditioning (HVAC), electrical system, suction system, compressed

Table 11.1. General Pressure Relationships and Ventilation Requirements for Certain Hospital Areas

Area Designation	*Pressure Relationship to Adjacent Areas*	*Minimum Air Changes of Outdoor Air per Hour Supplied to Room*	*Minimum Total Air Changes per Hour Supplied to Room*	*All Air Exhausted Directly to Outdoors*	*Recirculated within Room Units*
Operating room	P	5	25	Optional	No
Emergency operating room	P	5	25	Optional	No
Emergency examination and treatment room	E	2	6	Optional	Optional
Delivery room	P	5	12	Optional	No
Nursery unit	P	5	12	Optional	No
Recovery room	P	2	6	Optional	No
Intensive care	P	2	6	Optional	No
Patient room	E	2	2	Optional	Optional
Patient room corridor	E	2	4	Optional	Optional
Isolation room	E	2	6	Yes	No
Isolation room-alcove or anteroom	E	2	10	Yes	No
Examination room	E	2	6	Optional	Optional
Medication room	P	2	4	Optional	Optional
Pharmacy	P	2	4	Optional	Optional
Treatment room	E	2	6	Optional	No
X-ray, fluoroscopy room	N	2	6	Yes	No
X-ray, treatment room	E	2	6	Optional	Optional
Physical therapy and hydrotherapy	N	2	6	Optional	Optional
Soiled workroom or soiled holding	N	2	10	Yes	No
Clean workroom or clean holding	P	2	4	Optional	Optional
Autopsy	N	2	12	Yes	No
Darkroom	N	2	10	Yes	No
Nonrefrigerated body holding room	N	Optional	10	Yes	No
Toilet room	N	Optional	10	Yes	No
Bedpan room	N	Optional	10	Yes	No
Bathroom	N	Optional	10	Yes	No
Janitors' closet	N	Optional	10	Yes	No
Sterilizer equipment room	N	Optional	10	Yes	No
Linen and trash chute rooms	N	Optional	10	Yes	No
Laboratory, general	N	2	6	Optional	Optional
Laboratory, media transfer	P	2	4	Optional	No
Food preparation centers	E	2	10	Yes	No

Table 11.1 (Continued)

Area Designation	Pressure Relationship to Adjacent Areas	Minimum Air Changes of Outdoor Air per Hour Supplied to Room	Minimum Total Air Changes per Hour Supplied to Room	All Air Exhausted Directly to Outdoors	Recirculated within Room Units
Warewashing	N	Optional	10	Yes	No
Dietary day storage	E	Optional	2	Optional	No
Laundry, general	E	2	10	Yes	No
Soiled linen sorting and storage	N	Optional	10	Yes	No
Clean linen storage	P	2	2	Optional	Optional
Anesthesia storage	E	Optional	8	Yes	No
Central medical and surgical supply					
Soiled or decontamination room	N	2	6	Yes	No
Clean workroom	P	2	4	Optional	Optional
Unsterile supply storage	E	2	2	Optional	Optional

P = Positive N = Negative E = Equal

air and compressed gas system, fire safety system, plumbing system, steam systems, building structure, and others.

Although the clinical engineer is not expected to be an expert in all of these systems, he must understand the basic ingredients of each system, methods of distribution, and theories of operation.

Analysis by relationships

The fourth method of facility analysis involves analysis by interrelationships. If one accepts the thesis that the hospital is a systems problem, then these interrelationships are critical. The three foregoing analyses involve very narrow, vertically constructed views of the hospital involving the flow of people and materials, affinity relationships, and the utility and support systems. This last method of analysis is concerned with the relationship of these particular components. The center of this analysis should be the patient, whose well-being is linked to many components (see Fig. 11.5). Countless examples exist of patient and staff injury and death from improper facility interfaces. Such problems as cross-piping of anesthesia gases and inadequate electrical supplies are not unusual occurrences, but arise periodically throughout the health-care system.

To summarize the concept of facility analysis, the clinical engineer should be prepared to evaluate facilities and facility prob-

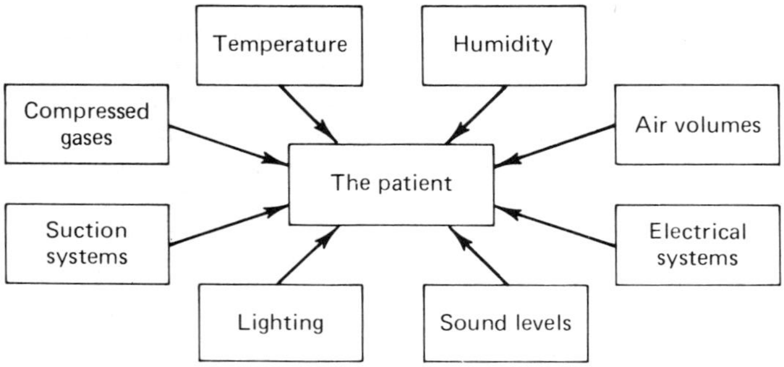

Figure 11.5 The patient is the focal point of utility and facility interfaces.

lems by any one or a combination of the four methods suggested above: by function, by layout, by utility and facility support system, and by interrelationships. Clinical engineering problems often overlap into the facilities area, and analysis by these methods will be useful in solving clinical engineering problems.

11.6 Facility Considerations in Clinical Engineering

Specific facility considerations in clinical engineering by the various methods involve a substantial body of knowledge. The relevant engineering principles represent all of the engineering disciplines as well as many of the arts practiced by health-care professionals. It is difficult to accept the old cliche that "medicine is an art, not a science" when we consider the health-care environment of bricks, mortar, copper wire, and piping.

Operating rooms

The operating room is considered a controlled environment, where control relates to all aspects of the environment including cleanliness, lighting, temperature and humidity control, electrical environment, support systems like gas and suction, ventilating characteristics, and building structural features such as floors.

Air exchange and room pressurization. In considering air exchange and room pressurization, Table 11.1 indicates that all operating rooms should have a positive pressure in relation to adjacent areas with a minimum of 25 total air changes per hour. The newest standards provide some allowance for recirculation of air.

Ventilating systems may be classified as either dilution ventilation or exhaust ventilation systems. Dilution ventilation is a design concept that is based on the simple notion that "the solution to pollution is dilution." Many potentially hazardous fumes, vapors, and gases may be rendered harmless by diluting them below a predetermined level. A concentration of isopropyl alcohol vapors, for example, should be diluted with 17,500 cu ft of air per pint evaporated to insure concentrations below currently accepted standards (*Industrial Ventilation*, page 2-2, 1976). Calculations leading to this solution are based upon threshold limit values (TLVs) established by the American Conference of Governmental Industrial Hygienists (ACGIH, 1976).

The filtration required is outlined in Table 11.2. Filter bed No.1 should be located upstream of the air-conditioning equipment and filter bed No. 2 should be downstream of the supply fan and any recirculated spray water systems and water-reservoir-type humidifiers. Although the cleanliness of incoming air is highly important, other considerations need to be recognized. Airborne contamination is by far a minor cause of infections in comparison to the primary cause of contact contamination. This is not to downgrade the need for adequate air control in operating rooms, but to put the issue in perspective.

To summarize, the requirement for 25 air changes per hour in operating rooms appears to be the standard for the foreseeable future. It is expected, however, that the demand for energy conservation and the recognition that airborne microorganisms are a minimal cause of infection will result in a push for greater amounts of recirculated air in operating rooms (Ulrich et al., 1976). The study conclusions were as follows: (1) Air of low viable particle count can be maintained in operating rooms during surgery when a turbulent system is employed. (2) As the rate of air flow increases, the viable

Table 11.2. Filter Efficiencies for Central Ventilation and Air Conditioning Systems in General Hospitals

		Filter Efficiencies (Percent)	
Area Designation	*Minimum Number of Filter Beds*	*Filter Bed No. 1*	*Filter Bed No. 2*
Sensitive Areas*	2	25	90
Patient care, treatment, diagnostic, and related areas	2	25	90†
Food preparation areas and laundries	1	80	-
Administrative, bulk storage and soiled holding areas	1	25	-

*Includes operating rooms, delivery rooms, nurseries, recovery rooms, and intensive-care units.

†May be reduced to 80 percent for systems using all-outdoor air.

particle count decreases. An optimal rate of 25 changes per hour was determined for systems. (3) Recirculated air at the optimal rate maintains lower viable particle counts than 100 percent outside fresh air used once and discarded. (4) High-efficiency filters removing 95 percent of particles down to 1 μm also remove viable particles in a recirculating turbulent system effectively. (5) During the period of open surgery, the viable particle level within an operating room can be maintained at one per cubic foot or less.

In 1976 the American College of Surgeons defined three air microbiologic classifications to enable surgeons to specify the air quality desired. These air cleanliness standards are as follows:

Class 1 microbiologic cleanliness

Viable microbiologic airborne particle count not to exceed one particle per cubic foot of air with a minimum sample of 30 cu ft of air.

Class 5 microbiologic cleanliness

Viable microbiologic airborne particle to average more than one and up to five per cubic foot of air with a minimum sample of 30 cu ft of air.

Class 20 microbiologic cleanliness

Viable microbiologic airborne particles to average more than five and not to exceed 20 per cubic foot with a minimum sample of 10 cu ft of air.

In addition to incoming air, several other considerations are important to the cleanliness of the operating room and absence of free-moving particles. For example, the obvious reason to require positive pressure in the operating room is that, when the surgical doors are open, movement of air will be from within the surgical suite outward to the corridor, thus preventing the movement of dust, dirt, and other microorganisms from the less controlled environment into the operating room.

Methods for maintaining positive pressures in operating rooms are clearly within the plant engineer's control, since the volume of air entering the operating rooms is determined by the speed of the supply fan, adjustment of dampers in the duct system, pressure drop across the coils, etc. Hospital blueprints and plans will indicate the number of cubic feet of air per minute that systems are designed to input and remove from the operating room. It is obvious that, to maintain positive pressure, the volume of air entering the operating room per unit of time must be greater than that which leaves the operating room.

Figure 11.6 is a typical blueprint of an operating room. Note the indication of the volume of incoming air and exhaust air. Using a velometer, we can, at the face of the incoming duct, measure air volumes and make calculations to determine the actual amount of air being delivered. Similarly, the same can be done for the exhaust air. Stating the incoming air in volume and the size of the room in cubic feet, we can calculate the number of air exchanges occurring in a given hour. For example, if the incoming air is 2000 cu ft per min (× 60), the incoming air in a given hour is 120,000 cu ft. If the operating room measures 20 × 30 × 10, or 6000 cu ft, 20 air changes per hour in that particular operating room will take place. If the exhaust for that particular room is 1200 cu ft per min, we can expect to find a positive pressure existing within that room.

Laminar flow systems. Though the proponents of laminar flow systems for protection against wound infection in operating rooms suggest that high-volume air exchanges, such as 6000 cu ft per min, will result in reduced wound infections in operating rooms, Laufman (1973) suggests the following: Airborne contamination plays a minor but exceptional role in the incidence of surgical infections. The evidence indicates that airborne contamination assumes importance as a source of wound infection only (1) when air-handling systems become grossly contaminated; (2) an otherwise well-installed and well-maintained effective air-handling system is

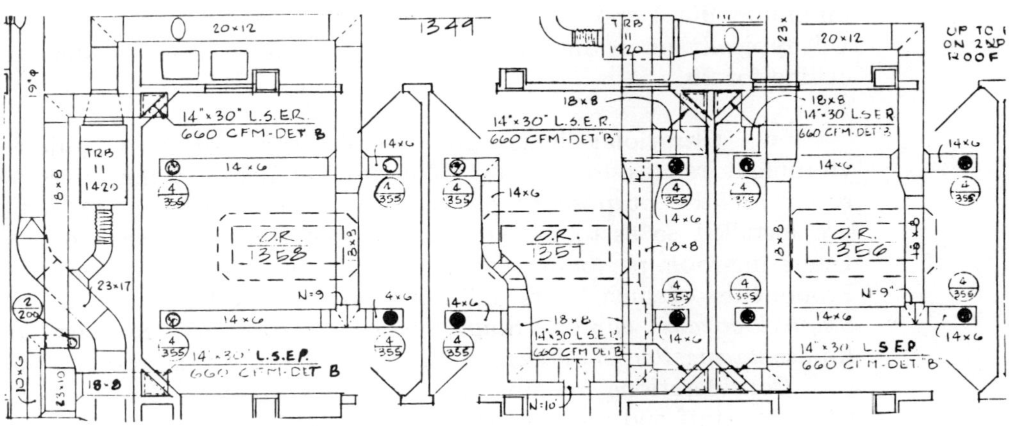

Figure 11.6 A typical mechanical drawing for an operating room.

abused; or (3) when a patient's resistance is markedly lowered by disease or drugs or when a large foreign body is implanted.

All available valid evidence supports the view that, in the presence of a properly installed and properly maintained conventional air-handling system, the superimposition of a laminar flow system, although it does reduce ambient particle counts, has no additional effect on lowering the incidence of wound infection rates. Given this evidence, we must conclude that laminar flow is not necessary for prophylaxis against wound infection. Other studies have indicated that, although laminar flow units are not imperative for the prevention of airborne contamination, there may be a place for "superasepsis" (Beck, 1976), and for "ultraasepsis" for laminar flow.

Other experience suggests that, where laminar flow is combined with isolated systems such as hoods, uniquely styled gowns, and protective mechanisms, there is indeed a marked reduction in operating-room airborne bacterial concentrations and contamination of sterile surfaces, including the wound. The lack of clarity surrounding laminar flow suggests a conservative approach to the acquisition of laminar flow systems.

Air filtration, residual gases. Case study No. 1 presented earlier in the chapter, which deals with the problem and hazard of residual anesthetic gases, has as one of its solutions the maintenance of an adequate air exchange in the operating room. Also, the need for comfort and fresh air in an enclosed environment such as the operating room is self-evident. Rooms must be designed so that there will be adequately filtered air entering and exhausted in such a way as to produce a positive pressure in the operating room. The filtration is achieved by high-efficiency particulate air filters (HEPA minimum 90%) and the number of air exchanges occurring hourly will typically be in the range of 12 to 17, with the newest standard calling for 25.

Ultraviolet lights. One further aspect of the OR air environment relates to the use of ultraviolet lights in the operating room for the purpose of destroying microorganisms. Such lights are installed so as to produce 25 $\mu W \cdot cm^{-2}$ in the operative field. Operating-room personnel wear eye shields, and the skin is protected. Proponents of such lights indicate substantial kill rates of *E. coli*, *Staphylococcus aureus*, *Staphylococcus epidermis*, *Proteus sp.*, *Klebsiella sp.*, *Pneumococcus sp.*, and other resilient microorganisms.

Considerations relating to ultraviolet concern the need for adequate staff and patient protection during surgery, and the actual effectiveness of ultraviolet radiation as a means of minimizing surgical related infections. The effectiveness continues to be a matter of study and evaluation.

Temperature and humidity. In terms of the operating-room environment, attention has been given primarily to the control of infections and airborne microorganisms. The matter of temperature and humidity control in the operating room is also important. Current standards for operating rooms call for their environment to be in the temperature range of 70 to 76° F with a minimum relative humidity of 50 percent and a maximum relative humidity of 60 percent. The National Fire Protection Association (1973a) specifies that the ventilation system must incorporate humidity equipment and control to maintain 50 percent relative humidity. Although it is not sufficiently reliable for complete dissipation of electrostatic charges, this humidity level does reduce the hazard of electrostatic spark discharges under many conditions. It also facilitates the control of airborne bacteria. The NFPA standards further state that the temperature to be maintained in the operating room should be chosen on the basis of the well-being of the patient and operating team. It is recommended that the equipment provided for maintaining room temperature be provided with controls for selecting any desired temperature within the range of 68 to 75° F.

The problem of temperature and humidity controls in the operating room continues to be a matter of concern. Current JCAH standards require that the humidity reading in each anesthetizing location be recorded daily.

Figure 11.7 is a drawing of a typical air supply system for an operating room, using 100 percent fresh air with the capability for heating, cooling, and humidifying. There is no recirculation of air.

Compressed gases. Compressed gas outlets and suction outlets are standard operating room fixtures (Sec. 7.8). The installation of compressed gas systems is performed in accordance with the current NFPA nonflammable medical gas systems standard (1974). Oxygen and nitrous oxide are brought into the operating room through copper tubing or standard weight brass pipe. The gas content of pipe lines should be readily identifiable by appropriate labeling that names the gas. The labeling should be done with metal tags, stenciling, stamping or adhesive marking that cannot be easily removed. This labeling should appear on the pipes approximately every 20 feet and at least once in each room and story traversed by the pipeline. The compressed gases will emerge in the operating room in the form of hoses with quick disconnects that can be attached to a gas anesthesia machine, volume ventilators, etc. These quick disconnects are configured so that gases like oxygen and nitrous oxide, for example, cannot be reversed. However, in initial installations, it must be made absolutely certain that the gas content of systems is certified, since cross-piping could have occurred further back in the system.

The main oxygen supply line should be provided with a shutoff

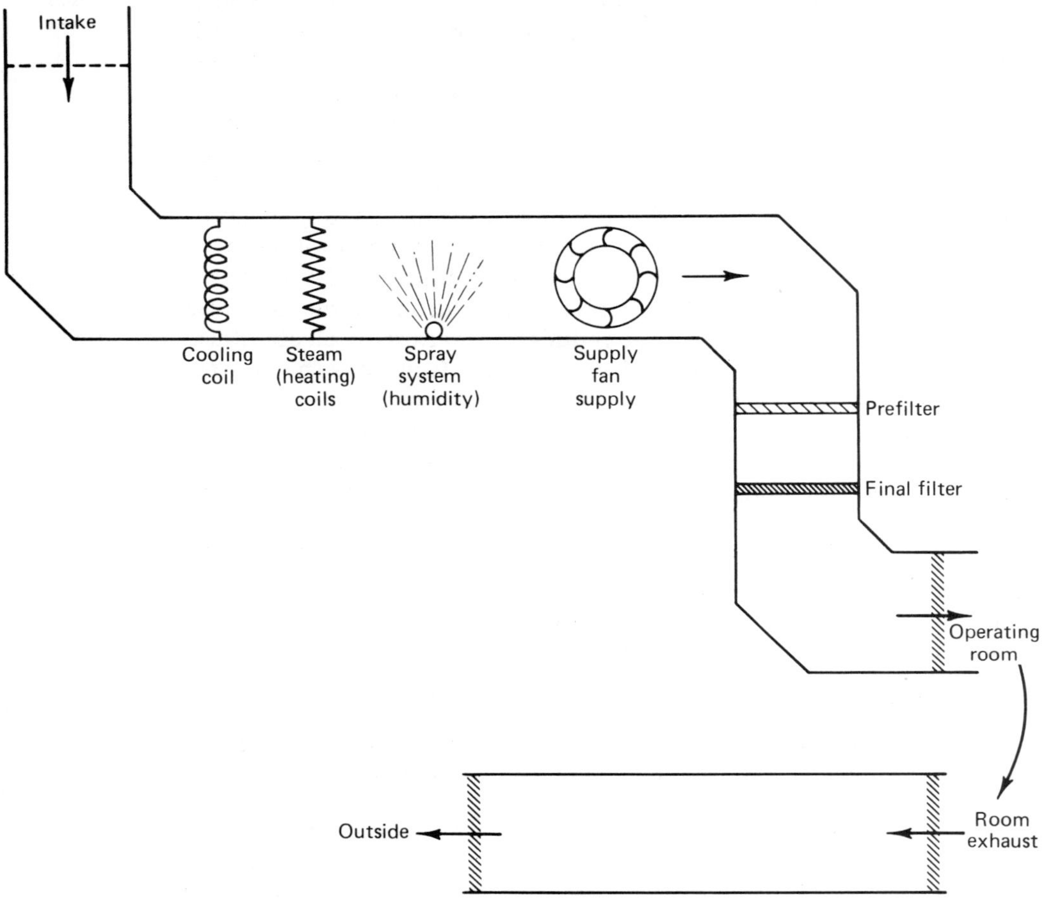

Figure 11.7 A typical drawing of a standard heating, ventilation, and air conditioning (HVAC) system for an operating room.

valve located for easy access in an emergency. Also, shutoff valves should be placed outside or inside each anesthetizing location on each nitrous oxide and oxygen line. These systems should be checked thoroughly for leaks prior to use after initial installation, and periodically thereafter. In one instance, it was found that a "hot spot" in the hallway, where nitrous oxide measured at 100 parts per million, resulted from a leaky connection in the room control valve system that probably had been present since the original installation.

Suction. Ordinarily the operating room has an installed vacuum system which is a piped medical-surgical vacuum system for the purpose of patient drainage, aspiration, and suction. Typically, the medical-surgical vacuum system is piped throughout the hospital in a planned network to locations where patient suctioning may be required, such as intensive-care units and patient rooms. A system usually consists of two or more vacuum pumps that operate alternately, with each pump able to maintain a minimum required vacuum of 75 percent of the peak calculated demand. The piping typically should be of the type K or L copper tubing, and vacuum gauges and manometers should be permanently installed for the vacuum system where adequate levels of vacuum can be established and controlled.

The suction system also requires shutoff valves. These should be labeled appropriately and should not be closed except in an emergency. These valves should not be located within easy accessibility to the general public.

The clinical vacuum installation should be installed in accordance with the requirements of the Compressed Gas Association pamphlet P–2.1 (1967); HEW minimum standards call for two oxygen outlets and two vacuum outlets per operating room.

Electrical environment. The electrical environment of the operating room and the conductive floor are topics that should be considered together since the isolated power system that is standard for the operating room was originated because of the installation of conductive floors.

The installation of conductive floors corresponded to the advent and use of ether and cyclopropane, which are highly flammable and explosive. Conductive floors are designed to bleed off the buildup of static electricity as people and equipment move through the operating-room suite. Earlier, bleed off of static electricity was accomplished by the use of drag chains on equipment, then by the use of conductive wheels and conductive shoe covers for operating-room personnel.

The criteria that apply to conductivity testing and requirements for conductive floors are contained in NFPA standards (1973a). However, with the increased use of halogenic agents and a decrease in use of flammable anesthetics, the requirements for conductive floors appear to be diminishing, and it is not at all uncommon to find newer hospitals building operating rooms without conductive floors. This trend is expected to continue into the foreseeable future.

However, the use of isolated power systems and line-isolation monitors in operating rooms is expected to continue. Because patients are vulnerable both to macroshock and microshock hazards in the operating room, an isolated power system and the line-isolation monitor are valuable. A typical line-isolation monitor and isolated power system are shown in Fig. 11.8.

An isolated distribution system serves only a single operating room. The choice of electrical outlets for the operating room previ-

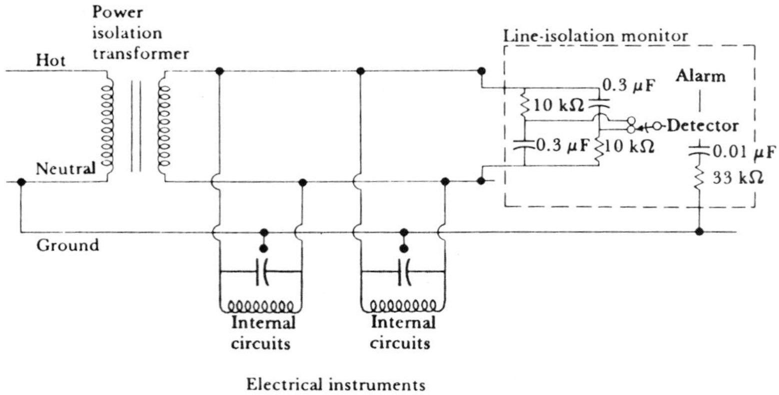

Figure 11.8 Power-isolation-transformer system with a line-isolation monitor to detect ground faults. From J. G. Webster (ed.), Medical instrumentation: application and design, Houghton Mifflin, 1978. Reprinted by permission.

ously was limited to twist-lock explosion-proof plugs; however, as the explosion hazard has now been minimized, the receptacles must either be rated for an explosive atmosphere as outlined in NFPA 56A or located above the five-foot level. The latter currently is receiving considerable attention since it is a more economical approach.

Central supply. Current trends in the handling and processing of sterile materials, instruments, and supplies center around the activity of the hospital's central supply services. Although procedures vary, all instruments and supplies requiring sterilization generally are brought to central supply for processing. The design and layout of the central supply function is critical since the flow of air, materials, and people will have a direct bearing on infection control in the hospital.

Air exchanges and room pressurization. Table 11.1 requires that the central medical and surgical supply area have specific pressure relationships established for the soiled, clean, and supply storage areas, with specified minimum air exchanges. These pressure relationships are of critical importance since any kind of reverse pressure would result in air moving from the soiled or contaminated room into the clean workroom, carrying with it contaminated particles, microorganisms, etc.

Utility systems support. Facilities support for central supply generally centers around adequate air exchanges and proper pressurization, adequate steam supplies for sterilization equipment, and adequate venting of areas where ethylene oxide is in use as a sterilizing agent.

Ethylene oxide in concentrations above the 90-percent level is generally considered flammable and explosive and should be vented. At any level of concentration, it is considered toxic. Ethylene oxide is a colorless, flammable gas at room temperature and pressure, and vaporizes at 51°F. The maximum recommended allowable concentration of ethylene oxide in the working area outside the sterilizers for a daily eight-hour exposure is 50 parts per million. The use of this agent was under intensive study at the time of this publication by the Environmental Protection Agency, and it is anticipated that the venting of ethylene oxide will become an increasingly important future consideration in the design and layout of central supply rooms.

In regard to steam sterilization and utility support, Fig. 11.9 reflects the space and service requirements for a specific steam sterilizer. Adequate steam pressure and steam flow are two critical considerations for the operation of the steam sterilizers. Since steam originates in the boiler of the hospital, it generally is the responsibility of the hospital engineer to deal with maintenance and engineering matters relating to utility support of sterilizers. However, questions occasionally arise concerning the adequacy of sterilization and whether or not sterilization is being achieved. Although sterilizers typically are not considered medical instrumentation within the area of the clinical engineer, it seems advantageous for the clinical engineer to understand the basic principles of steam sterilization.

Bacteria can be destroyed by heat which coagulates or solidifies their protoplasmic mass. The temperature at which this protoplasmic mass will coagulate varies with the amount of moisture present. Lewith experimented with proteins using egg albumen, which is very similar to bacteria protoplasm (U.S. Army, 1965). He discovered that when there is an abundance of moisture present the albumen, or protoplasmic mass, coagulates at a relatively low temperature. With very dry albumen, extreme heat of up to 338°F fails to cause the protoplasmic mass to coagulate. Heat and moisture, then, when applied for the proper period of time, are elements essential to the destruction of bacteria. Naturally, boiling water and moist steam under pressure are the easiest and most convenient ways that these agents can be applied.

The thermal death points of many bacteria have been carefully studied. These studies and repeated experiments have revealed that nonspore-forming bacteria such as those that cause typhoid, tuberculosis, pneumonia, etc. can be destroyed by exposure to water heated from 125 to 160°F for periods of 10 min. All vegetative forms of nonspore-bearing bacteria can be destroyed by boiling in water for a time period ranging from 10 to 20 min. Experts have been unable to agree on the minimum time required to destroy most resistant pathogenic microorganisms. Their conclusions range from 230 to

SYBRON|Castle

3112 Steam Sterilizer

SPACE AND SERVICE REQUIREMENTS*

SINGLE DOOR CABINET

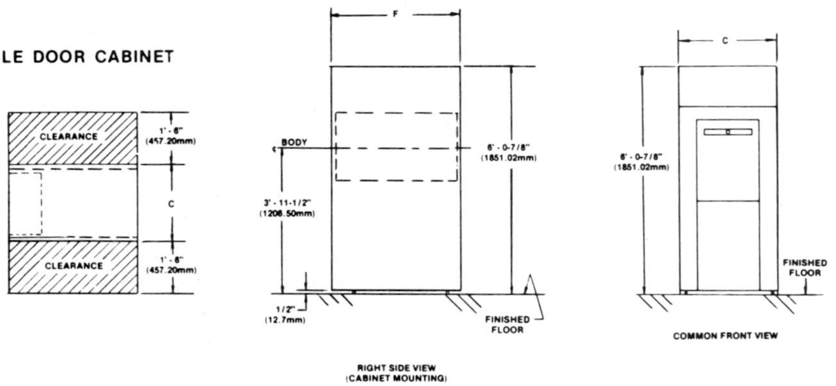

SINGLE DOOR RECESSED

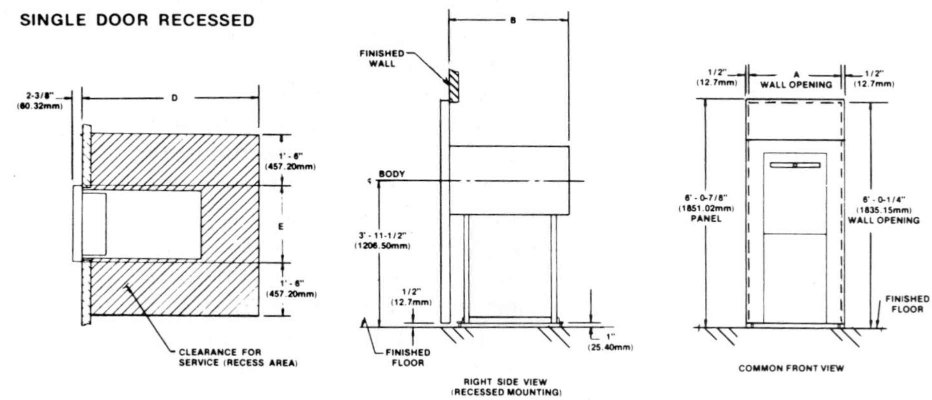

*Data for general use only.
For final information see Castle detailed installation drawings.

CHAMBER SIZE	DIMENSIONS						WATER PEAK FLOW RATE	STEAM PEAK FLOW RATE	OPTIONAL RECORDER 1 PHASE 120V	ELECTRIC STEAM GENERATOR 20KW. CONTROL CIRCUIT 1 PHASE, 120V, 5 AMPS	
	A	B	C CABINET	D	E	F				HEATER CIRCUIT	WATER RATE
16" x 16" x 26" (406mm x 406mm x 660mm) DOUBLE SHELL	2'-1" (635mm)	2'-2-5/8" (736.6mm)	2'-2" (660.4mm)	3'-8-5/8" (1131.47mm)	2'-2" (660.4mm)	2'-5-7/16" (747.71mm)	5 GPM (18.9 liters/min.)	96#/HR (43.5 KG/HR)	5 AMPS	208V. 3Ø, 48 AMPS 230V. 3Ø, 53 AMPS 230V. 1Ø, 91 AMPS	1.1 GPM (4.1 liters/min.)

Castle Company/Division Of Sybron Corporation/1777 E. Henrietta Rd., P.O. Box 23077/Rochester, N.Y. 14692
Phone 716-271-6060

Figure 11.9 Utility requirements for a typical steam sterilizer. Castle Company, Rochester, New York.

248° F of moist heat, and from 5 to 25 min. Current base sterilizers use moist steam placed under pressure between 15 and 20 lb per sq in. This insures a moist heat of 250 to 259° F which exceeds the highest temperature recommended by authorities in the field.

Present-day sterilizers, in addition to their function as basic plumbing systems with steam pressure, are equipped with electrical and electronic controls for timing the length of sterilization, automatically terminating the sterilizing cycle, controlling periods of time in which vacuums are drawn on the chamber, and controlling the drying process.

Radiology

The clinical engineer should know radiation principles and safety (Sec. 7.10) so that he can work in the radiology area. The number of computer interfaces required in radiology departments is growing, and thus, he needs to know about controlled temperature and humidity environments.

Facility considerations concerning radiology begin with the structure of the radiology department itself. The overhead tube systems in modern X-ray equipment constitute a dynamic load of about 3000 pounds which is supported from the ceiling. Older institutions have found it necessary to install structural steel within the ceilings in order to gain support from other permanent structural walls and columns. Another method of achieving this support has been to install direct support from the floors of the X-ray rooms up to mounted overhead ceiling railings. This problem of support for the overhead tube system brings into focus the fact that X-ray equipment for the most part is bulky and exceptionally heavy. In many instances, the floor loading capacity must be considered in the installation of equipment. Facility and utility requirements for support of one particular X-ray system are illustrated in Table 11.3, which demonstrates the substantial facility considerations that apply to radiology.

The standard document concerning radiation protection is NCRP Report No. 49 (1976), which specifies the amount of lead shielding necessary for various levels of X-ray capability. In addition, shielding is necessary for doors, X-ray film pass boxes, and other areas. Electrical requirements for the X-ray department differ substantially from those for some other parts of the hospital. When we consider that an X-ray unit may operate at 150,000 V at 800 mA, the power demand required of the system (120 kVA) is evident. For this reason, it is not uncommon to find X-ray equipment operating from three-phase power systems at 440 V. Furthermore, the length of conduits and conductors in relation to the power transformer should be as short as possible to prevent electrical voltage line loss. And finally, because X-ray equipment operates at very high power

Table 11.3. Examples of Some Facility and Utility Requirements for One X-ray System

Facility Considerations for Installation of Computer-Interfaced Radiology Equipment
1. Floor space
2. Floor loading capacity
3. Electrical power:
Voltage, single or three-phase
Current, size of breaker
Tolerance of line voltage variation
4. Water (heat exchanger)
Maximum and minimum gallons per minute
Proper pressure (psi)
Maximum incoming temperature
Adequate drain capacity
5. Room temperature and humidity control
6. Room air flow in CFM
7. Incoming air filtration
8. Room pressurization
9. Lead shielding of all required areas
10. Adequate fire protection (Halogenated, etc.)

demands for very short periods of time, we must consider the possibility of simultaneous exposures on the power system and their end result. Therefore, when analyzing electrical environments for X-ray equipment, we must determine the availability of the proper voltage and current, how much line voltage variation is acceptable, and which systems operate from the same power distribution system.

In addition to electrical system requirements, we must also consider that growing numbers of special procedure rooms are present in the radiology department. In those rooms where heart catheterization is likely to take place, the possibility of microshock is a matter of concern. For this reason, either equipotential grounding systems or isolated power systems must be in place. The electrical environment is extremely important where catheterization is done because a direct pathway has been created to the heart through a conductive catheter and this pathway could be the possible source of fibrillation caused by leakage currents (Sec. 7.4).

An important support equipment item in radiology is the automatic film processor. The utility requirements for automatic X-ray film processors include plumbing systems, drain systems, and methods for silver recovery. These matters should not be overlooked, and although an X-ray film processor would not be considered a clinical engineering responsibility, questions periodically arise concerning the quality of radiographs and the function of radiographic equipment when in fact the problem may be the X-ray film processor.

Probably the fastest growing department in terms of instrumentation is the hospital clinical laboratory. The growth of equipment in terms of both quantity and complexity has been the pattern in the laboratory over the past decade. Moreover, growing environmental concerns surrounding methods of disposing of used reagents, problems of contaminated materials, and the exhaust of noxious fumes and odors all point to an increasing need for facility and utility support. Clinical laboratories increasingly need stabilized line voltages from which to operate computer-interfaced diagnostic systems. Even very small line voltage variations can affect the accuracy of diagnostic information furnished by the laboratory. This problem may be intensified as energy conservation activity by larger metropolitan areas results in reduction of line voltage by as much as 5 percent during peak demand periods.

In addition, laboratories traditionally have lacked a sufficient number of outlets, and it is not at all uncommon to find four or five items of equipment plugged into the same outlet by means of a multiple connector.

A primary electrical consideration should be that all equipment in the laboratory is grounded. With the combination of metal surfaces, fluids, and electrically operated equipment, the potential for an electrical accident is high. For this reason, special attention should be given to periodic evaluation of the electrical environment in the laboratory.

Because the laboratory uses highly toxic and dangerous liquids such as acids, the need exists for a strong safety program. NFPA standard 56C spells out the general storage and safety requirements in the clinical laboratory (NFPA, 1973b). Air supply and exhaust must be balanced to provide a negative pressure relative to the surrounding hospital areas. Highly infectious or radioactive materials must be exhausted through 99.7-percent filters before discharging into the atmosphere. Exit corridors should not be used as plenums to supply or exhaust air from laboratory areas. Fume hoods must have face velocity ventilation adequate to prevent the backflow of contaminants into the room.

In the presence of cross-drafts or the rapid movement of an operator working at the face of the hood, the specified minimal face velocity of the fume hood is generally considered to be 100 ft per min. However, under unusual conditions, greater velocities in excess of this flow may be required.

Certain instruments may be useful for evaluating ventilation systems for proper operation. Since ventilation is an air-flow process, anemometers (hot wire, swinging vane, etc.) are useful for flow measurement. For clinical engineering purposes, the most useful instrument for cursory evaluation is the smoke tube. This device emits a puff of smoke which may be observed as it is moved by the ventilation system. For low-velocity systems, the smoke trail may be timed to estimate air-flow velocities. More generally, the smoke trail is a useful means to locate dead areas in a room, areas where flow imbalances may be present, or areas where the ventilation system is simply not operating.

The blood bank should have an electrical power source that is automatically connected to the emergency generation system in the event of a power failure. Moreover, the blood bank should be so equipped with an alarm that if the temperature within the blood bank exceeds the maximum allowable temperature, the alarm will sound in a location that is occupied on a 24-hour-per-day basis, such as the switchboard or emergency room.

Special-care units (SCU)

An area that has received substantial attention from the standpoint of clinical engineering has been the special-care unit. Typically a special-care unit may be an intensive-care unit (ICU), a coronary-care unit (CCU), or variations of these. In such SCU units patient care is highly instrumented, and patients are highly vulnerable to infection, to hazards of microshock, and a range of other potential equipment problems. The SCU has emerged only within the past fifteen years, and corresponding problems have arisen in facility equipment interface, new code proliferation, and requirements for facility design innovations and new approaches to patient care.

In considering the SCU from a facility point of view, we must address the overall environment, including such mundane factors as noise levels, light intensity, location of the central nursing station, and other practicalities. The development of SCUs by architects generally has been a step-by-step learning process, and, as a result, a substantial number of SCUs today exist throughout the nation with major design deficiencies or environmental flaws.

In special-care units, the patient is placed in an environment that for the most part is lighted 24 hours a day and has substantial levels of staff activity at all times. One expert has stated that patients who are placed in special-care units, for every one day of hospitalization, live through three full days of experience. Because the staff changes every eight hours and because light, activity, and noise remain continuously high, patients have difficulty differentiating time spans and can suffer what has been called "ICU psychosis." For this reason, special consideration should be given to facility interfaces by the clinical engineer.

Every effort should be made to lower sound levels in the SCU, although minimization of noise is difficult when patients are not in separately divided rooms, when ventilators and suction systems are used, and staff conversations must take place. Since the level of noise coupled with monitor alarms and audible signals can be distressing to patients, hospitals should strive for acoustically absorbing materials between rooms, separate patient rooms with closable doors, and floor covering that will absorb traffic sounds. Another point of view suggests that sound and noise are reassuring to the patient. Obviously there is some compromise in this matter.

Natural illumination for each patient is a must, although continuous artificial illumination is necessary for nursing stations. The design should provide adequate artificial illumination for the nursing station but should not disturb the patients at night. Table 11.4 demonstrates the general illumination levels recommended for the hospital. Note that the level for the intensive-care areas is specified and is higher than that for patient's rooms.

Perhaps the greatest environmental problem that clinical engineers have faced during the past several years in SCUs has involved the use of isolated electrical power systems. The basic concept has been that patients who are in SCUs may be monitored or paced with transvenous catheters and therefore are vulnerable to the hazard of microshock.

For this reason, many SCUs today are equipped with isolated power systems that provide for an alarm system to indicate when leakage currents from electrical devices in the unit or in a given room have exceeded the preestablished level. An isolated power system in the SCU is not significantly different from that in the operating room, and each is designed to meet the same general requirements. That is, if excessive leakage currents are generated, an alarm will indicate this with an audible sound and a red light.

From a practical point of view, isolated power systems and ground fault detectors present minimal problems. Weekly testing should be performed with a simulated load that triggers the alarm. The threshold of the alarm should be periodically tested and reset if necessary. It is of particular importance that the clinical engineer, when performing these tests, ensure that all nursing staff in the unit know that the tests are being conducted. Preferably, tests should be conducted only in those units in which patients are not being monitored. If a patient is in the room who is not being monitored, he should be alerted that the alarm will be tested and that the test is not a matter of concern. The nursing staff should understand the function of the ground fault detector and isolated power system, and have a planned course of action should the system go into alarm. Frequently these alarms occur during night hours, and an inexperienced nursing staff may have difficulty making decisions about what is to be done. In each SCU, therefore, the policy and procedure

Table 11.4. Recommended Levels of Illumination for Health Care Facilities (abstracted from *Lighting for Health Care Facilities,* CP-29, Illuminating Engineering Society, 1978. For a complete listing and information concerning the application of these levels see the complete report.)

Area	Illumination Levels on Tasks*	
	Footcandles	Lux
Cardiac function lab	100	1100
Dialysis unit		
For patient, see *Patients' rooms*		
For medical procedures	200	2200
EKG and specimen room		
General	30	320
On equipment	50	540
Examination and treatment rooms		
General	50	540
Local	100	1100
Inhalation therapy units	30	320
Laboratories		
Specimen collecting	100	1100
Tissue laboratories	200	2200
Microscopic reading room	50	540
Gross specimen review	200	2200
Chemistry rooms	100	1100
Bacteriology rooms		
General	100	1100
Reading culture plates	200	2200
Hematology	100	1100
Medical records room	100	1100
Nurseries		
General	30	320
Observation and treatment	100	1100
Nursing stations		
General	30	320
Desk for records and charting	70	750
Corridors, day	20	220
Corridors, night	3	32
Medication station	100	1100
Obstetric delivery suite		
Labor rooms		
General	20	220
Local	100	1100
Delivery area		
Scrub — general	200	2200
General	200	2200
Delivery table	2500	27000
Resuscitation of infant	500	5400

Table 11.4. (Continued)

Area	*Illumination Levels on Tasks**	
	Footcandles	*Lux*
Postdelivery recovery area		
General	30	320
Local	100	1100
Substerilizing room	30	320
Clean-up room	30	320
Occupational therapy		
Work area, general	30	320
Work tables or benches, ordinary	50	540
Work tables or benches, fine work	100	1100
Patients' rooms		
General	10	110
Observation by nurse	2	22
Critical examination	100	1100
On floor at night	0.5–1.5	5.4–16
Reading	30	320
Toilets	30	320
Physical therapy departments		
Gymnasia	70	750
Tank rooms	70	750
Treatment cubicles	50	540
Pulmonary function laboratories	100	1100
Radiological suite		
Diagnostic section		
General	10	110
Waiting area	10	110
Radiographic/fluoroscopic room, Special procedures	200	2200
Film sorting	200	2200
Barium kitchen	100	1100
Radiation therapy section		
General	10	110
Waiting area	10	110
Isotope kitchen, general	100	1100
Isotope kitchen, benches	200	2200
Computerized radiotomography section		
Scanning room	10	110
Equipment maintenance room	100	1100
Surgical suite		
Operating room, general	200	2200
Operating table	2500	27000
Scrub room	100	1100
Instruments and sterile supply room	30	320
Clean-up room, instruments	100	1100
Anesthesia storage	20	220
Substerilizing room	30	320

*These levels are target maintained values, unless otherwise indicated. Initial values are higher than the maintained values listed above due to such factors as losses from dirt accumulation and lamp lumen depreciation.

 manual should contain specific directives as to action in the event the ground fault detector/isolated power system alarms.

Codes that relate to isolated power systems and SCUs, as formulated by the NFPA, are constantly undergoing change and upgrading. Requirements that relate to the electrical environment of the SCU can be found in the NEC and other specific NFPA publications.

An alternative measure to insure electrical safety in SCUs is the equipotential grounding system by which all metal surfaces and electrical ground points in the room be bonded to a single common ground point. This method simply precludes the possibility that a difference of potential can develop between ground points because of ground loops and subject a patient to potentially hazardous leakage currents. Both the isolated power system and the equipotential grounding method are acceptable under the current national standards and codes (some states have specific requirements), and each approach has its individual merits.

Just as in the operating room, the special-care unit also has specific requirements for pressurization and air changes. Table 11.1 shows that the intensive-care unit must have a positive pressure with air changes totaling six per hour at a minimum. The obvious reason for maintaining positive pressure in the intensive-care unit is to prevent the movement of undesirable particulate matter from adjacent areas into the unit, and thereby lessen any risk of cross-infection. Some discrepancy exists in current codes for isolation rooms in intensive-care units. The HEW minimum standards call for equal pressure, while the JCAH calls for a negative pressure for isolation rooms. One reason for this conflict is that occasionally it will be necessary to maintain reverse isolation, as in the case of a burn patient where the intent is to prevent his becoming infected, as opposed to a patient who might be a source of infection.

Other facility interfaces that are required in the SCU are the oxygen and suction systems generally provided in the direct wall connection, complete with regulators and quick disconnects. Oxygen normally is used routinely in the coronary-care unit for recovering patients with mild coronary infarctions and other cardiac ailments. To minimize the cardiac workload while increasing the oxygen concentration in a patient's blood, oxygen is normally supplied from a central oxygen supply system that provides an even flow of oxygen at approximately 50 psi. Most hospitals are equipped with alarm systems to indicate if the oxygen supply system drops below a minimum preestablished level. Shutoff valves are situated throughout the institution, as are suction shutoff valves. These shutoff valves should be protected in such a way that they cannot be turned off by accident; however, experience shows that the valves too frequently are placed in high-traffic areas and are readily accessible to the public. Therefore, periodic inspection of the valves is of great importance. Figure 11.10 shows a central oxygen system.

The central vacuum system in SCUs is for the obvious purpose of removing mucus or solid food particles to minimize internal obstructions. On other occasions it may be necessary to use this suction for thoracic drainage and the evacuation of serous fluids from the lung cavity. The central vacuum system that provides vacuum to the SCU and the entire hospital involves two or more vacuum pumps that are duplexed alternately for the system. The pumps are normally capable of supplying 75 percent of peak demand and are designed to maintain approximately 12 to 15 in. Hg at each individual outlet. The system outlined in Fig. 11.11 includes a receiving tank, optional engineer alarm systems, conventional risers and shutoff valves, and the obvious bedside outlets. Normally, the SCU is equipped with more than one suction outlet since there may be a need for simultaneous suction. Fittings are provided to accommodate the bottle in which fluids are collected to prevent aspiration directly into the suction system.

Compressed air is required in the special-care unit for pneumatically operated equipment or for intermittent positive pressure breathing. The system must provide moisture-free air at 50 psi. The system is operated by central air compressors, an air dryer which dehumidifies compressed air, and the conventional risers and shutoff valves. The compressor tank should be periodically "bled" to insure that the accumulated condensation is drained from the system and not blown out into the air lines. Figure 11.12 shows the general outline of the compressor system.

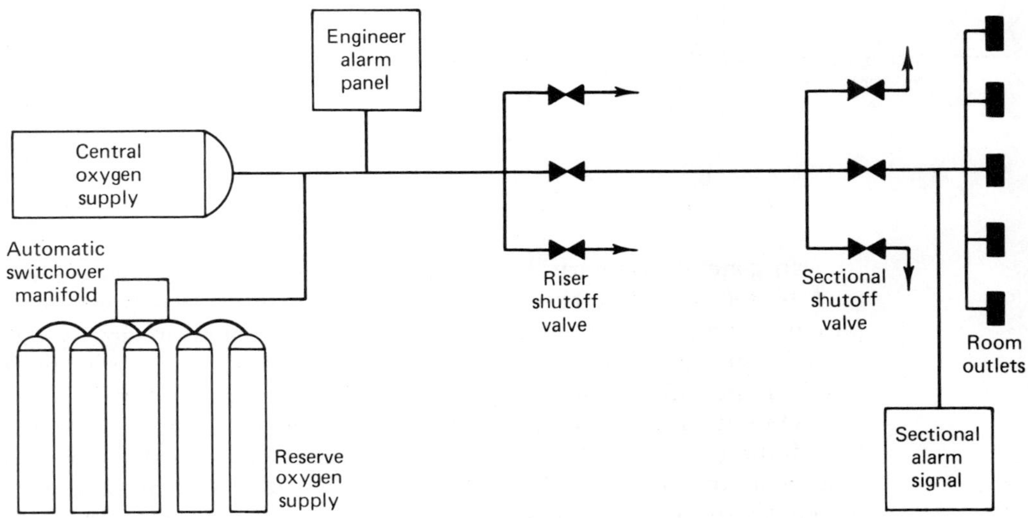

Figure 11.10 A typical central oxygen system.

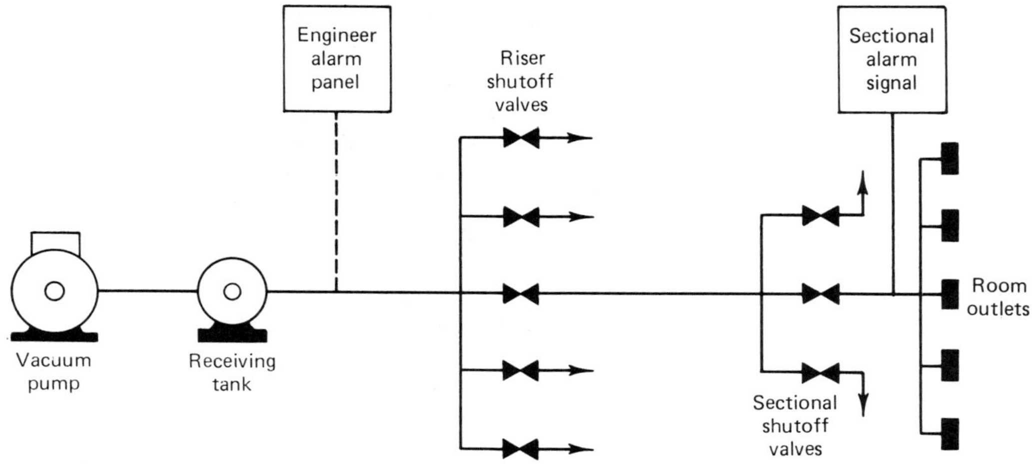

Figure 11.11 A typical central hospital vacuum system.

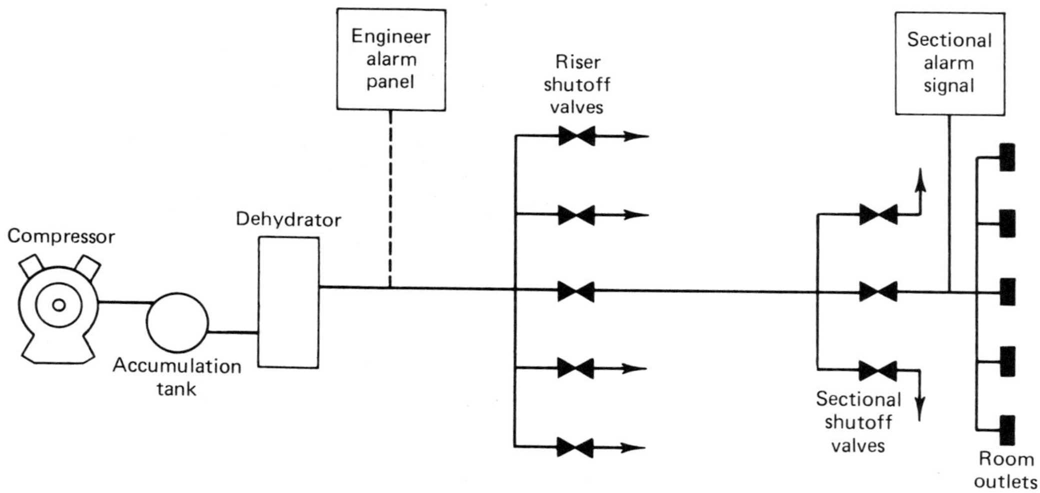

Figure 11.12 A typical hospital compressed air system.

In general summary, Fig. 11.13 shows what the engineer can expect to see as a facility systems interface in the conventional SCU patient room. It includes all of the basic ingredients previously discussed: suction, compressed air, oxygen, electrical ground fault monitor and safety system, and a single environmental interface.

One final point concerning the SCU is the obvious need for that function to be supplied by the emergency electrical generation system in the hospital. All outlets and electrical systems should be linked to the emergency generator to insure continuous supply of electrical power.

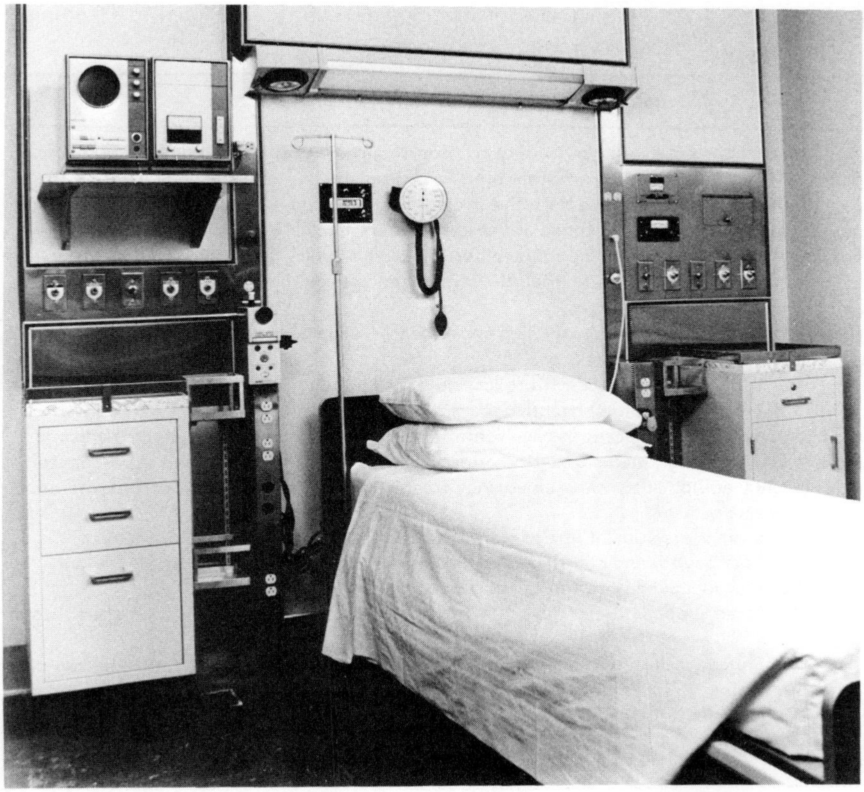

Figure 11.13 A typical special-care room with all usual systems. Square D Company, Oshkosh, Wisconsin, Modular Medical Walls Brochure.

Other services

The emergency room has utility system needs similar to those of the operating room and the SCU. The unit must have available emergency electrical power in the event of power failure. It must have oxygen vacuum supplies for use with critically injured patients, and special illumination in the form of overhead OR-type lights in trauma rooms. Table 11.5 shows that the emergency room requires special planning and analysis that must be undertaken when considering the layout and design of functional services within the hospital.

The nursery has a controlled environment with access limited to nursery personnel. This minimizes outside environmental effects to protect vulnerable infants. The gown, washing, and surgical mask procedures result in cleanliness of the environment and control of outside contact within the nursery. Table 11.1 shows that the

Table 11.5. An Extract Demonstrating the Utility and Facility Planning Necessary for an Emergency Room

Minimum Requirements for Emergency-Care Patient Services

The extent of the emergency services to be provided in the hospital will depend upon community needs and availability of other organized programs for emergency care within the community. Hospitals having a narrative program calling for a minimum level of emergency services shall provide at least the facilities indicated in items (1), (4), and (9) below. Other hospitals shall provide all of the following to the degree called for in the narrative program.

(1) An entrance at grade level, sheltered from the weather, and with provision for ambulance and pedestrian access.

(2) A reception and control area conveniently located near the entrance, waiting area(s), and treatment room(s).

(3) Public waiting space with toilet facilities, public telephone, and drinking fountain.

(4) Treatment room(s). Handwashing facilities shall be provided in each room or shall be conveniently adjacent to each room. The rooms shall contain cabinets, medication storage, work counter, medical suction outlets, X-ray film illuminators, and space for storage of emergency equipment such as emergency treatment trays, defibrillator, cardiac monitor, and resuscitator.

(5) Storage area out of line of traffic for stretchers and wheelchairs.

(6) Staff's work and charting area(s). This may be combined with reception and control area or located within the treatment room.

(7) Clean supply storage. This may be a separate room or located within the treatment room.

(8) Soiled workroom or area containing clinical sink, work counter and sink equipped for handwashing, waste receptable, and linen receptacle.

(9) Patients' toilet room convenient to treatment room(s).

— Department of HEW

nursery unit has a positive pressure in relation to all adjacent units, and has 12 minimum total air changes per hour supplied to the room. Recirculation of air into nurseries is generally discouraged unless the minimum filtering requirements are met. In general, nurseries must have the ability to isolate infants either through a separate isolation room or isolating-model infant incubators. Besides the requirement for positive pressure, other specific requirements exist for temperature and relative humidity. Table 11.6 specifies a temperature of 75° F (24° C) and a range of humidity from the minimum of 30 percent to a maximum of 60 percent for nurseries.

Electrical service to the nursery must be a guaranteed electrical distribution system operating off of main power systems with back-up from the emergency generator. All electrical outlets in the nursery must be wired to the emergency generation system.

Respiratory therapy departments have minimal facility/utility requirements because most such therapy services are provided directly at bedside in the patient's room. However, consideration should be given to the provision of oxygen and compressed air since future

Table 11.6. Temperature and Humidity Criteria for Critical Hospital Departments

Area Designation	*Temperature* °F	°C	*Relative Humidity (%)* Min	Max
Operating rooms	70–76*	21–24*	50	60
Delivery rooms	70–76*	21–24*	50	60
Recovery rooms	75	24	50	60
Intensive-care rooms	75–80*	24–27*	30	60
Nurseries unit	75	24	30	60
Special-care nursery unit	75–80*	24–27*	30	60

*Variable range required

plans may be to provide outpatient treatment in the respiratory therapy department. When ethylene oxide sterilizers are used to sterilize respirator components, plastic tubing, etc., adequate means must exist for exhausting the ethylene oxide from the area and the sterilizer to the outside, since ethylene oxide is toxic and, in high concentrations, is flammable and explosive.

Physical therapy departments have a substantial amount of electrical equipment that comes in direct contact with patients, in combination with metal surfaces and water, thus presenting the highest possible potential for electrical safety hazards. One recommended approach to this problem is the installation of ground fault interrupters (GFIs) (not to be confused with line isolation monitors) for the purpose of protecting patients in the event that a fault occurs. If a fault occurs and leakage current is detected by the ground fault interrupter, the whole circuit will be shut down much as a circuit breaker disengages a circuit.

Other obvious facility interfaces within the physical therapy department apply to water supplies and plumbing systems in support of the variety of hydroelectrical devices operated.

Summary. This chapter has outlined some considerations essential to the clinical engineer's understanding and knowledge of hospital physical plant facilities and utilities. It was not our objective to present in-depth engineering detail, since obviously that would require a separate textbook, but rather to familiarize clinical engineers with the varied facility considerations that are intrinsic to the healthcare setting.

Note the wide range of engineering disciplines involved and the complexities of the interrelationships that this chapter has highlighted. The challenge for systems engineering is nowhere more pronounced than in the hospital.

11.7 Study Questions

11.1 Describe one advantage and one disadvantage of conductive floors in operating rooms.

11.2 List three major components of a computerized axial tomography scanner.

11.3 Present three solutions to the problem of residual anesthetic gas in anesthetizing locations.

11.4 List three examples of problems that demonstrate the requirement for facility interfaces.

11.5 Describe four requirements that may be generated by the purchase of a new item of medical equipment.

11.6 What is the conflict between regulations and cost containment?

11.7 Define *functional analysis* and give an example of its benefit.

11.8 What is meant by *analysis by layout?* Summarize its benefits.

11.9 Define *support systems analysis* and give an example of its benefit.

11.10 Define *analysis by relationships* and give an example of its benefit.

11.11 Name and define the two classifications of ventilating systems.

11.12 List three recommendations for the air quality of operating rooms.

11.13 Decide whether or not you would recommend a laminar flow system and explain why.

11.14 Why are ultraviolet lights installed in the operating room? What new requirements do they add?

11.15 Why should temperature and humidity be controlled in the operating room?

11.16 List two safety features recommended for compressed gas lines.

11.17 Explain under what circumstances conductive floors are required in operating rooms.

11.18 Describe the room air-pressure requirements for central supply and the reason for them.

11.19 Give two problems associated with the use of ethylene oxide.

11.20 Why do sterilizers use steam? Why do they use pressure?

11.21 List five facility considerations for installation of computer-interfaced radiology equipment.

11.22 Describe the ventilation requirements of the clinical laboratory.

11.23 List means of reducing ICU psychosis.

11.24 Describe testing protocol for isolated power systems.

11.25 List four facility system interfaces found in a conventional SCU patient room.

11.26 List five minimum requirements for emergency-care patient services.

11.27 Describe the purpose of a ground-fault interrupter.

11.8 References

American College of Surgeons (1976). OR air cleaniness standards. Jan. Bulletin.

American Conference of Governmental Industrial Hygienists (1976a). Industrial ventilation: a manual of recommended practice.

American Conference of Governmental Industrial Hygienists (1976b). TLVS: threshold limit values for chemical substances and physical agents in the workroom environment.

Beck, W. C. (1976). Special air systems: the case for laminar flow. *Med. Instrum.* 10:275-276.

Compressed Gas Association (1967). Pamphlet P-2.1, Standard for medical-surgical vacuum systems in hospitals. New York.

Gordon, R., G. T. Herman, and S. A. Johnson (1975). Image reconstruction from projections. *Sci. Amer.*, 233(4):56-68.

HEW (1975). Minimum requirements of construction and equipment for hospital and medical facilities. DHEW Publication No. (HRA) 76-4000. Washington, DC: U.S. Government Printing Office.

Illuminating Engineering Society (1978). Lighting for health care facilities. CP-29. New York.

Jacoby, S. W. (1977) Personal communication. Air Test Labs, Inc., Rochester, NY.

Laufman, H. (1973). Current status of special air-handling systems in operating rooms. *Med. Instrum.*, 7:7-15.

Lindheim, R. (1971). Uncoupling the radiology system. Chicago: Hospital Research and Educational Trust.

Martin, J. T., ed. (1974). Handbook of hospital facilities for the anesthesiologist. Park Ridge, IL: American Society of Anesthesiologists, Inc.

National Council on Radiation Protection and Measurements (1976). NCRP Report No. 49. Structural shielding design and evaluation for medical use of X-rays and gamma-rays of energies up to 10 MeV. Washington, DC.

National Fire Protection Association (1973a). Standard for the use of inhalation anesthetics (flammable and nonflammable). NFPA 56A. Boston.

National Fire Protection Association (1973b). Safety standard for laboratories in health-related institutions. NFPA 56C. Boston.

National Fire Protection Association (1974). Standard for nonflammable medical gas systems. NFPA 56F. Boston.

Ulrich, J. A., W. Cribbs, and G. S. Michaelsen (1976). Recirculation of air in operating rooms. *Med. Instrum.*, 10:282-286.

U.S. Army (1965). Principles of sterilization. AMEDS Publication No. 50401. Washington, DC.

12
Educational Responsibilities

KENNETH C. MYLREA AND FRED J. SAVAGLIO

12.1 Learning Objectives

Upon completing this chapter you should be able to:

1. Describe the major components of a clinical engineering education program.
2. Design a clinical engineering continuing education program.
3. Provide BMET continuing education.
4. Develop an in-service training program on medical equipment operation.
5. Develop instructional materials suitable for use in in-service training programs.

12.2 Introduction

This chapter describes the educational responsibilities connected with clinical engineering. We provide either the aspiring or practicing clinical engineer with basic guidelines and principles required to successfully develop a meaningful educational program for himself, his subordinates and other members of the health-care team with whom he must interact.

12.3 Clinical Engineering Education

Clinical engineering is an evolving field that currently includes practitioners with varying levels of formal education. Job activities also vary widely and include work with medical equipment as well as other activities involving engineering in health care. Because of this diversity in personnel, activities, and goals, we will approach the subject of clinical engineering knowledge requirements in a general way. That is, we will describe the knowledge requirements for a variety of in-hospital clinical engineering activities.

Knowledge requirements

To practice clinical engineering, you must have a solid technical background and an understanding of the unique features of health care. Technical expertise is most important and should not be sacrificed or short-changed while acquiring other necessary knowledge and skills. Clinical engineers usually have minimal interaction with other engineers and need solid engineering competence. Cook (1978) has reported the results of a survey of skills required for clinical engineering practice.

The pyramid in Fig. 12.1 shows a stratification of knowledge requirements for clinical engineering. Certainly you can practice clinical engineering without obtaining the entire pyramid of knowledge. However, after acquiring the technical base, you need the additional knowledge and skills shown in Fig. 12.1. Additional requirements in clinical engineering are the ability to interact with people and the initiative to promote your capabilities in an environment where few other personnel understand or appreciate your skills.

The practicing clinical engineer must know the limits of his knowledge and ability. The diversity of engineering and clinical/medical information required make it virtually impossible for any

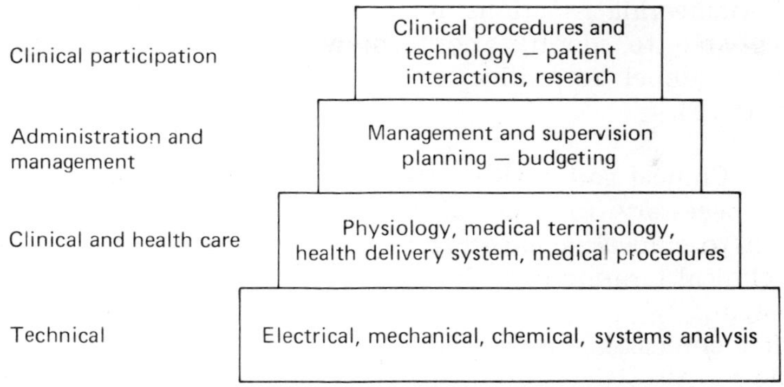

Figure 12.1 Pyramidal concept of knowledge requirements for various levels of clinical engineering activity.

individual to be competent in all areas. Thus, it is essential to know yourself and your limits and to be able to call for assistance as appropriate.

Technical requirements. The clinical engineer serves as the hospital's technical resource. Although he needs to be technically competent, he may find that within the hospital, appreciation of his technical expertise is limited to maintenance and repair on medical instrumentation and providing safe and reliable equipment (Boxerman, 1977).

As the hospital's technical representative, it is essential that the clinical engineer know the regulatory requirements that involve technical aspects of the hospital (Chap. 4). Also, to the extent possible, he should be involved in the development of technical codes, regulations, and standards which form the basis for these regulatory standards. In the past, hospitals have had meager representation on technical code and standard setting bodies. An important activity for clinical engineers is involvement in this process to ensure that the full interests and concerns of health care are represented and to assist the hospital administration in their proper interpretation.

Clinical engineers deal with medical instrumentation and computer technology applications in health care (Chap. 5). Some hospital instruments already contain computers or microcomputers and this trend will undoubtedly continue. Thus, knowledge of microcomputers, digital electronics, and signal-handling capability is necessary (Webster, 1978).

Two additional skills important in clinical engineering and not unique to any specific branch of engineering are: (1) the ability to understand and analyze systems involving equipment, personnel, and health care; and (2) an analytic problem-solving capability. Few clinical engineering functions are purely technical and the engineer must be able to identify nontechnical problems and work with clinical personnel to evolve meaningful and acceptable solutions (Oakes, 1975).

Clinical and health care. Clinical and physiological knowledge is necessary for clinical engineering. A functional systems approach to clinical engineering requires an understanding of physiology, clinical terminology, the practice of medicine, and the health-care system.

Most clinical engineering education programs include a course in physiology. Also, there are many available reference texts (Selkurt, 1976; Guyton, 1977; Vander et al., 1973). Medical terminology can be learned on the job or as part of formal course work. A self-learning text is valuable in this endeavor (Smith and Davis, 1972). Useful information on medicine and clinical practice for engineers is contained in Jacobson and Webster (1977).

Administration and management. As shown in Chap. 13, clinical engineers usually find themselves operating as managers. They may become directors of divisions or departments and may thus find themselves competing with other service and clinical departments for space and funding. To be successful it is necessary to understand and adopt methodologies, justification techniques, and economic analyses typical of hospital administration. The hospital administrator probably will not understand engineering or engineers and will want your program described using techniques and terminology with which he is familiar. Preparing and justifying your budget from the point of view of the hospital administration will be essential (Chap. 2, 3, and 13).

The administrative role will probably include management of technical and clerical personnel. Many books are available in this subject area (Odiorne, 1965; Carroll and Tosi, 1973; Morrisey, 1970).

Another administrative function is the organization and operation of the clinical engineering program. In the hospital setting this often requires evolving new ideas, alone or with others. This leads to interaction with various interest groups to obtain concurrence and funding to undertake and carry out the work. These administrative qualities together with the ability and insight to make decisions are characteristics that are necessary for clinical engineering.

Clinical participation. Advanced clinical engineering requires an ability to teach and demonstrate equipment applications in a positive and effective manner. You may be required to point out difficulties and explain proper procedures to responsible individuals in the presence of patients, without causing problems or creating antagonism. The clinical engineer also is involved in continuing education and in-service programs for a variety of medical, nursing, and technical personnel. In all of these tasks communicative and interactive skills are useful if not essential.

In the future it is likely that clinical engineers will be more involved in clinical procedures and in teaching and demonstrating skills involving sophisticated medical instrumentation. This will require an extensive knowledge of both clinical procedures and technology. For example a qualified clinical engineer may be the best person to teach procedures involving computerized scanners, catheterization and data collection, computerized clinical laboratory equipment, etc. These activities represent an aspect of clinical engineering that has not yet been approached by many individuals in the field, but which we believe represent one true realization of clinical engineering potential.

Some clinical engineers work on specific research projects in the hospital. Meaningful involvement in research includes participation in identifying research activities, data analysis, and publication of results. Depending on the type of research there may be extensive requirements for engineering-oriented expertise involving designing, fabricating, testing or acquiring instrumentation, as well as systems integration and extensive data analysis.

12.4 Formal Training Programs

There is a wide range of formal training programs in clinical engineering. These include two- and three-year training at technical schools and community colleges, bachelor's and master's degrees, and Ph.D. programs. Also, within each category, there is a variety of educational goals, program durations, degree names, and specific emphases of the programs.

Biomedical equipment technician (BMET) training programs stress the practical aspects of medical instrumentation and usually include extensive laboratory work. Junior colleges and technical schools throughout the country offer supplemental or complete BMET training programs (Altman, 1977). The above reference also includes a list of state and local BMET associations.

A 1975 survey showed that 24 institutions were offering a bachelor's degree program in biomedical engineering (Jendrucko, 1976). Most bachelor's degree graduates in biomedical engineering go on to graduate or medical school. The content of a program should be carefully reviewed to determine the extent of traditional engineering material as well as the biomedical and interdisciplinary subjects. Clinical engineering activities in the hospital place primary importance on engineering capability.

Educational Responsibilities

A variety of master's and Ph.D. educational programs in clinical and biomedical engineering are also available (Schwartz and Long, 1975). Early programs in biomedical engineering concentrated on graduating Ph.D.s with emphasis on theoretical modeling of physiological systems. More recently practical instrumentation aspects of biomedical engineering have gained a foothold and many master's degree programs with clear emphasis on clinical engineering are available (Oakes, 1976).

Some programs offer the master's degree in clinical or biomedical engineering while others consider these as options with the degree in a traditional discipline, usually electrical engineering. Most master's programs for clinical engineers are approximately two years in length and include a hospital internship.

Ph.D. training is most useful if the individual contemplates extensive involvement in research. This training normally requires coursework in both technical and biomedical areas plus a dissertation that usually generates new information in the selected area of research.

12.5 Continuing Education for the Clinical Engineer

By definition, the clinical engineer's continuing education program must keep pace with the constantly changing medical/clinical environment. Much like the practicing physician, the engineer must sustain a working knowledge of established clinical practices while concurrently exploring new techniques and technologies. Even established clinical instrumentation applications produce new information and new methods of diagnosing or treating patients. For example, ECG waveform monitoring in coronary intensive care is an established clinical instrumentation technique. However, only recently, through the use of computerized arrhythmia monitoring, has the significance of quantitative assessment of the frequency and types of arrhythmic waveforms been demonstrated. Thus, it is likely that the majority of the continuing education of the practicing clinical engineer will take place on the job.

On-the-job training

This continuing education responsibility can be conscientiously addressed by recognizing the educational opportunities provided in three categories of the clinical engineer's on-the-job duties.

The first category deals with the solving of specific equipment problems and the crises they sometimes produce. During the time that a particular equipment problem occurs, the hospital staff is usually keenly interested in what caused the problem, how it was corrected, and what can be done to prevent a repeat of the situation. At this point the engineer's attention should focus on the entire clinical setting to determine if the problem was associated directly with a failure in the equipment or the way it performed in the particular application. This analysis will provide information required to establish criteria for reviewing new equipment designs and in defining improved techniques or methods for reducing the occurrence of the problem. A log book will prove useful for later reference and for checking for trends in equipment malfunction (Chap. 9).

A second opportunity for continuing education lies in the wealth of information that can be learned through new equipment purchases (Chap. 8). Some engineers make it a practice to establish unsolicited, scheduled programs of equipment reevaluation in order to gather the latest information on new equipment. In addition to the availability of technical experts from the companies whose products you are reviewing, manufacturers provide a wealth of literature in the form of clinical studies, product specifications and design theory. Also, a search of medical and technical journals allows the engineer to become familiar with the specific technical aspects of the clinical equipment application (Sec. 15.6).

The third category of on-the-job continuing education is that of observation and participation. The clinical engineer should routinely observe clinical equipment being used in practice. This is particularly important when new clinical procedures are being implemented.

New procedures must be documented with clinical protocol and reviewed by the medical staff. Participation in the drafting of clinical protocols provides the engineer with an opportunity to study the clinical application of equipment. The byproduct of this observation process and participation in protocol development is an improved understanding of equipment interaction with the patient.

Additionally, a wealth of educational opportunities are provided by attendance at clinical in-service education programs conducted by clinical departments in the hospital. Furthermore, department directors and hospital in-service directors are aware of in-service classes being conducted at nearby hospitals. A listing of valuable clinical settings for the clinical engineer to observe either at his or other hospitals is:

1. Arrhythmia monitoring.
2. Bypass or open-heart surgery.
3. Cardiac catherterization.
4. Cardiac defibrillation/cardioversion.
5. Cardiopulmonary resuscitation.
6. ECG stress testing.
7. On-line blood pressure monitoring.
8. Renal dialysis.
9. Respiratory therapy (set-up).

Professional affiliations

Clinical engineering, like other professions, provides the opportunity for sharing of experiences and ideas with a large cross-section of members through local, state, and national societies. Listings of pertinent societies and organizations and their areas of interest are found in Chaps. 14 and 15.

Professional societies typically sponsor annual meetings with multiple presentations of selected topics. The annual meetings provide a simple means of presenting technical papers and publishing condensed abstracts. The abstracts are printed in a seminar proceedings booklet which in itself is a simple means of reviewing the latest ideas being exchanged in the field. A clinical engineer should require that his employer sponsor attendance at a minimum of one such meeting per year.

12.6 Continuing Education of the BMET

The structure and emphasis of continuing education for the biomedical equipment technician (BMET) must follow a thorough, objective assessment of both the job and the BMET's capabilities (Altman, 1977). We assume that the clinical engineer will have direct managerial responsibility for the BMET. However, in some organizations the BMETs may be assigned to a separate support group. The latter case is mentioned to emphasize the underlying principle that must be used by the clinical engineer in assessing the direction and content of the BMET's continuing education needs. That principle, simply stated, is "Does the BMET have sufficient knowledge of the design concept for each piece of equipment to correctly assess whether it is functioning properly?" Keeping the foregoing question in mind let's consider three objective criteria used in BMET job and capability assessment and hence the selection of a continuing education program.

The first criterion is technical skill capabilities, particularly in the area of equipment troubleshooting. In addition to determining the BMET's troubleshooting capabilities in electronic, electromechanical, mechanical and chemical systems, the engineer must ascertain the BMET's understanding of the operation of basic instrumentation subassemblies such as power supplies, operational amplifiers, RF circuits, digital circuits, and optical systems. Depending on the types of equipment the BMET will be expected to repair, a program of both formal and on-the-job education must be defined to achieve proficiency in the system, instrument, and subassembly areas.

The second criterion may be categorized as interpersonal and verbal communication skills. This skill area applies to in-hospital communications as well as the ability of the technician to solicit support from manufacturers and parts supply vendors. Clearly these skills become of less concern if the BMET is to do only bench repair work. However, for the BMET who will be required to directly respond to equipment repair calls, the following skills must be assessed: (1) the ability to accurately obtain a description of equipment problems, the mode of use during the suspected failure and the urgency of the situation; (2) the ability of the BMET to be sensitive to the patient-care situation existing at the time he arrives to review the problem; (3) the capability of the BMET to accurately communicate his assessment of the problem to the person who initiated the call. Continuing education to improve these capabilities is often best addressed by a combination of situations—where the BMET can be exposed to the typical war stories of experienced individuals—and by careful, controlled introduction into new equipment support situations.

The third criterion involves attentiveness to recordkeeping and written communication. The medical-hospital field places an extremely high importance on accountability through careful recordkeeping, precise execution of established written procedures, and exact descriptions of results. A technician who is not at ease with written communications must be given the opportunity to develop these skills through formal technical writing classes. Classes in medical terminology are also helpful to the technician who is not conversant with clinical applications. The engineer can promote the development of these skills by requiring participation in technical report writing.

The obvious question at this point is "What about continuing education in additional areas of medical instrumentation application?" In our experience the benefit of offering the BMET the opportunity to broaden his capabilities into new areas is often a motivational issue. As a minimum, the first and second criteria must be concurrently addressed in order to be assured that the new interest is properly matched with a new skill.

12.7 In-service Training for Allied Health Personnel

Introduction

The previous sections of this chapter have dealt with educational responsibilities of the clinical engineer as they pertain to himself and those who work for him. This section addresses another educational responsibility: that of teaching hospital staff how to use clinical equipment effectively and safely. Training programs developed by the clinical engineer must compete with an immense scope of nontechnical, medical and administrative training required by nearly all members of the clinical support staff. A reasonable estimate of the portion of time available in the in-service schedule of a typical nurse, for clinical equipment training, is about 10–15 percent (Noble, 1971). Considering the in-service activity of most nurses, this translates to approximately one to two hours per year. Those who specialize in the field of in-service education are quick to illustrate the classical problems of getting sufficient time, interest, and materials to provide effective, high-quality programs (Wiley, 1976). Consequently, the clinical engineer must develop an opportunistic approach to achieving an appropriate level of in-service training. The primary motivations for nurses are increased awareness, added knowledge, new experience, and familiarization with resources.

An essential goal of the in-service program is to achieve safe and effective utilization of equipment in the care and treatment of patients. The clinical engineer typically has the role of teaching the proper use of new and existing technology. He must also assist the hospital in the proper selection, operation, and maintenance of new equipment (Dodson and Latimer, 1976). He must perform this basic teaching role within the context of in-service education programs. Education of the hospital administrative and medical staff in understanding the clinical engineer's role in technology application is also essential if hospital patients are to receive safe, effective, and economical care.

The task of sustaining the clinical engineer's role in prepurchase evaluation (Chap. 8) and selection of new technology is a continuing educational process. The most effective method for accomplishing this is to routinely disseminate information on new technology development to appropriate administrative/medical staff. By simply bringing pertinent information on existing and new technology to the attention of concerned hospital staff, the clinical engineer will become recognized as the hospital's technical resource. A measure of effectiveness for an educational program is an increased interaction between clinical engineers and the physician and nursing staff.

Establishment of an equipment control program (Chap. 9) will increase the understanding of the clinical engineer's role by the hospital's administrative and medical staff.

Development

The clinical engineer must assemble from a number of possible sources (see Sec. 12.8) the specific materials and programs that fit the needs of his particular institution. In-service needs of a given institution should be defined, in part, through direct input from the staff for whom the programs are being designed. The most often asked question of those for whom the in-service was developed is "Why didn't you ask me what I needed?" Following is a list of typical subject areas that should be in the clinical engineer's repertoire of in-service programs.

1. Blood pressure transducers and calibration.
2. Defibrillators/pacemakers.
3. ECG electrode application and artifact reduction.
4. ECG monitoring.
5. Electrical safety.
6. Electrosurgery use and safety.
7. Neonatal monitoring.
8. Ventilators.

The clinical engineer should use these and other topics to survey the nursing staff as to their interests in continuing education.

The next and most often repeated development hurdle is related to cost—both direct costs of materials and the time required to assemble and present the various programs. This cost element is usually a concern of the hospital administration. Arguments used to justify these costs are as vital to development of the program as the materials *per se*. There are three arguments that can be used in this context.

First is the cost of equipment maintenance. Fifty percent of emergency repair calls involve operator error rather than failure of equipment. This directly translates to a potential of 50 percent reduction in maintenance cost if the operator is properly trained and this type of call is eliminated.

The second argument involves liability. Operator's use or misuse of equipment is as much a factor in a potential liability suit as is the failure of the equipment itself. The third argument involves the maximization of the use of investments made in equipment. Too often particular features of a given piece of equipment, which significantly affect its initial cost, fall into disuse simply because of a lack of sufficient training.

The next step in in-service program development is coordination with the in-service director of the institution. Close coordination with this individual will permit various segments of the equipment program to be combined with medical in-service programs. As an example, electronic aspects of ECG monitoring can be taught at the same time as the clinical and patient-care aspects of the procedure.

Acquisition costs of in-service materials and labor costs to provide the in-service training can be reduced by making optimum use of manufacturers' programs developed for promotional or post-sale training on new equipment. The cost of such programs will often by borne by the manufacturer. Courses and course materials can be included with the request for quotation for the equipment (Chap. 8). The same situation applies to operator's manuals and other such documentation. At the request of the clinical engineer, the manufacturer will often make company experts available for in-service programs rather than the often less-informed local sales personnel. In-service programs can be accomplished during the time new equipment is being evaluated if the clinical engineer carefully structures the material to be presented.

A final resource for program development material is professional organizations and commercial organizations (Sec. 12.8; Chap. 15).

Structure

Figure 12.2 illustrates several elements that comprise a comprehensive in-service program structure. Each element is described briefly in the following text.

Orientation. Each new hospital employee must go through a formal orientation program. The clinical engineer's component of orientation should include basic guidelines on electrical safety, practical work on the routine use and setup of clinical equipment, procedures for notifying equipment repair and support personnel, and a digest of available in-service programs and other equipment practices and policies.

Routine scheduled rounds. The clinical engineer should establish a routine schedule of visits to each clinical department for which he is responsible to see if he can answer questions or perhaps observe equipment use problems. Problems may not be considered important enough to call for help, but they may be symptomatic of a training need. Often, visits can occur during regularly scheduled departmental staff meetings and the engineer should be invited to participate. The clinical engineer must remember to schedule his visits during all three working shifts. This is clearly a necessity.

Lectures. The bulk of the formal material of an in-service program will be presented during scheduled lecture periods. Advice on the presentation of lectures is simple to state and seldom followed; be organized, use audiovisual aids (Sec. 12.8), do not talk down to the audience, stimulate the audience, allow for questions, and keep the presentation brief.

Clinical problem solving. The best time to conduct an effective in-service program is when a particular problem is fresh and

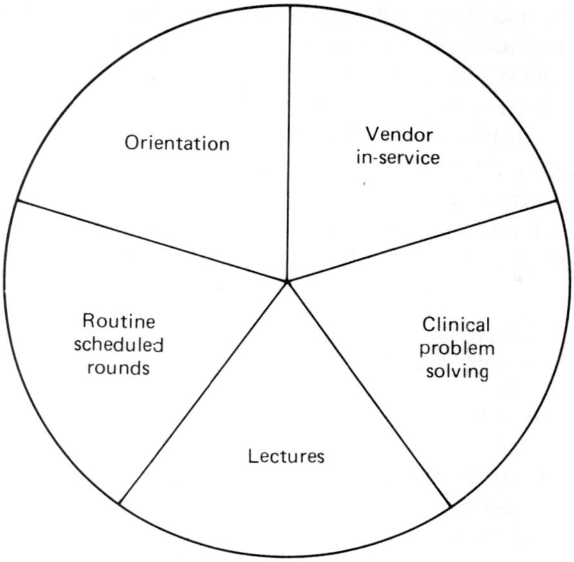

Figure 12.2 Elements of an in-service education program.

prevention of reoccurrence is of prime importance. You may generate mountains out of molehills, using good judgment, to drive home a particularly significant point. Repeated lectures on the proper application of physiological electrodes are never as effective as witnessing the actual solution of a noise or artifact problem.

Vendor in-service training. As previously mentioned, the opportunity for in-service education through postsale and promotional programs presented by equipment vendors and manufacturers can be a valuable component of the overall in-service program. However, these should be carefully planned and structured by the clinical engineer. It is also necessary to provide for follow-up programs which can be given to personnel who do not participate in the initial in-service orientation. For example, a vendor in-service session for a new ICU will need to be repeated as new nurses are assigned to the unit.

Evaluation

Evaluating the effectiveness of an in-service program can be accomplished in three ways. First, the emergency maintenance cost and categories of equipment problems can be monitored. High call rates for operator errors indicate the need for retraining or in-service for employees. A second method is through the use of a standard set of questions asked periodically during visits to the clinical department. The questions should be probing but not demanding. In the same manner, a standard physical checklist related to equipment handling or setup can be established and reviewed. The third method is the use of an evaluation form periodically sent to the staff. This form should list specific items in each of the subject areas covered during in-service programs that the staff should have completed. Design the evaluation forms with care. Do not exceed one page for the form. Present questions in terms the staff will understand. Use simple yes or no answers on each item. The burden of responsibility for low scores should clearly rest with the clinical engineer.

12.8 Internships and Work Study

Internships are a valuable part of clinical engineering training. The health-care environment is sufficiently unique that an engineer needs to develop an understanding of its local activities, interactions, and constraints, in order to function efficiently and effectively. Additionally, clinical engineers are likely to find that they are practicing solo, in contrast to, for example, the medical equipment industry where a variety of engineering skills is usually available through coworkers or supervisors. A hospital internship will also be valuable for those engineers considering employment in the biomedical equipment industry since they can gain an understanding of how and by whom their products will be used.

The purpose of an internship is threefold. First, a student gains experience in the health-care environment and the practice of medicine. He should become familiar with the patient-care environment and gain confidence and maturity regarding his engineering skills and his ability to interact with medical personnel. Second, the hospital may obtain additional engineering support at low cost (Sec. 6.5). The intern must, however, be assigned to a practicing clinical engineer. Third, he will learn the caveats and constraints of the environment.

The internship should not be viewed as a passive exposure to the health-care environment. Instead, it should be an active involvement in the day-to-day activities of clinical engineering with some measure of the pressure and frustrations involved. It should include functional interactions with various hospital personnel as well as participation in clinical settings such as staff rounds, and nursing functions. Many types of internships have been developed, and a number of problems and opportunities exist. A review of several approaches is available (Oakes, 1976).

Study program

Internships should be scheduled during the latter part of educational programs. In general, greater didactic work prior to the internship allows higher level and faster learning experiences in the hospital environment. If possible, prior coursework should include physiology, medical instrumentation, hospital operation, clinical engineering, and management.

Initiating and operating an internship requires appropriate planning, administrative approval, funding if necessary, and supervision of the intern. Planning includes factors such as duration, payment, insurance coverage, and activities and evaluation of the intern (Cook et al., 1978). Duration should be determined on the basis of what the hospital expects from the intern as well as the educational responsibilities. If the intern is paid by the hospital, the duration should be sufficient to allow a reasonable period of productive activities in addition to the exposure, observation, and learning. We recommend that the hospital pay the intern at least a minimal amount and expect him to function at least partially as a hospital employee. This arrangement often alleviates the problem of insurance coverage since most hospitals have policies that cover their employees.

Activities planning and methods of evaluation should be determined jointly with the faculty representative from the educational institution. Activities should be planned to provide the following:

1. Practical experience with a variety of medical equipment, including maintenance, repair, and calibration.
2. Instruction in the constraints and requirements surrounding the use of technology in medical care.
3. Involvement in clinical engineering problems requiring a systems appraisal, interpersonal interactions, and judgmental decision making.
4. Acquisition of knowledge, ability, and confidence concerning operation of a clinical engineering program.

These objectives must be balanced off against the needs of the hospital and the responsibilities assigned to the intern by the clinical engineer. A cost-effective approach, in which both the intern and the hospital benefit, is more desirable.

Responsibility for a significant clinical engineering project provides a good learning experience for an intern and a means for evaluation. The project should be selected by the student with faculty and clinical engineering inputs. It should require engineering knowledge and skills to complete, but should be primarily interdisciplinary and require extensive interpersonal interactions and exposure to the hospital environment. The students should prepare a project proposal for review by the clinical engineer, his faculty supervisor, and a clinical preceptor if required. Verbal and written progress reports are essential.

Evaluation of the intern should be the joint responsibility of the faculty advisor and supervising clinical engineer. Close involvement with a major project will provide a measure of feedback concerning the student's progress. Additionally, simple, periodic reports from the clinical engineering supervisor in the hospital can provide feedback regarding skills in areas such as equipment specification, selection, inspection, PM and repair; interaction with nursing and medical personnel; solution of interpersonal relations problems; and interest and dedication to clinical engineering. The faculty advisor should provide meaningful feedback and remedial suggestions at specified intervals throughout the internship.

Supervised learning. When the intern's skills have been evaluated and remedial training identified, it is necessary to identify a responsible individual to carry out the supervised learning. This could be the supervising clinical engineer, another engineer, or a senior technician. The individual can then work with the intern to develop problem definitions noting the specific skills required for their solution. In this way, the intern is fully aware of the skill that he is to acquire, and the task for which he must develop and use his skills is defined.

Supervised learning should provide the intern with a range of experiences encompassing technical jobs, interpersonal interactions, and clinical exposure. Emphasis should be given to areas where skills are lacking or weak.

Project responsibilities. As the capabilities of the intern increase, he can move from specific jobs into self-contained projects. The projects can be simple or complex, but should be assigned completely to the intern. If the work requires many of his recently acquired skills, this will afford him an opportunity to reinforce past learning while developing new skills. In the process of moving from problems to projects, it is important that the intern feel some of the normal pressures of clinical engineering. Projects should also require a measure of decision making so that the intern can move from defined, single-solution problems into multiple-option, undefined, people-intensive problems.

Projects assigned on a day-to-day basis must be different from the major internship project described under the study program above. In general, the student should not begin his main project until he has acquired most of the skills necessary to carry it through essentially on his own. Although the main project is a learning and skill-development experience, it should be viewed more as his opportunity to use and demonstrate his abilities and a means of evaluation. If properly selected, the main project will clearly reflect his abilities as a clinical engineer.

Functional responsibilities. The next step from project responsibility is functional responsibility for a specific area in the hospital, a type of equipment, or a portion of the clinical engineering program. This aspect of employee training is included for the sake of completeness since it is unlikely that interns will be in the hospital long enough to assume such a responsibility. An example of a functional responsibility is to oversee all clinical engineering aspects of an intensive-care area. This includes equipment acquisition, in-service training, preventive maintenance, and emergency repairs.

12.9 Training Materials and Methods

A clinical engineer's responsibility includes a planned, continuing education program for himself and his subordinates. He also must provide technically related, in-service, and continuing education to nurses, other hospital employees, and medical staff. Training and materials for continuing education of clinical engineering personnel were discussed in Sec. 12.4. This section will discuss the availability and preparation of materials used by clinical engineers in meeting their responsibility for in-service education to the hospital's non-technical personnel.

Training materials

The expansion of technology-oriented in-service education programs along with requirements for them from JCAH (Sec. 4.7) has resulted in a wide variety of commercially available prepared programs. A directory of such programs is available (Arnold and Paula, 1977) and should be reviewed before selecting training materials or undertaking development of specific lectures or materials.

Continuing education for nurses and other professionals has been the subject of sessions at the annual conferences of the Alliance for Engineering in Medicine and Biology and the Association for the Advancement of Medical Instrumentation (Sec. 15.8). Keeping abreast of the developments in this area is a responsibility of the clinical engineer.

We will use electrical safety as an example to discuss available audiovisual materials and the clinical engineer's potential contribution in this area. Electrical safety is of concern to medical and nursing personnel and is clearly in the domain of clinical engineering. To the engineer, it is not a complex problem, although it draws on a good deal of his background and education. The material must be explained to hospital personnel so that they understand the concepts of safety and so that appropriate behavioral changes will occur, resulting in a safer environment for the patient and employee.

All hospital employees should be exposed to information on electrical safety. This can be done most effectively during orientation of new employees with a short, high-information-content presentation. Since it must be repeated for each orientation group, it is an appropriate use for a packaged audiovisual presentation. A good slide-tape presentation for this purpose is available (Arnold and Paula, 1977).

The next step in electrical safety education is a more extensive presentation to nursing personnel working in heavily instrumented areas, such as operating rooms and intensive and cardiac care. Reference material is available in the literature (Mylrea and O'Neal, 1976) suitable for distribution to nursing personnel. A second option is to use this material along with a series of animated slides for a presentation by the clinical engineer. These slides should be simple and carry a message that is significant and easily remembered. Examples of slides prepared to convey the concept of electrical resistance and how it relates to people are shown in Fig. 12.3. A one-hour illustrated lecture covering electricity and electrical safety in the hospital is available on videotape from Microcampus at the University of Arizona in Tucson, AZ.

In the selection or presentation of technical material a clinical engineer should consider the background and training of the audience, the duration of interest span, and the specific information that will produce the desired behavioral changes in the listening group. Selection and presentation of materials is an important clinical engineering function which can have a significant effect on equipment-related problems in the hospital. A useful text for teaching instrumentation to nurses is available (Hoenig and Scott, 1977).

Preparing educational materials

In-service education will require preparation of written, verbal, and visual presentation materials. Since the mechanisms for information transfer are different in each case, the purpose of the program must be clearly understood in order to properly select an appropriate medium or combination for a particular program (Grayson and Biedenbach, 1975). In-service education should be viewed as having a definite purpose: the transfer of information or skills that will in some way modify the behavior of recipients. This behavioral change should be your goal. Behavioral objectives, such as those at the beginning of each chapter in this text, will help the student know exactly what behavioral change is desired. Such objectives should be provided for each in-service presentation. From the point of view of the recipients, if the in-service education does not have a direct effect on their job, then they will probably not have time for it. This is certainly true with respect to nurses who want information applicable to the care of their patients.

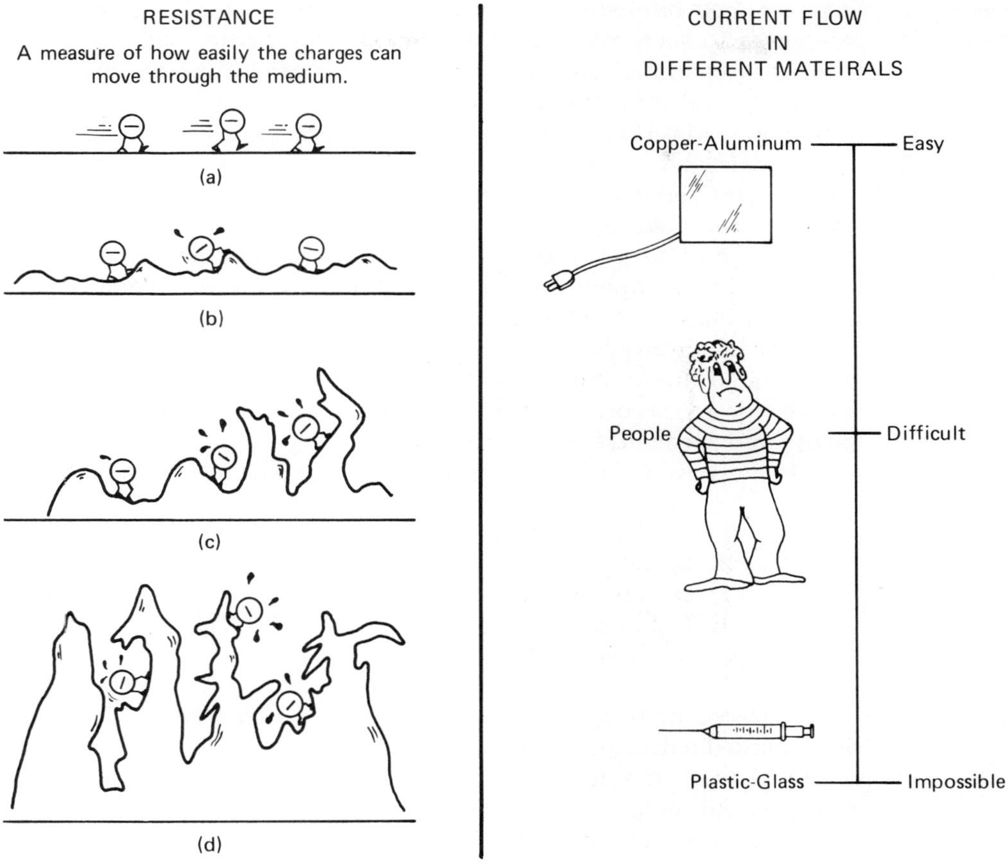

Figure 12.3 Examples of two visual displays designed to convey the concept of electrical resistance and how it relates to people.

Written communications. Written communication permits the rate of information transfer to be determined by the recipient. Sentences or paragraphs can be reread, the meaning of words can be ascertained, additional material can be referred to, and tests or procedures can be tried during the reading. Thus, the recipient retains considerable flexibility in the information transfer process. However, since the goal of the material is to effect a behavioral change, it is most important that the material be such that it require some evaluation or self-test by the learner. The major disadvantage of written material is that it requires a large degree of student motivation compared to an oral presentation.

In the area of continuing and in-service education, written material can be effective in two modes: first, totally new areas of information can be learned through books and journals. This type of learning requires considerable self-motivation and the clinical engineer may be involved only to the extent of having a file noting references to such material. A second use for written information is to delineate defined operations or procedures that may be performed irregularly and by different people. This second method is probably more appropriate for in-service education directed toward nurses.

Checksheet typewritten material that defines operational procedures may serve the goal of minimizing difficulties in the setup and operation of medical equipment. This will mean fewer calls for assistance and better clinical utilization of the equipment. For example, in the setup and operation of blood-pressure monitoring equipment, a list of things to do and check, in zeroing, calibration, and flushing could be made available. Also, simple trouble-shooting steps could be provided for common difficulties such as damped wave-forms, failure to zero, no output wave, etc. Information should be presented clearly and succinctly, for example:

If the wave-form appears damped, check for:

1. Air bubbles or kinks in the tubing.
2. Blood clot in the tubing.
3. Occlusion of the catheter.
4. Leaky transducer.
5. Incorrect catheter tubing.

Material of this type can be provided on both sides of a single sheet embedded in plastic and kept with the equipment. In the event of difficulties, the nurse can be expected to consult such a sheet which would serve as reminder either for setting up the equipment or for first-line trouble shooting. It is unlikely that she will read through an operations manual or several-page document to obtain the required information. Instead, she will change instruments and initiate a request for service.

In general, when preparing written information, you should consider the situation in which it will be used and the personnel who will be using it. Keep in mind the goal of effecting a change in activities or behavior and design the format and content accordingly.

Oral presentations. Verbal interactions on an informal or formal basis are an important aspect of clinical engineering. If necessary, the clinical engineer should consider training through a Toastmasters Club or other speaker training activity.

Unlike the reader, the listener cannot go back over information or look up word definitions. Abstract ideas and complex formulae are difficult to assimilate without visual aids. However, the verbal technique does have advantages, including the ability to stress information, emphasize points, and in most cases to provide an interactive learning situation. A combination of audio and visual material provides an opportunity to use strengths of both media.

An oral presentation should be relatively slow with emphasis on key points. The presentation should follow a logical sequence, building on material that is easy to understand and progressing to the more complex ideas and concepts. Use simple words to avoid any possibility that the listener may not understand the meaning. Only a few key points should be presented since too much information leaves the listener confused, with relatively less retention. Key points should be explained and emphasized through the use of visual material. Illustrations are an excellent means of transferring information that is difficult or cumbersome to verbalize.

A typical sequence in a presentation to explain the operation of a new piece of equipment would be as follows:

1. Introduction of the equipment and a clear statement of what the presentation is to accomplish from the point of view of the listener (behavioral objectives).
2. Explanation of equipment operation using appropriate terminology at a level understandable by the audience.
3. Explanation of procedures for set-up and calibration of the equipment.
4. Identification of potential problems and checklists to follow for finding the source of the problem.
5. Reenforcement of learning through a questioning technique.
6. Specific recommendations and summary.

The learning process is augmented if a short quiz is given at the end of the presentation. If you choose to do this, inform your audience of the quiz before beginning the lecture. This also provides some feedback to the clinical engineer regarding the success of the presentation.

Selection and presentation of visual material is an important aspect of conveying information. Audiovisual information should be well coordinated. If presented in slide form, a general rule of thumb is approximately two slides and 60 spoken words per minute. While this may seem slow, you should remember that concepts and ideas that are familiar to you may be new and difficult to the listener.

Visual material should be simple with each slide presenting only a single idea. An easy way to appraise the format of the slide is to draw the material on an 8½ by 11″ sheet and place this between your feet while in a standing position. With the correct amount of material, it would be easily viewed and read. Humor can be an effective tool in conveying information and stressing key points. However, the humor must be closely tied to the idea you are attempting to convey.

Preparation of your presentation can be simplified through the use of a storyboard mechanism, illustrated in Fig. 12.4. This simplifies the organization of your presentation and helps to insure reason-

SLIDE | ORAL PRESENTATION

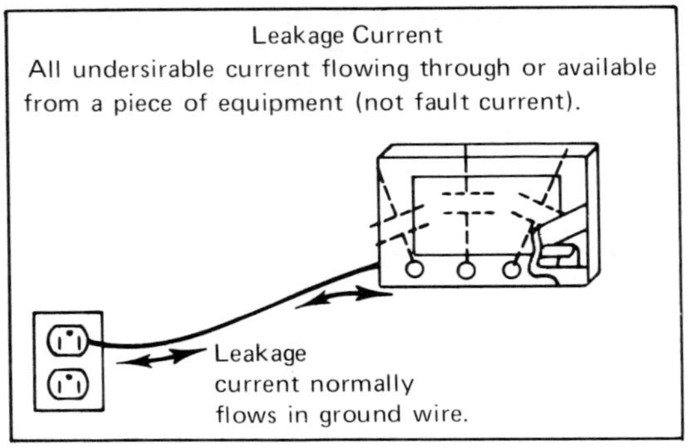

Leakage current is all undesirable current flowing through or available from a piece of equipment. This current flows when the equipment is operating properly and does not include fault current. The current results from capacitive and resistive characteristics which can be minimized but not completely eliminated. Leakage current normally flows in the third (grounding) wire back into the power system.

When an individual touches a piece of equipment a second path to ground is created and current can flow in either path. However, the ground wire has a much lower resistance and carries almost all of the leakage current from the equipment. Thus, a properly connected ground wire is an important safety factor for medical equipment.

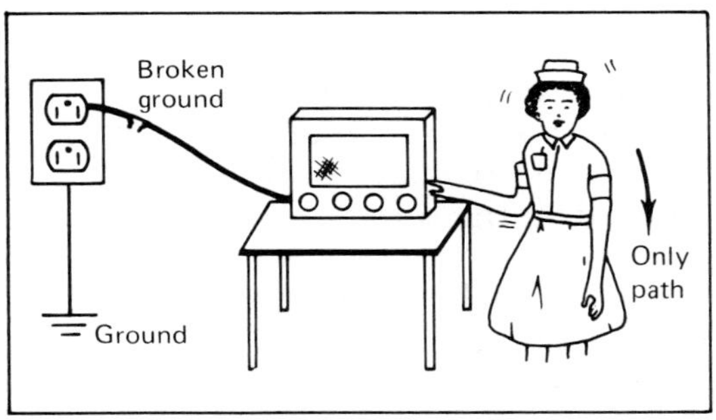

If the easy path to ground is open because of a loose plug or broken wire, the current will flow to ground through the person. This might cause only a tingling or a more severe shock. In any case, this is a dangerous situation and you should request that it be repaired. The equipment should not be used with patients.

Grounds are important.

DO NOT pull, twist or jerk the cord or plug.

DO call maintenance for loose or damaged plugs, frayed cords, any shock or tingling

Equipment grounds are important. Pulling, jerking, or twisting the cord or plug can break the ground and eliminate this safety. Treat equipment with tender loving care and call maintenance if you receive a tingling or shock, see broken loose or damaged plugs, or frayed cords.

Figure 12.4 An example of the storyboard method of preparing coordinated oral and visual material.

able oral content to augment each slide. It is preferable that you not read your presentation. However, knowing clearly what it is you are going to say along with each slide will result in a more meaningful presentation with a greater probability of effecting the desired behavioral change in the listener.

In a true learning situation, there is a tendency for the listener to take notes. This detracts from the material being presented and key points may be missed. Notetaking can be avoided by providing handouts that in effect summarize the presentation and provide a list of the key points and recommendations.

12.10 Study Questions

12.1 What are the basic areas that a clinical engineer must have competence in?

12.2 List the major technical requirements for clinical engineering practice.

12.3 List adminstrative skills that the clinical engineer must have.

12.4 Explain the statement: "The majority of a clinical engineer's continuing education will take place on the job."

12.5 What are the three classifications of on-the-job duties that may lead to continuing education opportunities?

12.6 What are the major aspects of a continuing education program for BMETs?

12.7 What is the role of the clinical engineer in in-service training for nurses?

12.8 Describe the five elements of an in-service education program.

12.9 What are methods by which a clinical engineer may obtain materials for in-service training, and what are sources for this information?

12.10 List the three purposes of a clinical engineering internship.

12.11 What activities should be included in a clinical engineering internship?

12.12 Develop an in-service lecture on electrical safety for nurses.

12.13 What is a behavioral objective for a lecture, and why is it useful to the student?

12.14 How do written and oral presentations differ?

12.15 What learning tasks are best suited to the use of written material?

12.16 What is the major disadvantage of written material as a learning tool?

12.17 What is a major limitation of oral presentations?

12.18 In what cases are oral presentations particularly useful?

12.11 References

Altman, B. J. (1977). The biomedical equipment technician. *J. Clin. Engr.* 1:16–23.

Arnold, B. E. and Paula, A. F. (1977). Directory of audiovisual aids for in-service training of clinical engineers, biomedical engineers and BMETs. Diamond Bar, CA: Quest Publishing Co.

Boxerman, S. B. and Hooton, C. J. (1977). Clinical engineering—the administrator's perspective. *IEEE Trans. Biomed. Eng.*, BME-24:57–59.

Carroll, S. and H. Tosi, (1973). Management to objectives: applications and research. New York: Macmillan.

Cook, A. M. (1978) Skills required for clinical engineering practice: results of a survey, *J. Clin. Eng.*, 3(3):279-283.

Cook, A. M., L. E. Rose, and B. H. Barkalow, (1978). Cost effective clinical engineering internships in community hospitals. *J. Clin. Eng.*, 3(2): 173-177.

Dodson, B. and B. W. Latimer, (1976). Strategies for clinical engineering through shared services. Battle Creek, MI: W. K. Kellogg Foundation.

Grayson, L. P., and J. M. Biedenbach, (eds.) (1975). Teaching aids in the college classroom. Washington, D.C.: American Society for Engineering Education.

Guyton, A. C. (1977). Basic human physiology. Philadelphia: W. B. Saunders.

Hoenig, S. A. and D. H. Scott, (1977). Medical instrumentation and electrical safety: the view from the nursing station. New York: Wiley.

Jacobson, B. and J. G. Webster, (1977). Medicine and clinical engineering. Englewood Cliffs, NJ: Prentice-Hall.

Jendrucko, R. J. (1976). The evolution of undergraduate biomedical engineering education and the professional opportunities for program graduates. *Ann. Biomed. Eng.* 4:17-24.

Morrisey, G. (1970). Management by objectives and results. Reading, MA: Addison-Wesley.

Mylrea, K. C., and L. B. O'Neal, (1976). Electricity and electrical safety in the hospital. *Nursing*, 76 (6):52.

Nobel, J. (1971). Philadelphia: Emergency Care Research Institute.

Oakes, J. B. (1975). Clinical engineering—the problems and the promise. *Science*, 190:239-242.

Oakes, J. B. (ed.) (1976). Proceedings of the conference on clinical engineering education. Publication 76CH1214-6 EMB. New York: Institute of Electrical and Electronics Engineers.

Odiorne, G. (1965). Management by objectives. New York: Pitman.

Schwartz, M. D., and F. M. Long, (1975). A survey analysis of biomedical engineering education. *IEEE Trans. Biomed. Eng.*, BME-22:119-124.

Selkurt, E. E. (1976). Physiology. Boston: Little, Brown.

Smith, G. L. and P. E. Davis, (1972). Quick medical terminology. New York: Wiley.

Vander, A. J., J. H. Sherman, and D. S. Luciano, (1973). Human physiology the mechanisms of body function. New York: McGraw-Hill.

Wiley, L. (1976). What if you didn't have in-service education. *Nursing*, 76: 65-68.

Webster, J. G. (ed.) (1978). Medical instrumentation: application and design. Boston: Houghton-Mifflin.

13 Management and Supervision

MICHAEL O. BRINKMAN

13.1 Learning Objectives

Upon completion of this chapter you should be able to:

1. List six reasons for the existence of management and supervision in a clinical engineering program.
2. Justify clinical engineering management based upon economic, quality assurance, and code compliance bases.
3. Explain basic theories of personnel management.
4. Describe typical clinical engineering program costs.

13.2 Introduction

Two of the most important aspects of the clinical engineer's job, and two for which he is generally least trained, are management and supervision. Although these two words can be defined to have essentially the same meaning, for our purposes *management* will be defined as *the direction or control of the total clinical engineering function* while *supervision* will be limited to *the overseeing of personnel within the clinical engineering function.*

Why are these two concepts so important? Normally a clinical engineer has a professional role within a framework that possesses no other technically qualified management people, i.e., technically qualified in clinical engineering. Because of this setting, he is looked to as the authority on clinical engineering and must be able to plan, organize, and direct that particular function. If he cannot, it is safe to say that that function will control him and that there will never seem to be enough funds or personnel to cope with the work that needs to be done.

Proper supervision is just as important, if not more important, when the clinical engineering group contains more than one employee. The efforts of the group are then dependent on two or more persons, and the group must act as a team. If the clinical engineer cannot properly supervise his subordinates, he will find his own workload immeasurably increased. With these introductory thoughts in mind, let us proceed to detailed discussions of how to manage and supervise. Although Chaps. 6, 8, 9, and 10 have dealt at length with managing outside service vendors, this chapter will primarily deal with management and supervision of a clinical engineering organization.

13.3 The Need for Management and Supervision

Before a clinical engineer can do an effective job of managing his operation and supervising his subordinates, he must understand why such actions are necessary. Six major reasons are given below to point out the necessity for proper direction of the clinical engineering function.

Economic justification and cost control

The word heard most frequently today in health-care circles is "costs." As pointed out in Chap. 3, the costs of health-care delivery are rising substantially faster than the general cost of living.

The reasons for this increase are many and varied, but all the talk has focused extraordinary attention on each health-care dollar that is spent. Whether the clinical engineer is in a hospital, a shared engineering service group, or another type of organization, he will be called upon to justify the dollars he wants to spend. Why is the task necessary? Does it need to be done in that way? Can it be done with outside labor, or is it cheaper to hire someone? How much cheaper? Have all costs been included? Can he juggle existing funds or people to get the new task done without increasing costs? And on and on. The clinical engineer must be prepared to answer these unspoken questions when he makes his plans known to his management. If he cannot hold his ground under fire with well-formed plans, he is unlikely to gain approval.

Once his plans are approved, the clinical engineer must be ready to control his costs. His management now expects him to carry out those plans at the price indicated. The clinical engineer who does not have a consistent track record of successful implementation will find it difficult or impossible to gain approval for subsequent programs.

Thus we have the first major reason for thoughtful and effective management on the part of the clinical engineer. Dollars are a limited resource; they will go to projects that are considered necessary and have a high probability of success.

Personnel morale and motivation

In the larger hospital setting or in the shared engineering service organization, the clinical engineer may be called upon to supervise other employees. Except for very large hospitals, hospital groups, or shared services, these other employees are usually biomedical technicians and/or clerical help.

Direction of subordinates significantly alters the overall management role of the clinical engineer. On the one hand, his task is to plan, implement, and control a program using his own talents and motivating only himself; on the other hand, his most overriding task becomes that of influencing others to perform the tasks he has planned in the way he has planned them without compromising the overall objectives in any significant way. Unless the clinical engineer can function as a leader and motivator, his department or group cannot achieve its goals in an efficient manner, the second major reason for effective management and supervision.

Quality control and program improvement

As clinical engineers learn during the course of their education, there is always room for improvement in any product. This is also true of any service. It can be made faster, more competent, more responsive to the consumer's needs, broader, less expensive, etc.

The clincial engineer must be constantly aware of the fact that he is supplying a service to someone, whether in the hospital environment or not. He and his group are there to solve someone else's problem. It may be an administrator's problem of complying with codes; it may be a nurse or operator's problem of getting a defective instrument operating again; it may be a physician's problem of using the right technique to obtain data on a patient. Whatever the situation, the clinical engineer is at the beck and call of others and must recognize that his function revolves around their needs.

Thus we have the third major reason for management. The clinical engineer must seek out and sense the needs for his services, must constantly match his existing services with those needs, and then must redesign and reshape his approach to meet the changing requirements of his users. He cannot afford the luxury of creating and implementing a program that does not serve their needs.

Compliance with codes

As explained in previous chapters, a significant portion of today's clinical engineering workload centers around compliance with local, state, federal, and voluntary codes. These codes are constantly being reviewed and revised by their originators with varying degrees of impact upon the health-care field.

It is usually the lot of the clinical engineer to monitor the revisions of the codes, interpret their meaning and intent, and apply them to his situation. The engineer has been trained to think in a logical fashion, but needs to apply good judgment and common sense to the problem of compliance with codes. It is primarily up to him to manage that compliance by determining the minimum required, and phasing the steps that need to be taken to cause the least disruption to the hospital's finances. Management of compliance requires careful planning and thoughtful implementation within the context of available technical and financial resources. This is the fourth major reason for the management function.

Liability exposure

The fifth major reason for management within the clinical engineering function concerns the liability exposure connected with the purchase, use, and repair of equipment. No one who reads the daily newspapers in the United States can be unaware of the liability problem that exists in the health-care field today. Hospitals, physicians, manufacturers, and nurses are constantly being sued for malpractice; product liability and clinical equipment play a role, albeit a variable one, in these suits.

It is up to the clinical engineer to manage the hospital's liability exposure in the areas of clinical equipment acquisition, use, and PM and repair. This will require attention to prepurchase evaluations, in-service training, service vendor control, supervision of internal PM and repair, and thorough but limited documentation of all such actions. Since it is relatively easy to overemphasize reactions to liability exposure and to overdo efforts to minimize this risk, management by the clinical engineer becomes doubly important. He is the hospital's balance for weighing the liability risk against the cost of reducing the risk and must therefore act in a responsible managerial fashion.

Input to administration

As mentioned previously, the clinical engineer provides a service capability supplying the needs of others within the overall organization. His most important interface is with administration since he or his group represent the organization's sole technical capability in the clinical engineering area.

Irrespective of questions asked by administration, the clinical engineer should regularly initiate advice to administration on such things as clinical equipment acquisition, code compliance, in-service training, PM and repair costs, departmental achievements and performance, incidents involving clinical equipment, etc. In a true sense, the clinical engineer must act as his own public relations man and ensure that his group's activities are made known to the proper administrators. This is the sixth major reason for him to perform a management function to ensure administrative awareness of the problems, needs, and accomplishments of the clinical engineering group.

13.4 Economic Justification

Make-or-buy analyses

Whether a clinical engineering program is in the design stage or is in operation and is being reviewed for cost-effectiveness, the clinical engineer must generate continual make-or-buy analyses to justify the existence of the program. These analyses compare the cost of providing the service internally to the cost of purchasing an identical service from an outside source.

Before explaining several methods of performing make-or-buy analyses, it is worthwhile to explore the concept of *identical service.*

No comparison can be valid that views apples and oranges as the same fruit, yet this is precisely what often happens in the clinical engineering field. *Identical service* exists only when all measures of the service are equal; for example, response time, availability, content and frequency of PM, mean time to repair, parts inventory and parts cost, technical competency, tools and test equipment, content and frequency of electrical safety testing, etc. If the services are not identical, then weighting factors must be used in the analyses to even out the differences.

Comparison with prior costs

One of the more accurate methods of performing a make-or-buy analysis of a new program is to compare the cost of a planned internal approach to the known costs of an existing vended approach. The analysis is begun by taking a full fiscal year's records of outside expenditures by individual department and instrument and separating them into the categories of parts, labor, and travel. The content of the vended service on each instrument is then reviewed to see if it contains all the necessary elements of electrical safety testing, calibration, PM, and repair. When these two audits are completed, the clinical engineer will have the complete up-to-date story of clinical equipment PM and repair.

If the hospital has not had any internal capability prior to the analysis, it should expect to see costs of each portion of the program expressed as a percentage of the total costs of the program falling somewhere near the following ranges and norms:

Item	*Percent of Total Costs*
Parts	40–60%
Labor and travel	40–60%
X-ray PM and repair	50%
Laboratory PM and repair	16%
Monitor PM and repair	8%
Central supply PM and repair	8%
Surgery PM and repair	6%
Respiratory therapy PM and repair	4%
Nuclear medicine PM and repair	2%
Other PM and repair	5%

The respective percentages of parts and labor and travel costs will shift as travel distances increase and as the usage of replacement X-ray tubes varies in a given year. These ranges and norms were developed by the author in approximately 80 hospital studies in various sections of the United States.

Invariably the clinical engineer will find that the present approach is lacking in one respect or another. Usually, the hospital purchases no clinical engineering services nor does it provide PM on all clinical equipment. When the engineer has identified these deficiencies, he must specify their content and then take steps to determine what their cost would be if they were vendor-provided. He can then add these additional costs to the historical costs to obtain a total cost figure for providing a complete clinical engineering program on a vended basis. Parts costs should be eliminated from the total and presumed to be constant for both internal and external programs. In some cases, it may be impossible to obtain certain services from outside vendors, especially in the more rural areas. These deficiencies would have to be taken into account in the ultimate comparison.

The other half of the analysis or comparison consists of developing an internal program that will deliver the same range of services. It begins with a determination of the number of annual man-hours required to provide engineering services, electrical safety testing, PM and calibration, and repairs. Engineering services are made up of such activities as prepurchase equipment evaluation, in-service training, PM and repair management, code interpretation, hazard alert processing, and design and implementation of documentation. Preventive maintenance, repair, and safety man-hours would be estimated by considering each individual instrument, its frequency and content of scheduled PM, and its reliability. Normally, equipment electrical safety testing is performed as part of scheduled PM while facility electrical safety testing (receptacles, beds, line isolation monitors, etc.) is handled by the Plant Services Department. The analysis by instrument should thus produce a total number of man-hours required by each unit each year.

The clinical engineer will gain a clearer picture of the total PM and repair and engineering program and be able to structure the proper organization if he separates his instrument workload by the type of work skills normally found among experienced biomedical technicians. These work skills would include X ray, laboratory, electronics, electromechanics, respiratory therapy, nuclear medicine, and diagnostic ultrasound. The latter two work skills would rarely be considered since most hospitals have less than 100 man-hours of annual labor in either category, and the work skills are highly specialized. Although the work skills of electronics, electromechanics, and respiratory therapy can occasionally be combined and handled by one technician with unusually broad experience, the laboratory and X-ray work skills usually cannot, and must stand on their own with sufficient man-hours to justify the employment of a technician.

When annual man-hours by work skill have been tabulated, the clinical engineer is then in a position to define the shape of the clinical engineering program. He can do this by dividing the estimated man-hours for each work skill by the number of man-hours equal to one full-time equivalent (FTE). For smaller hospital programs employing 1–4 people, perhaps 1250 man-hours should be used as equal to 1 FTE. For programs with 5 people or more, 1350 man-hours is suitable. Beyond 10 people further requirements can be made, but few hospitals can support groups of this size unless they band together. Where possible and desirable, the engineer should combine work loads of the different work skills before determining the number of people required.

Once the clinical engineer has determined the number of technicians and engineers he needs, he can put together an operating budget for his department to include the requirements for capital expenditures. Details of drawing up a budget are included in Sec. 13.5. With completion of this budget, the engineer is now in a position to compare his planned approach with the existing vended approach. The comparisons should be presented to answer the following basic questions:

1. Which approach is least expensive?
2. What are the capital requirements of each approach?
3. Do both approaches comply with regulations?
4. What are the advantages and disadvantages of each approach?
5. What approach is recommended and why?

Quite obviously, the foregoing approach is time-consuming and somewhat tedious. However, it can be performed in 2–4 weeks, and its results are difficult to contest. It should be used only when starting a new program, when considering adding a major new service to an old program, or when there is serious doubt about the cost-effectiveness of an existing program. The next section will deal with a simpler method of cost comparison for operating programs.

Comparison with outside purchase

An easier method for an accurate make-or-buy analysis comes into play with an operating program. Presuming that the program is using effective management information systems similar to those outlined in Sec. 13.5, it can easily calculate productivity and cost per hour. Since all time and materials spent on any given instrument or engineering activity are documented fully, it is then a simple matter to calculate total costs of labor and parts for any instrument, group of instruments, department, or engineering function (total labor hours × cost per hour + actual material costs = total cost). To be completely accurate, it would be necessary to use the cost per hour of the specific individual or individuals doing the work since there may be a wide variation in salaries and wages.

The comparative outside cost can be determined by formal or informal bid. The clinical engineer could begin by defining the task as it is performed by the internal group. He would then look for outside sources competent to do the same type of work and request bids. He must be careful to qualify the bidders by seeking references, information as to parts inventory and test equipment, technical documentation, PM format and frequency, and/or other criteria that seem fitting. The bid responses can then be compared with the internal program costs and content to determine which option represents the best alternative.

Comparison with equipment acquisition cost

One relatively quick method of estimating the cost-effectiveness of an internal program is to compute its total cost as a percentage of the selling price of the equipment it maintains. This method does not have the accuracy of the comparisons previously described, but it does have the distinct advantage of being easy to determine, especially if the clinical engineer's equipment control program includes data on the cost of each item that is maintained. All that is required is to divide the total costs of the program by the total acquisition cost of the equipment served. The resulting percentage would normally fall in the range of 8–10 percent. If it is significantly lower than this range, the content of the activities of the internal group should be questioned and reviewed. If it is higher than 10 percent, the difficulty can normally be traced to low productivity.

The clinical engineer must be cautious about using this method of economic justification if his program is narrow in scope. Realistic percentage annual costs on medical equipment can vary from a low of 4 percent to a high of 30–40 percent. If the majority of the equipment is lower priced electromechanical instrumentation, the percentage will be high. If it is composed of high-dollar items, the percentage will be quite low.

Comparison with other hospitals and other programs

The final method of economic justification available to the clinical engineer is to compare his program with that of other hospitals or other groups. Since these various entities are unlikely to have the same equipment complement or provide exactly the same range of services, the only valid means of comparison would be size, cost per hour, and productivity. The organizations would also have to be in geographical proximity since salaries and other costs do vary by region.

Size is important in this type of comparison because smaller programs necessarily have lower output and greater work content per productive employee. The degree of supervision is also greater because the supervisor is spread over fewer people. Clerical cost per employee is also higher. For example, a one-man program would be expected to have 5–15 percent less productivity and 5–10 percent greater cost per hour than a four-man program. The clinical engineer must expect and welcome examinations of this kind since it is the type of comparison most likely to be made by administrative managers of different institutions.

The literature contains numerous rules of thumb that have been put forward by various groups for determining the size of internal biomedical programs. The Veterans Administration has recommended that for staffing its hospitals there should be one clinical engineer for every 500 general medical or surgical beds, and, additionally, one clinical engineer for every 400–1000 domiciliary or neuropsychiatric beds. These guidelines also recommend another clinical engineer for each spinal cord injury unit or other major medical specialty.

The VA further recommends one biomedical technician for every 150 general or medical care beds and one BMET for each 400 neuropsychiatric or domiciliary beds. Alternatively, the VA states that personnel requirements can be based on the total cost of equipment; for example, one BMET for each $300,000 worth of technical medical equipment, one BMET for each home dialysis center, and one BMET for each $500,000 of radiological or nuclear medicine equipment.

Unfortunately, the VA's recommendations for staffing based on bed capacity do not take into account two major concerns: actual equipment complement and availability of funds. Studies of actual PM and engineering costs have shown a ratio of 10 to 1 in PM expenditures between two 250-bed hospitals. For that reason, we strongly recommend that staffing patterns for any institution be based on actual equipment complement and clinical engineering work load. In studies we performed in eighty hospitals throughout the United States, cost data have tended to indicate a staffing pattern for general-care hospitals with "normal" equipment complements of one clinical engineer or manager per 750–1000 beds and one BMET per 350 beds.

The VA's second alternative for staffing based on total cost of equipment is much more realistic, but neglects average expenditures for replacement parts. This rule assumes that $300,000 worth of technical equipment would have an absolute maximum annual repair cost of $30,000. On the average in nonnuclear and nonradiological departments, parts would make up 35–45 percent of that total, or approximately $12,000. The remaining $18,000 would be for labor and travel. Since the average BMET costs the VA $26,000 annually in labor and overhead, use of this formula would cost the government and the taxpayer $8000 more per man than the purchase of outside services. The same type of calculation holds true for radiological or nuclear medicine equipment. Here the cost is again about $30,000 per year, but now the parts portion would increase toward 50 percent of the total.

Other estimates on personnel needs come from a report of the program at the Downstate Medical Center in New York (Ben-Zvi, 1975). This report recommends an even higher level of staffing than the VA, suggesting an engineer, four technicians, one secretary, and one clerk/typist for a hospital of 200–300 beds. This staffing level does not presume any work on radiological or nuclear equipment, although they normally make up 50 percent of the work load.

The Downstate Medical Center does use this type of staffing; however, it is heavily oriented towards teaching and research, and significant grant monies are available to defray the cost of technical personnel. The average general-care hospital is not engaged in teaching or research and has no grant money available. Its role is to provide properly operating equipment at minimal expense with sufficient documentation to protect it in liability situations and satisfy regulatory bodies. As mentioned previously, the number of technical personnel varies directly with the amount of equipment to be handled.

In another evaluation of the 300-bed hospital, Topham (1976) determined that a total of 1800 man-hours would be required for PM and repair. This study also included no radiological or nuclear equipment and estimated PM requirements from procedures put forth by the VA and the Downstate Medical Center. Topham stated that this level of work could be accomplished by one well-trained, experienced person who would probably be a clinical engineer.

Since the recommendations of the VA and the Downstate Medical Center have already been discussed, we will not rehash them again. Instead we must deal with the mistaken concept that an individual works 2080 productive man-hours per year. He is paid for that amount of time, but the available time is far less. Vacation normally accounts for 80 hours per year, holidays for another 64 to 80, sick leave for an average of 40, coffee breaks for at least 100, meetings for 150 to 200, and so forth. Actual hands-on time in a real productive sense is normally restricted to 1250 to 1350 hours, especially in a one-man program. Attempts to plan productive levels at a significantly higher rate will normally be unsuccessful.

In looking at these three suggestions for staffing levels, it has been our intention to indicate the problems associated with quick and dirty rote formulas for determining required staff. If the institution is looking for an accurate approach to this problem from its clinical engineer, he can only do it by beginning with the proper foundation, equipment complement, and building his program from that base.

13.5 Cost Control

Budgeting and typical program costs

When the clinical engineer puts together a program for his employer, the major element of that plan must be a budget. Administrative managers need to know what the annual operating costs of a clinical engineering program will be, what accounts the funds are earmarked for, and how much money (capital) will be needed initially to purchase test equipment, office equipment, and parts. Typical elements of an internal clinical engineering budget would be:

1. Salaries (engineers, technicians, clerical).
2. Overtime.
3. Fringe benefits (FICA, FUTA, group insurance, etc.).
4. Space.
5. Tools (less than $100 in per unit cost).
6. Depreciation (office and test equipment).
7. Training costs.
8. Technical documentation.
9. Telephone/pager.
10. Office and operating supplies (forms, postage, pencils, etc.).
11. Equipment repair and calibration.
12. Travel.
13. Dues and subscriptions.
14. Miscellaneous.

If the clinical engineer and his program serve multiple hospitals, then automobile expenses and perhaps liability insurance costs must be added. Vacation, holiday, and sick leave are included in salaries.

Based on 20-odd studies performed by the author in conjunction with the Hospital Council of Southern California, it is generally impossible to make predictions about the type of clinical engineering program a hospital should have based merely on its number of licensed beds. The program will be completely dependent upon the hospital's clinical equipment complement modified by its specialty status if it is a teaching or research institution. In our studies, we found that hospitals would have to spend 0.8 to 1.0 percent of their total expense budget on the average to provide compliant clinical engineering programs. Since California requires more PM and safety in certain areas than JCAH or other states, this range should be lowered for other locales. It should be used only as a guideline and a check for the more accurate studies based on equipment inventory.

Rather than attempting to present target budgets for small, medium, and large sized hospitals and further confusing this issue, Table 13.1 provides target budgets for one-man, four-man, and eight-

Table 13.1. Target Budgets for Various Clinical Engineering Programs (These 1977 costs will probably increase about 9 percent per year because of inflation.)

	One-Man Program	*Four-Man Program*	*Eight-Man Program*
Salaries	$19,000	$72,000	$131,000
Overtime (5% est.)	-	2,700	5,500
Fringe benefits (16% est.)	3,000	11,500	21,000
Space	500	1,000	2,000
Tools	300	800	1,600
Depreciation	1,100	2,000	3,000
Training costs	500	1,500	3,000
Technical documentation	500	1,000	2,000
Telephone/pager	1,000	2,000	3,000
Office and operating supplies	1,000	2,000	3,000
Equipment repair and calibration	300	500	1,000
Travel	300	500	700
Dues and subscriptions	1,000	2,000	3,000
Miscellaneous	500	1,000	2,000
Total	$29,000	$100,500	$181,800
Productive hours	1,250	5,000	10,300
Cost per hour	$23.20	$20.10	$17.65

man programs. The total number of full-time equivalents required for a given program would be determined by the man-loading analysis discussed in Sec. 13.4.

The one-man program would have a clinical engineer or senior BMET plus the part-time use of clerical help (40 percent). Proposed salary for the engineer or senior BMET would be $16,000 annually. The program would normally provide complete engineering services as previously defined in Sec. 13.3 plus equipment electrical safety, PM and calibration, and repair on all electronic and electromechanical clinical equipment. The program would need approximately $8000 worth of test equipment, $800 worth of office and shop equipment, and $3–5000 in parts inventory. Dues and subscriptions would cover subscribing to the more useful periodicals plus belonging to a local shared engineering service for technical support if one is available. We include travel to cover one trip to one professional seminar per year. We estimate productivity to be 65 percent.

The four-man program would consist of a clinical engineer/manager ($18,500), an electronics technician ($15,000), an electromechanical technician ($13,500), an X-ray technician ($17,000), and a clerk/typist ($8000). This program would handle the same items as the one-man program plus all X-ray equipment.

The program would need approximately $13,000 worth of test equipment, $3000 worth of office and shop equipment, and $8–

10,000 in parts. We estimate productivity to decrease to less than 50 percent for the manager and to increase for the technicians to 75 percent.

The eight-man program would be composed of one clinical engineer/manager ($22,000), two electronics technicians ($15,000, $14,000), two electromechanical technicians ($13,500, $12,500), two X-ray technicians ($17,000, $14,000), one laboratory technician ($15,000), and one clerk/typist ($8000). This program would handle all clinical engineering services and electrical safety, PM and calibration, and repairs on all clinical equipment except nuclear medicine and diagnostic ultrasound. The program will need approximately $18,000 worth of test equipment, $6000 worth of office and shop equipment, and $15–17,000 in parts. Productivity will drop further for the manager while the technicians will remain constant at approximately 75 percent.

The three budgets that have been presented would be fairly typical for programs with the number of men indicated. There has been no attempt made to correlate these budgets with hospital bed capacity because clinical engineering programs are solely dependent upon the amount of clinical equipment a hospital possesses and not upon its number of licensed beds. A clinical engineer would develop his program from the equipment inventory by using the methods described previously in Sec. 13.4. After he has developed his organizational structure, he can then draw up his budget, beginning with salaries and fringe benefits. It should be recognized that salaries will have some regional variation with significantly lower rates being paid in Maine and most of the southern states.

A budget is used both as a means to plan future expenditures and as a guideline to measure costs once it has been formally adopted. It can obviously be altered at any time, but most managements will not accept major variations or requests for major budgetary changes unless the changes can be justified in one way or another. Thus, a clinical engineer can not just draw up his financial plan, get it accepted, and then shove it in his top drawer. He must monitor his performance according to his plan, and keep his management aware of his program. He must realize that his budget and those of other departments have been reviewed, pared, rereviewed, possibly pared again, and finally adopted as the organization's financial plan for the upcoming fiscal year. The purpose of the whole budgeting process is to match income with expenditures to ensure that funds will be on hand to cover costs as they are incurred. If some department overspends, another must underspend, or the organization will find itself out attempting to borrow money to make ends meet. The plight of New York City in 1976 is a prime example of what can happen when a budget is totally ignored.

We have pointed out that a clinical engineer must monitor his performance according to his plan. This entails accounting for all operating expenses and comparing them with budgeted costs, making only those capital expenditures that have been approved, and controlling the purchasing of spare parts for inventory. The next six subsections will deal with methods for accomplishing these objectives.

Accounting reports

Every operating department of a hospital or other organization normally receives its own accounting report on a monthly basis which recounts its monthly and year-to-date performance with respect to its budget. The report's format is determined by the format of the budget and usually is arrived at in discussions between the head of the operating department and his counterpart in cost accounting.

It is up to the clinical engineer to see that the clinical engineering program receives its own accounting report, and that the accounts contained in the budget and on the report are sufficiently detailed to provide him with the information necessary for him to manage his function properly. Section 13.5 has presented 14 common accounts that provide the necessary information. However, many hospitals do not allocate space or telephone expenses, and these categories may not appear on the accounting report. Other organizations may lump together salaries, overtime, and fringe benefits into one account. Still others may lump all nonpersonnel-related expenses into one category. Such reports make it difficult to manage his expenditures, so the clinical engineer must stand his ground and insist on a useful report.

A common complaint voiced by clinical engineering programs is that their accounting reports make no sense, and any resemblance between the report and the budget is purely coincidental. This may be true and, in many cases, is true. At that point, the engineer must put on his manager's hat and trudge off to the accounting department to get the report straightened out to his satisfaction. He must continue this process, month-by-month, until he routinely receives correct reports. He has no choice inasmuch as he cannot effectively manage without them.

The accounting report is just one of the tools the engineer uses to provide him with the data he needs to manage his department. It tells him the total amount of dollars he spent and the accounts in which the expenses were incurred. It does not tell him what services he was able to render with that money, nor how well he or his men performed.

The work order/service report

Another tool is the form that is used to record work performed by members of the clinical engineering department. This form has many names and many uses; for our purposes, we will call it the work order or service report and we are primarily concerned with its cost-control value. Figure 13.1 shows a sample of an uncomplicated work order form. Note that the reverse side of the form could be used for documenting electrical safety measurements or other information if so desired.

The work order has two major functions as a cost-control instrument. One is to record the time and materials put in on each task; the other is to relieve inventory by the cost and quantity of materials used.

Monthly management report/clinical engineering

Using the work orders as his source documents, each month the clinical engineer can record the total hours reported by each departmental employee. He can then take the total time paid to each employee, divide it into the total time reported, and determine the productivity of that man. By summing the total number of productive (reported) hours worked by each of his technicians and himself and dividing them into total expenditures reported on the accounting report, the engineer can also determine his cost per hour for the department. Figure 13.2 shows a sample of a form that he can use to present this data in a simple fashion on a monthly basis.

Productivity for the clinical engineer should vary between the limits of 25 and 66 percent, depending upon whether or not he has any technicians reporting to him. In a one-man department, his productivity should approach 66 percent or slightly more. This productivity is equivalent to 1250 hours and is arrived at by taking the number of hours in a working year (52 weeks × 40 hours = 2080 hours), subtracting an allowance for vacation, holidays, and sick leave (200 hours), and then multiplying the computed available hours (1880) by 66 percent. The remaining hours would be consumed by budgeting, accounting, correspondence, daily breaks, phone conversations, supervision of the part-time clerical support, etc. As the department grows, more and more of the engineer's time will be lost to supervision, correspondence, meetings, and the like until he puts in approximately 25 percent of his available time (470 hours) on reportable activities.

Technicians' productivity should fall in the range of 65–66 percent (1250 hours) to 75–80 percent (1350–1400 hours). Usually the technician will have less available time since he has fixed sick leave in excess of the five or less days assumed for the clinical engineer. The clinical engineer should ensure that the work order reporting system that he selects does not allow for the reporting of a full day's work on one sheet. If it does, invariably all employees will report eight productive hours even though such a number is unat-

CLINICAL ENGINEERING WORK ORDER

Date	Location	Model No.	Serial No.	Control No.

Equipment Description	Time Reported	Requestor	Phone Extension

Type of Work Required ☐ Repair ☐ PM ☐ Recall	Next Srv. Date	Maint. Cycle

Description of work
to be performed ______________________________

____

____

____

Description of work performed
(if different than above) ______________________________

____

____

LABOR	Date	Employee Name-No.	Time-Start	Time-Stop	Reg. Hrs.	O/T Hrs.	Remarks

MATERIAL

Stock No.	Qty.	Descrip.	Unit Cost	Total Cost	Stock No.	Qty.	Descrip.	Unit Cost	Total Cost

Figure 13.1 Clinical engineering work order.

MONTHLY MANAGEMENT REPORT

CLINICAL ENGINEERING

For month of______________19___

Employee	Time Paid	Time Reported: Clinical Engineering	Time Reported: Preventive Maintenance	Time Reported: Repairs	Total Time Reported	Productivity

Total Departmental Expenses $____________ Total Material Cost $________

Total Departmental Hours ____________ Current Inventory $ ________

Departmental Cost/Hr. $ ________ /Hr.___ Avg. Response Time______ Hrs

Average Productivity__________ % ___ Mean Time to Repair______ Hrs

Number of Recalls__________

Figure 13.2 Monthly management report—clinical engineering.

tainable. Design the system so that the employee reports his activities by the job and is not forced to account for his unreported time. Only a system of this nature will produce reasonably accurate reported hours.

The Monthly Management Report also contains three other entries. The first is an entry of average productivity for the department for the month. This figure will give the clinical engineer a standard to start with when gauging the performance of any individual employee. If the employee is above that level, he is above average. If he is below that level, his performance is subpar and the engineer should take steps to see what the problem is. Again, the engineer must exercise managerial judgment and look at the data in light of other factors that may affect the employee's performance. Data and figures don't necessarily lie, but they always tell only as much as they can. Unfortunately, that may not always be the whole story.

The second entry, total material cost, provides the clinical engineer with his parts cost. When he adds that cost together with his operating expenses (essentially total labor costs), he arrives at a total cost for his efforts. Note that parts cost is not included in the calculations for cost per hour. Total material cost is obtained by summing the individual material costs recorded on each work order for that month.

Another entry is for current parts inventory value for the department. This is obtained by taking last month's inventory value, adding to it the value of the parts purchased during the month, and subtracting the total material cost. The department should have a dollar goal for its inventory and should strive to keep it as low as possible. A good way to establish such a number would be to stock a two month's supply of every item that turns over fairly rapidly, and one of each item that has a relatively long lead time for procurement (90 days or more) and historically fails every six months.

Parts inventory should be a major concern of the clinical engineer in his managerial role. Since parts may not normally be included as part of the operating budget, but instead charged to the department in which they are used, everyone tends to dismiss them. In addition, parts purchased for inventory are considered to be an asset until they are used. Thus they do not appear as an expense on anyone's budget. It is relatively easy to lose control of these inventoried parts and allow the inventory to grow rapidly. Such a problem was one partial reason for the recent demise of a shared clinical engineering group.

There are many different methods and systems available commercially for maintaining a perpetual inventory. We will not recommend any single method, but any system that the clinical engineer uses should be capable of telling him what he has, where it is, what it cost him, when he last bought it, from whom, how many he has on hand, how many he uses, when and how many he should reorder, and what the price breaks are. Whether on computer, file card, or in a logbook, the inventory should be readjusted for usage at least once a week to avoid stock-outs. The objective of inventory management and control is to keep as low a stock as possible consistent with the downtime needs of the ultimate users.

In Sec. 13.7 we will discuss the final three entries on the Monthly Management Report—average response time, mean time to repair, and number of recalls.

13.6 Personnel Morale and Motivation

When a clinical engineering program progresses beyond the one-man stage, the engineer's most important task becomes that of supervising. He is no longer able to handle the task himself, but must rely on the supportive efforts of others. How well he can motivate his subordinates to provide their best efforts will go a long way toward determining whether his department will be successful or just one of many others that plod through each day.

As anyone who has ever held a job knows, there are both good and bad supervisors. Unfortunately, most subordinates and workers feel that bad supervisors predominate, and they tend to spend a great deal of their day complaining about their lot in life. How can the clinical engineer strive to avoid the "bad supervisor" label? This section will provide some food for thought that will help the engineer in his role as leader of his group.

The work climate

Work should be fun! Does that sound like a startling departure from the norm or a way-out idea? Perhaps it is, but it shouldn't be. Work should be enjoyable for all concerned, from the owner right down through the lowliest employee. In organizations where this is true, output is higher, quality is higher, and everyone is concerned about his or her performance as well as about what the customer receives. Even in organizations that have an unfavorable overall climate, it is possible to have a department that has the right attitude. What's more, that department will stick out like an oasis in the middle of a parched desert. The key is the supervisor and how he directs his department.

So how does the clinical engineer develop such a nice-sounding group? He begins with himself and his own attitude. Does he enjoy going to work and running his department? If he doesn't, his subordinates won't either. If he gripes about his superiors and treats them with disrespect and disdain, his subordinates will do likewise, and they'll add him to the list of bad guys.

Thus the first secret of successful supervision is to provide an enjoyable climate in which to work. The engineer sets the tone with his own attitude. He must be willing to work at being a supervisor if he expects to be a good one and to enjoy the respect of his subordinates, his peers, and his superiors. He must be genuinely interested in his job, in the quality of services he provides to others, in seeing that everyone he serves (no matter how obnoxious) receives 10 percent more than that to which he's entitled. He must be interested in his people and their problems. He must be fair in his dealings with them, conscientious in praising them as well as constructively criticizing them, and firm when his authority is challenged. Above all, he must enjoy doing all these tasks, because if he doesn't, he is either in the wrong position or working for the wrong employer.

None of these skills come naturally although the clinical engineer may have talents that make it easier (or harder) for him to be a supervisor. It takes effort and perseverance and willpower to develop the skills and make them into habits. But once the habits are developed, they become ingrained, and the engineer will find that day-to-day supervision comes much more easily. The succeeding subsections will give additional thoughts on specific ways to supervise personnel.

Theories of supervision

In the early and middle '60s, two theories of managing workers were put forth by McGregor (1960). These were called Theory X and Theory Y. Theory X was the traditional approach to the worker and presumed that people generally dislike work and have to be driven to perform. Theory Y, on the other hand, presumed that people have a need to work and want to have some responsibility and success. In a true sense, Theory X presumes workers to be children that have to be constantly watched and directed. Theory Y presumes everyone to be adult and mature.

Experiences with Theory Y during the '60s and early '70s have indicated that a supervisor has to be careful how he applies this theory in practice. It is true that Theory Y is backed by impressive evidence and makes more sense as a way of treating people. However, not every worker is mature or developing, and the seasoned supervisor or manager must be careful how unstructured he makes the job for each employee.

The secret of managing or supervising people successfully lies in the manager's ability to understand the motivations and desires of each of his subordinates. The manager must know how much responsibility to give each one, how much direction needs to be provided, and how much security the individual needs. Ultimately, the manager is attempting to realize the maximum amount of creative energy in each employee without exhausting the supply or allowing it to go untapped.

Such results do not occur overnight, nor can they necessarily be predicted in advance. By trial and error, patience and diligence, the supervisor can gradually zero in on each employee's needs and capabilities and provide a framework in which the employee can function to the best of his ability and achieve job satisfaction. What the clinical engineer must be careful to do is not to exceed the limits of any employee. Retrenchment from that position is a traumatic and trying experience for the employee.

Let's look at the clinical engineering program and try a practical application of these theories. In the case of technicians who are providing electrical safety, PM, and repairs, Theory X would be a climate in which the technicians' every move would be subject to direction by the engineer. Extremes would include such things as working alongside the technician beyond his training period and telling him precisely how to fix or maintain an instrument, or perhaps handling all communications with other departments and discouraging technicians from any thoughts of their own.

Theory Y at its limit would suggest recruiting an employee, handing him his tools when he starts the job, and telling him what equipment is his responsibility. In this situation, the technician would be left free to determine his own PM procedures, the frequency of PM, the response time to repair calls, the scheduling of his own time, the ordering and stocking of parts, etc. Many technicians are not accustomed to these tasks and wouldn't know how to react without guidance and direction.

The happy medium lies somewhere in between. What is the technician's technical prowess? Can he originate PM procedures on the equipment with which he's familiar? Can he schedule his time to handle his workload, or does he sit around for an hour agonizing over what has priority? Does he know how to troubleshoot logically? Is he aware of the funds that are available for parts purchasing? The engineer's management function is to answer these questions and others like them, size up his technician's growth potential, and train him to make the most of his ability.

We must stress one last concept. The manager or supervisor must be sincerely interested in the welfare of his subordinates. They are not just inanimate tools to be picked up and put down as needed. They are living tools that can create other tools to make a manager's life that much easier.

A common complaint voiced by many workers is that supervisors haven't told them what to do and how to do it. Taking a portion of this complaining with a grain of salt, it is still obvious that this is a major problem with many organizations. Workers are recruited and hired to fill specific roles but no one takes the time to sit down with them after they're on the new job and tells them what is expected. It's presumed that they know because they were hired for their background and experience.

The clinical engineer should take steps to ensure that everyone who reports to him knows what he or she is supposed to do and how the engineer would like it done. This includes other engineers, technicians, clerks, and secretaries. It requires that the engineer sit down with his people on a periodic basis to review what they're doing and how they're doing it. If they're not doing things as they should and don't know how, then it's up to the engineer to train them or to arrange for such training with other sources. Brief reviews ought to be held on a monthly basis while major reviews should be done on a formal basis every six months. These latter reviews are covered in greater detail in a later subsection.

Training of personnel can be an expensive and unrewarding activity, especially if it is on a formal basis with outside sources at a service school. These schools are generally expensive in terms of tuition, travel and living expenses, and technician or engineer time or salary. Unfortunately, the results are not always productive because of the nature of the school. Usually, the person attending has not worked on the instrumentation and thus has no real questions to ask. Instead he sits in a lecture-type situation, is deluged with facts that can't be comprehended at the rate they're presented, and returns to his home base knowing 50 percent of the instrument or system, if that. To top off the situation, his superior can't understand why he's uncomfortable with the unit(s) when he's just received some very expensive training.

How can this situation be combatted? Assuming that the equipment is already handled by the hospital, an alternative exists. This would involve the clinical engineer setting up on-the-job training programs for new employees. It requires that the department live with little or no productivity from inexperienced employees for the first three months of their employment while they learn how the department does its work and why. Essentially, the engineer must assign a new employee either to himself or another qualified person within the department. The twosome will then work together as a team on those instruments or services that the employee will be expected to handle in the future. The experienced employee must act as a teacher during the training timespan and must seek to impart as much hands-on experience to the new employee as he can. Three months is a relatively good target timespan for this initial training; some will take more and some less.

Expanding services to a new technical area is quite another matter. In this case, no one within the organization knows the equipment or service and some outside training becomes necessary. To guard against returning from such a session with only a minimal amount of training, the clinical engineer should sit down with the employee who will attend and draw up a list of objectives that should be met by the session. The employee then knows what he needs to pick up at the session and can insist on more of the instructor's lecture or private time to obtain all that he needs to know.

Although we have been dealing primarily with technical training, we should not neglect training in customer relations, forms, policies, and procedures. Customer relations is an extremely important issue in that fixing the instrument in the clinical engineering field is not necessarily synonymous with pleasing the user. Each employee in the department needs to be taught to respond quickly, to tell the requestor of the service what is happening and why, and to give more than 100 percent. The secret of good customer relations is generally found in good communications with good follow-through. Tell the requestor when you'll be there, then be there. Tell him when you'll be finished and then be finished. Don't try to avoid dealing with the user, and don't try to con him. He already had a problem when he called you; don't give him two.

Completion of forms is another important area that needs some degree of orientation or training. Don't presume that forms are self-explanatory; they're not. Technical people don't like to document what they do, and they'll try to complete only half the form if they can get away with it. Insist on perfection and keep giving the incomplete forms back to the responsible party until they are done properly.

In short, the clinical engineer should not presume knowledge on the part of any employee no matter how skilled or experienced that employee may be. He should instruct each employee in what is and is not expected and then should take pains to ensure that performance is as required.

The employee review

All organizations normally have structured formal reviews on a periodic basis. Usually, the period is one year; sometimes it is every six months. The latter is preferable, and it is recommended that the clinical engineer institute formal reviews every six months in his department if his organization is on an annual basis. Too much can happen in a year's time, especially with bad work practices, so a six-month period is preferred.

Most organizations use a formal review document which must be completed by the engineer and perhaps the employee he is reviewing. The document will deal with the employee's work habits, his ability to get along with others, his productivity, his attitude, his ambition, his responsibility and trustworthiness, his technical ability, his managerial ability, his performance, and his potential. It will also usually include recommendations for wage and salary increases, especially if the cycle is annual. Another reason the six-month cycle is recommended is to avoid the situation in which the employee is given a small or no increase because of poor performance, yet he has never been told his performance is lacking. He now must wait an entire year for another chance at a raise even though his finances may have been disrupted by sharp rises in the cost of living and/or by increases in his home responsibilities. All the wind is taken out of his sails, and his attitude and self-esteem plummet. A good supervisor will not let this situation develop. He will stay on top of his people, will let them know when and how their performance is lacking, and will prepare subordinates if the next raise won't be forthcoming. In other words, the good supervisor reviews his people constantly, and corrects or rewards their behavior whenever the occasion demands.

Professional development

Another aspect of personnel morale and motivation is contained in the concept of professional development. Both clinical engineers and biomedical equipment technicians have voiced their desire for recognition as professionals by the health-care community. Of course, this type of treatment must be earned by the individual involved, but the clinical engineer can help the development. Certification is certainly one means of enhancing status, but there are others that may be just as effective though more unusual. Does the employee have difficulty in verbal communication with others? Why not encourage him to enroll in a course specializing in that field and have the organization underwrite the tuition if successfully completed; or what about high school or college level English courses for the individual who cannot communicate in writing; or perhaps the organization will fund tuition for work towards a degree; or pay for dues to professional societies; or fund trips to seminars of interest or to national meetings that concern the employee's specialty. All of these actions will help the employee to develop both as an individual and professionally, and should be encouraged by the engineer and the organization. They will help to create the attractive work climate that was previously discussed.

13.7 Quality Control and Program Improvement

A major task of any manager is to take continual stock of his operation to determine whether or not it is meeting its performance objectives. A clinical engineer must do the same with his program. This section will discuss ways of monitoring that performance and setting new goals for improvement.

Perception of quality by other departments

Commercial organizations often use marketing research firms to contact the organizations' customers and find out just what the customers' attitudes are towards the organizations' products or services. In a significant number of cases, the organizations find that the customers have attitudes totally at variance with what the organizations thought. This may also be true in the clinical engineer's setting.

The clinical engineer should seek to avoid such problems by having regular contact with the departments or others he services. The contact should be face-to-face, and the engineer should make it plain that he wants straight talk from the user. As pointed out early in this chapter, his sole function is to serve the needs of others. If he is not doing that, he needs to know it, and to know it from the people involved. He is a problem solver and complaints are just another type of problem that require a little more careful treatment.

In dealing with negative comments about himself or his department, the engineer should try to listen only, and refrain from commenting unless the complainer wants him to take some sort of action. It is important to get the other side of the story before attempting to make any judgments or commitments. The customer may always be right, but he sometimes needs a little gentle education to understand his position.

Positive comments should be passed on to the rest of the clinical engineering department, especially if they involve good performance on the part of any particular individual. It is always good for morale and for quality to hear good comments about what you're doing.

Complaints should receive careful study to determine both short- and long-term solutions. Let's take an example that might be typical. A technician in a particular department points to an instrument and states that he called four hours ago for service, but no one has showed up to fix it. He needs it immediately to finish up the tests for the day. This complaint would raise the following questions:

1. Is your normal response time less than four hours?
2. If not, did he indicate that the call was an emergency?

3. Did anyone commit to immediate response?
4. If they did, why wasn't the technician called back?
5. Who received the call?
6. Was it assigned?
7. Is anyone available for immediate response now?

The answers to these questions would point the manager in the right directions for both types of solutions. The most obvious short-term solution is to get someone there to repair the unit, or do it himself if it's simple. The long-term solution may be any one of a number of things. The clinical engineer may need to get a letter out to all his users to tell them his normal response time and what can be expected in an emergency. Then again, his organization may have dropped the ball by fumbling the call. If that happened, he must find out if it's a hole in his system or just human error. Perhaps his groups needs some reeducation on the quality that's expected. No matter what the ultimate solution, the clinical engineer is correctly using his consumer to shape the services of his organization. Their answers will guide him as to how he must improve his program to meet their needs.

Availability

An important facet of any clinical engineering program is its availability. When a one-man department is created as the initial step in a clinical engineering program, quite often that man's availability is set at 8 hours per day, 5 days per week, 50 weeks per year. All other hours are handled through the backup of the manufacturer or other service vendors. As a department grows, it begins to offer a greater availability, usually by being on call. Ultimately it may put a full-time man or men on the second shift to perform PM and cover emergency repairs.

The engineer must determine the proper level of his department to match his customers' needs. He will then adjust its capabilities based on the comments his users make in response to his quality assessment efforts.

Response time

Response time is another measure of the department's performance. For our purposes, it is defined as *the time from call receipt to the time when a service individual arrives on the scene.* Customarily, an outside service vendor promises response within 24 hours. More is expected of the inside group, but how much more? Should the group respond immediately (15 minutes) to all repair calls? Should it stagger its response such that life-support equipment receives immediate response, but suction pumps are not handled until somebody can fit it in? These are the types of questions the engineer must answer in setting his policies and in revising them to improve his services. However, he should recognize that totally instant response carries a higher price tag since it will have a marked impact on productivity. Common sense should be the prevailing guideline for setting response times.

Mean time to repair

Some clinical engineering programs have made it a practice to respond quickly to all calls, pick up the instrument if it's portable, remove it to the shop area, and then take three days to a week or more to get it repaired and returned to the user. The delay may be because parts stock is inadequate, or because the department is short-handed, or for other reasons. No matter what the reason, this type of service would not be considered suitable by most users.

Mean time to repair is thus an important measure of a department's capability on emergency repairs. The clinical engineer must define what he intends to provide; for instance, the department may elect to have a goal of repairing 90 percent of all equipment failures within 24 hours. Whatever the goal, the engineer must be prepared to revamp it to meet the real needs of his users. Downtime on a necessary unit in a rural hospital is much more critical than downtime on the same unit in an urban area where more alternatives exist.

Number of recalls/rerepairs

One measure that can tell a clinical engineer much about his department's performance and problems is the number of recalls or rerepairs. The department should strive for none, but should realize that some will occur, especially when intermittent problems are involved or there is a rash of calls to handle in a short time. Recalls are possible indications of a number of things: poor training, incompetence, sloppy work, rushed work because of workload, or intermittent failures. The latter problem is not really a problem; intermittents will occur and will be difficult to solve. The other four causes are indicative of deficiencies that the clinical engineer must correct by management action.

Poor training is obviously correctable by better training. Here the manager's task is to determine what change in training will get the job done properly. Incompetence can only be corrected by termination. The engineer's difficult task here is to decide what is incompetence and therefore not correctable, and what is poor training and therefore correctable.

Sloppy work may also be grounds for eventual termination.

The clinical engineer must seek the cause for the poor work and attempt to correct it through sessions with the individual involved. If the problem cannot be solved, termination will have to be considered.

More often than not, an increasing level of recalls is a sign that the group does not have enough man-hours available to cope with the workload that it has. When this occurs, members of the group or department start to rush their work and take less pains in checking an instrument so they can get to the next call they have to make. They then wind up fixing one or two problems and leaving another. The solution to this cause is to add additional manpower if the problem persists or to purchase more outside help for a time if the problem is felt to be temporary.

Setting goals and objectives

In planning or revising his services, the clinical engineer must set goals. In the cases of electrical safety testing, PM and calibration, and engineering services, the goals will be stated in terms of such things as frequency and content of the service. Criticisms of these services would normally be that they are either not getting done, or what's being done is inadequate.

The same cannot be said of repairs. Here the program's services are highly visible and much is expected. The user has a definite problem, and is looking for a reasonably quick solution. He wants his unit fixed, he wants it done in a reasonable length of time, and he doesn't want it to fail again in the near future. The engineer must respond to these needs by specifying availability, response time, and mean time to repair, and he must be prepared to revise his specifications if they prove unacceptable. Each user will have his own needs, and the program will have to be structured to meet them within the framework of the financial resources that are available.

Implementing goals and objectives

Implementation of goals and objectives is fairly straightforward. The clinical engineer's specifications are used in conjunction with the equipment inventory to determine staffing levels for the department. Staffing levels are used to draw up the budget. Once it is approved, it is then used as the guideline for departmental action.

The clinical engineer needs to develop some means to measure the performance specification he has set. To a degree, this is easy in that his staffing level should determine whether or not he can supply the coverage he has specified. Response time, mean time to repair, and number of recalls can all be determined from the work order and reported on the Monthly Management Report (Sec. 13.5).

This work can be handled by the clerk/typist when the monthly report is issued.

13.8 Compliance with Codes

Chapters 4 and 9 have described in some detail the various types of codes and regulatory bodies that have an effect on the clinical engineering function. Without repeating here the information contained in those chapters, we will briefly explore the concept of codes from the standpoint of the clinical engineer's management role.

Although there are many regulations that affect the hospital's clinical engineering program, the clinical engineer will find that he is primarily concerned with three: those of the JCAH, those of the state in which he functions (such as California's Title 22), and those of the NFPA. The latter are important because they are incorporated into so many other federal, state, and local codes. These three sets of regulations are important because they impact the clinical engineer's day-to-day activities. They specify the content and frequency of his PM program, they touch on his documentation, they tell him how and when to conduct electrical safety testing, they tell him to tag defective devices, they tell him to perform incoming inspection, they tell him how often to calibrate his test equipment, they specify where manuals should be kept, and they tell him how some medical departments should be constructed and operated. There is some overlap between standards and occasionally a conflict will exist between contradictory sections. The engineer must then use his common sense and good judgment to handle the problem.

Not all of the regulations have the force of the law, and thus the clinical engineer has some leeway in his actions. For instance, the standards of the JCAH are considered to be optimum achievable standards, a goal for which the hospital strives. Thus the surveyors of JCAH will hold the hospital responsible for meeting other regulations which are law, but will stop short of refusing accreditation if the hospital meets the basic laws, but does not come up to the level of the standards of JCAH.

The engineer's task in the confusing regulatory maze is to identify those rules with which his employer must comply, and separate out those rules that are not yet mandatory. He can then apply prudent common sense to the latter to come up with a program that achieves eventual compliance at a pace the employer can afford.

In some cases, the clinical engineer will identify a regulation or standard that does not make any sense and involves significant expenditures with no tangible benefits. When these situations occur, the engineer should document his findings and report to his superiors in an effort to marshal forces to overturn the regulation. This has been done quite successfully in the recent past with the NFPA, JCAH, and some states, and is of great interest to the American Society for Hospital Engineering, the technical arm of the American Hospital Association. The clinical engineer should become involved in the standard-setting process as much as time permits, and provide a practical input to the standard setters.

13.9 Liability Exposure

A recurring theme at national meetings and seminars throughout the country is the concern of clinical engineers about their liability exposure when an incident occurs that may involve them, their department, or program. This concern is legitimate, especially in light of malpractice suits in the health-care field. Section 14.3 will consider the risk and responsibility of the clinical engineer at some length. This chapter will deal briefly with the practical actions the engineer has at his disposal.

Is personal malpractice liability insurance a practical alternative for the clinical engineer? In today's insurance market and insurance climate, the answer has to be "No." The clinical engineer would normally be covered under the liability insurance policy or policies written by his employer. Although he could be sued individually for personal negligence in a given situation, the cost of insurance to cover this contingency runs to thousands of dollars if the coverage is complete. This is just not economically practical for the average clinical engineer with today's salary ranges. However, it may be a necessity for the clinical engineer in private practice, that is, if he can obtain the insurance at all.

The clinical engineer's answer to the malpractice-suit craze lies in careful attention to his operation. He must ensure that his program is thoroughly researched, well-planned, carefully implemented, legally compliant, and above all, well-documented. Nothing serves as well to remove a clinical engineer and his employer from a suit than documentation that graphically depicts the history of the department. It's when the engineer cannot reconstruct his department's actions that he and his employer are in trouble. The guideline is thus to be thorough and complete.

13.10 Input to Administration

Earlier chapters have discussed the clinical engineer's interfaces with other hospital departments and with administration. Here we will consider the management aspects of the relationship with administration. The discussion will center on the clinical engineer's obligation to keep his superiors informed about his area of responsibility.

Inputs on equipment acquisition

The engineer's primary responsibilities in equipment acquisition are to see that the organization does not buy unsafe equipment and that it does not purchase a Cadillac where a motorcycle would do. He should also guard the hospital from purchasing unreliable equipment, or equipment that will not fit or cannot be operated in the particular environment. The clinical engineer must learn to couch his criticisms in a tactful and diplomatic fashion, but he must ensure that they are heard. He must also ensure that his inputs are made whether solicited or not.

Inputs on compliance with codes

Administration needs to know what codes exist, whether or not the hospital is complying with the codes, what it will cost to become compliant, and how much time is involved. The clinical engineer is the one to provide these inputs and should do so as new codes take effect. Again, in his role as a manager, he should sense the needs of administration and provide the inputs without being told to do so.

Other inputs to administration

The previous two subsections have given examples of just two areas in which the clinical engineer must exercise his initiative as a manager and keep administration aware of what is happening with the clinical engineering program and its sphere of activity. Comparable inputs also need to be made on such topics as:

1. Electrical safety.
2. In-service training.
3. Equipment control.
4. Liability exposure considerations.
5. Maintenance and repair costs.
6. Departmental performance.
7. Departmental objectives.
8. Departmental growth.

If a monthly report format exists, the engineer should consider including brief and concise remarks about those areas in which changes have occurred in the last month. If there is no such thing as a monthly report, the engineer should consider starting one so that he has a formal mechanism for alerting his superiors to his problems and accomplishments. He should also consider special project-type reports for any changes of a major nature, and be careful to make recommendations for action rather than tossing the decision ball to his boss. The engineer has the technical expertise and is the one most qualified to suggest a solution. If funds are not available, he can then work with administration to establish alternatives.

13.11 Summary

We have seen that management and supervision are two very important aspects of the clinical engineer's job function. Although these responsibilities include management of outside vendors, we have focused on the internal managerial roles that a clinical engineer may have to play.

Our purpose has been to direct the engineer's attention beyond the technical side of his work. We have sought to show why planning is needed, and how budgets can be developed and controlled. We have offered some tips on how to motivate and supervise personnel, and talked about means of measuring people and departmental performance. Hopefully, we have provided the engineer with some simple and easy guidelines for running the department.

We want to stress two thoughts again in closing. One is that management and supervision require labor to make them work, but there is great satisfaction in having a well-run, cost-effective, problem-solving clinical engineering group. The second is in your relationship with your superior. If you can make his job easier, you've gone a long way toward becoming an effective manager.

13.12 Study Questions

13.1 Distinguish between the terms *management* and *supervision*.

13.2 List four major reasons for proper direction of the clinical engineering function.

13.3 Why must the clinical engineer determine the costs of his program?

13.4 Give two examples that show that the clinical engineer depends upon serving the needs of others.

13.5 List three ways that the clinical engineer can manage the hospital's liability exposure.

13.6 List three subjects on which the clinical engineer should initiate advice to administration.

13.7 List three examples of measures of service that are used to define the concept of *identical service*.

13.8 Name the two hospital departments that have the highest preventive maintenance and repair costs.

13.9 List three basic questions to be asked in a make-or-buy analysis of a new clinical engineering program.

13.10 List three methods of comparing costs in a make-or-buy analysis of a new clinical engineering program.

13.11 List two methods used by the VA for estimating clinical engineering staff needs.

13.12 List six elements of an internal clinical engineering budget.

13.13 What information is and is not found on an accounting report?

13.14 List four items found on a clinical engineering work order.

13.15 How is the monthly management report used to control inventory value?

13.16 Differentiate between two opposing theories of managing workers.

13.17 What are two methods of training of personnel?

13.18 How often should each employee have a structured formal review? Why?

13.19 List two methods of promoting the professional development of employees.

13.20 List four measures of the performance of the clinical engineering department as perceived by the customers.

13.21 What three bodies set regulations that most concern the clinical engineer?

13.22 Should a clinical engineer obtain personal malpractice insurance? Why?

13.23 List five topics that the clinical engineer should cover in his monthly report to management.

13.13 References

Ben-Zvi, S. (1975). Preventive maintenance program, final report, Publication PM-5. Chicago: American Society for Hospital Engineering.

Brinkman, M. O. (1973). What to do when the equipment warranty runs out. *Modern Hospital.* 121 (December):74-77.

Brinkman, M. O., and R. H. Drue, (1977). Biomedical maintenance—What's it worth to you? *Hospital Topics* March/April.

Drucker, P. F. (1973). Management: tasks, responsibilities, practices. New York: Harper and Row.

Drucker, P. F. (1964). Managing for results. New York: Harper and Row.

Drue, R. H. (1973). Equipment maintenance: How to tell whether you need an in-house staff, an outside contractor or both. *Modern Hospital* 121 (December):70-73.

Laird, D. A., and C. Eleanor (1957). The techniques of delegating. New York: McGraw-Hill.

Marting, E. and H. F. Merrill, (1958). Developing executive skills. New York: American Management Association.

Maslow, A. H. (1965). Eupsychian management. Homewood, IL: Richard D. Irwin.

McClelland, D. C. (1969). Motivating economic achievement. New York: Free Press.

McGregor, D. (1960). The human side of enterprise. New York: McGraw-Hill.

Ridgway, M. G. (1976). Designing (and implementing) a complete biomedical engineering support program for your hospital. Presentation at AHA Institute. October.

Topham, W. S. (1976). The cost of a clinical engineering department. *J. Clin. Eng.* 1:24-26.

14
Legal, Professional, and Ethical Aspects

RICHARD J. JENDRUCKO and RICHARD E. FAVORITI

14.1 Learning Objectives

Upon completing this chapter you should be able to:

1. Explain the legal basis for clinical engineering liability.
2. Describe the processes of certification and registration.
3. List the major professional and technical societies involved in clinical engineering.
4. Describe the criteria for making ethical decisions.

14.2 Introduction

In the course of the past ten years, the issue of product liability has reached crisis proportions. Those who engage in the design, development, manufacture, marketing, or distribution of a product now find themselves the focus of attention in a consumer-oriented climate directed toward increased legal exposure. For the clinical engineer (like most professionals) this exposure is a real one, whether he is an independent contractor, consultant, or employee.

The profession of clinical engineering stands out as centrally focused on the overall mandate of engineering, service to mankind. As a professional, a clinical engineer is expected to act in a responsible manner at all times, to work for the advancement of the profession and to maintain personal competence through registration, certification, membership in technical and professional societies and regular participation in technical meetings.

Clinical engineers in their various job functions are called upon to make decisions on courses of action, the results of which often directly affect the quality of life for other individuals. Associated with the authority for making such decisions is the ethical responsibility for making the best or most correct decision given a set of circumstances at hand.

14.3 Legal Issues in Clinical Engineering Practice: Defectiveness in General

Product liability begins with an item, commonly referred to as a "product." The product is allegedly defective because it failed to meet the reasonable expectations of a consumer or user and, as a result, was the proximate cause of an injury to that person. The term *product* normally implies the existence of an article or device. However, almost anything—a building, health-care facility, procedure or system—may fall within the concept of a *product* and, hence, within the rapidly expanding area of liability.

A defect may show up in various ways. The two more prominent theories advanced are defective design and failure to warn.

Liability will generally be assessed against those individuals or entities responsible for a product considered to be defective because of its design and/or development. This could include the specifications or blueprints for the product, its composition, and the process used to manufacture, test, or inspect it. The engineer is held to a standard of designing or developing a product that is safe for its intended use. He should take into consideration those uses that, although not intended, may be reasonably anticipated. Many courts will examine whether or not a better and/or safer design is feasible. Even if an individual designs or develops a product in accordance with standards accepted by the industry as a whole, he may not avoid liability if that design or development creates an unreasonable risk of harm.

In recent years, lawsuits based on design defects have increased. When the design of a product is found to be defective, an engineer's liability may be more extensive than where an isolated occurrence produces a defect to a single item. A design defect may affect every product manufactured on the basis of the same design and, if serious enough, may necessitate a recall or some other emergency procedure.

The second general area of increased exposure in products liability is the failure to give adequate warning. In other words, this failure to give adequate warning constitutes a defect. Like the design cases, it is important to consider the likelihood of an accident and the degree of the danger created when the product is put to a foreseeable use in the absence of a warning. In addition, we should also consider the feasibility of adopting a more effective warning.

The first issue arising under the duty to warn is whether or not there is such a duty, considering the product involved and the danger anticipated. It is generally agreed that there is no duty to warn of a danger that is obvious to the user. Similarly, it is generally accepted that there is no duty to warn against the abnormal use of a product. Nevertheless, if the obvious danger or the unusual use is foreseeable, there is the strong possibility that a court might conclude that there is a duty to warn. This would depend upon the product's utility; the severity of the danger; the economic ease in providing a warning; and the reduction in the danger with the warning having been made.

One of the more difficult problems under this first issue is the duty to warn about dangers that cannot be determined prior to use. It is difficult to envision a warning that could protect a user or consumer from a danger that is not known. Courts nevertheless appear to have extended this responsibility to those instances where scientific knowledge has developed to such a point whereby there is a suspicion of danger. In those cases a court may require that a warning be provided in terms as specific as possible under the prevailing circumstances. This depends upon the severity of the danger, the value of the product to society, and the practicability of an effective warning.

If a court concludes that the duty to warn exists, the next issue is to decide the extent to which that warning must be communicated. Courts generally consider that the duty to warn extends to those persons reasonably expected to use the product. Whether or not this means that the warning must be transmitted to every potential user depends upon the danger being guarded against and the relevant ease or difficulty in accomplishing the task. In those situations where the product's danger is not extreme and it is reasonable to expect that the immediate user will communicate the warning to other probable users, some courts have relaxed the duty to warn by not requiring that every potential user be put on notice.

In addition to considering whether there is a duty to warn and to whom the duty is extended, the engineer must also consider the physical adequacy of the warning, its prominence, clarity, and completeness, as well as the method of transmission. Courts have regularly held that a manufacturer is bound to utilize the most effective method in communicating a warning. Compliance with statutory, regulatory, and/or industry-wide standards, will not, in themselves, necessarily avoid liability.

Even in those situations where a warning has been given and it is deemed adequate, we must remember that the duty to warn is a continuing one. As technology advances, new discoveries are made, and new dangers uncovered, most courts hold that there is a continuing duty to alert users to these discoveries.

Legal theories

Given a defect, whether it be in the design, failure to warn, production, marketing, or retailing aspects of the product, the task then becomes one of stating the defect in terms of a legal theory recognized in a court of law. In general, products liability acknowledges three acceptable theories: (1) negligence; (2) breach of warranty; and the most current, (3) strict liability.

The injured party has the burden of establishing his claim through one or more of the above theories. Although he could conceivably pursue a recovery under all three theories, by far the simplest and most popular concept in terms of procedure and burden of proof is strict liability. Breach of warranty is more contractual in nature and, because of its burdensome requirements and pitfalls, it is the least chosen of the three remedies.

In terms of negligence, courts have stated that an individual is held to the standard of exercising due care in the design, manufacture, distribution, and marketing of a product. If the failure is material, the person responsible will be held liable to any individual who is foreseeably injured as a proximate result of that negligent conduct. Due care for an engineer includes such matters as a careful evaluation of the concept or plan to be implemented, materials to be used, process of manufacturing, testing, inspecting, and the need for safety measures. Where there is an anticipated danger that is unavoidable, the engineer may have a duty to warn and/or provide direction. The fact that the user has the opportunity to inspect and/or test for defects will not necessarily relieve the engineer of his duty to exercise reasonable care.

The concept of negligence is one of having to prove the failure to exercise due care, i.e., to prove fault on the part of the prospective defendent. Depending upon the circumstances and the product involved, this may be a very difficult task. This burden has been one of the principal motivations for the development of strict liability. In order to recover in a strict product liability action in terms of a defect, the injured party must prove that: (1) the damage resulted from a condition of the product; (2) the condition was an unreasonably dangerous one; and (3) the unreasonably dangerous condition of the product existed at the time it left the defendant's control. Unlike negligence, there is no requirement that fault be established in order to justify recovery under strict liability.

Under both negligence and strict liability an individual may also be answerable in damages when he misrepresents a material fact regarding the character or quality of the product sold.

Negligence and strict liability are tort concepts under product liability. A tort is a legal wrong committed upon a person or property independent of any contract. The third concept, breach of warranty, is provided by the Uniform Commercial Code and is contractural in nature. Warranties are either implied or expressed. Each type has its own specific requirements and characteristics. Oftentimes it is difficult and awkward to establish these requirements in situations other than the commercial contractual setting. Thus the trend in product liability has been to gradually move away from breach of warranty as a viable approach to recovery.

Implied warranties arise by operation of law, separate from any provision in the contract, and are one of two types: (1) implied warranty of merchantability; or (2) implied warranty of fitness for a particular purpose.

The implied warranty of merchantability is based upon the justified expectation of the average consumer and simply means that a merchant warrants that the product he sells is fit for its ordinary purpose. In contrast, the implied warranty of fitness for a particular purpose contemplates a situation where the seller knows of a particular or special purpose for which the product is required and the buyer relies upon the seller's skill or judgment in selecting that product.

An express warranty does not arise by operation of law but rather is created by the parties. It is an affirmation of fact or promise, description, sample or model of the product, which becomes a part of the basis for the transaction. In such a situation, the seller is said to warrant that the goods conform to those representations.

Legal defenses

Separate and apart from attacking the basic elements of proof required by the injured party, such as establishing the defect, e.g., the duty to warn, or the need for a safety device, etc., the amount of recovery may or may not be affected by the injured party's conduct at the time of the occurrence and how that conduct is legally characterized.

Contributory negligence has been defined as the injured party's failure to exercise due care to discover the defect or guard against its possible existence. The result of this failure is the injury for which the plaintiff now seeks recovery. Although contributory negligence is a complete bar to recovery in a product liability case based upon negligence, courts generally consider that such conduct will not bar recovery in a strict liability action.

In contrast, assumption of risk has been stated to exist where the plaintiff knows of the defect creating the danger; realizes the full extent of the danger involved; and, with that awareness in mind, nevertheless voluntarily and unreasonably proceeds to encounter the danger. If established, assumption of risk will be a complete bar to recovery in both negligence and strict liability actions.

Courts are divided as to whether the foregoing tort defenses are available in cases where breach of warranty is pursued. Nevertheless, there are certain contractual defenses available, such as the exclusion or modification of warranties or the failure to give notice of a breach of warranty within a reasonable time after discovery. If proven, each may bar recovery or limit the amount of recovery to the injured party. Despite their apparent existence, courts have restricted their applicability and generally, have developed approaches that have seriously diluted their effect.

Further reading

The foregoing analysis presents a surface treatment of the subject of product liability. For a review of the subject and some of the issues raised refer to the following pertinent articles: Hoenig, "Product Designs and Strict Tort Liability: Is There a Better Approach?" *Southwestern University Law Review*, 8:109–137 (1976); Twerski, et al, "The Use and Abuse of Warnings in Products Liability —Design Defect Litigation Comes of Age," *Cornell Law Review*, 61: 495–558 (1976); Vetri, "Products Liability: The Developing Framework for Analysis," *Oregon Law Review*, 54:293–314 (1975); Fischer, "Products Liability — The Meaning of Defect," *Missouri Law Review*, 39:339–362 (1974); Wade, "On the Nature of Strict Tort Liability for Products," Mississippi Law Journal 44:825–851 (1973); Keeton, "Product Liability and the Meaning of Defect," St. Mary's Law Journal 5:30–39 (1973).

14.4 Professionalism

Engineering, as a professional discipline is focused on service to mankind. As a "learned profession," engineering implies a set of required characteristics of practitioners (Beakley and Leach, 1972):

1. Knowledge and skill in specialized fields above that of the general public.
2. A desire for public service and a willingness to share discoveries for the benefit of others.
3. Exercise of discretion and judgment.
4. Establishment of a relation of confidence between the engineer and client of the engineer and employer.
5. Acceptance of overall and specific codes of conduct.
6. Formation of professional groups and participation in advancing professional ideals and knowledge.
7. Recognition by law as an identifiable body of knowledge.

Beyond motivation to accept the above principles based on a sense of professionalism and responsibility, there are in many work environments, requirements for formal registration as a professional engineer. In this section we consider the background, objectives, and procedures for professional engineering registration.

History, objectives, and necessity of professional engineering registration

Professional registration procedures in the United States have been established for all professional groups including medicine, dentistry, nursing, law, architecture, accounting, and engineering. Beginning with the adoption of the first engineering registration law by Wyoming in 1907, all fifty states, the District of Columbia, Puerto Rico, Canal Zone, and US possessions now provide the mechanism for licensing to practice professional engineering. Professional registration laws have been developed and are enforced to protect the public from possible consequence of contracting services from incompetent, nonprofessional practitioners. Engineering differs somewhat from other professions in that all practicing engineers are not licensed. This results partly from the fact that there are a number of exemptions specified in the various state laws (Constance, 1963). In addition, individual engineers are often employed by organizations wherein they function in groups, frequently at a professional level quite removed from any direct contact with the public that consumes the products of their services. In such cases, the requirement for registration is not always clear. As a general guideline, however, whether or not an individual engineer must or should become professionally licensed depends on:

1. Laws of the state in which the engineer is practicing.
2. Regulations and requirements of any organization by which he or she is employed.
3. The extent to which registration aids one personally in his or her practice by public conformation of competence and adherence to a professional code of conduct.

Many convincing arguments have been made for the necessity of professional registration for engineers, even when this is not explicitly required by an employer (Ellis, 1972; Fine, 1972). In the specific case of clinical engineers, with the frequent personal responsibility to individual hospital or clinic patients, registration is highly desirable to support claims of competency should any legal issues arise in which the latter is questioned. Jarzembski and Rowley (1976) surveyed all the state boards of engineering registration and determined that registration is necessary for the practice of clinical engineering.

Registration requirements

Specific prerequisite requirements for registration as a professional engineer depend on the laws of the state or territory in which the engineer will be practicing. Table 14.1 provides a summary and comparison of the requirements and procedures for registration in the various states. Generally, there are two stages in the registrations process:

1. Certification as an Engineer-In-Training (EIT).
2. Registration.

Most states have formalized, through law, programs for certification of new engineers as "engineers-in-training." (Exceptions are noted in Table 14.1.) The chief objectives of the EIT Program may be summarized as follows (Constance, 1958):

1. To make the young engineer more conscious of his or her obligation as a member of the engineering profession and to acquaint him with the procedure, requirements, and long-range advantages of engineers' registration.
2. To permit him or her to take the first step toward registration by passing a written examination on theory and engineering fundamentals while the subject matter is still fresh in mind.

Table 14.1 Minimum requirements for engineering registration

This table, current as of January 1973, was developed from data furnished by the Examining Boards of the states and territories. Each Board retains the right to change these conditions at any time.

A.
- (1) Prerequisites are for professional engineering. Requirements for land surveying, in general, are the same but do vary in some states.
- (2) Valid EIT status, 10 year limit.
- (3) Land Surveyors work experience time is three years less than professional engineers.

B.
- (1) ECPD school (Accredited by Engineers' Council for Professional Development)
- (2) Approved school (Acceptable to Board)
- (3) Board approved school of science and engineering
- (4) 4 year engineering curriculum
- (5) Degree
- (6) High school

C.
- (1) 6 hour examination
- (2) 8 hour examination
- (3) 12 hour examination
- (4) 13 hour examination
- (5) 14 hour examination
- (6) 16 hour examination
- (7) Examination as required by Board (Board may waive portions of examination
- (8) Oral or written examination (based on application)
- (9) Personal interview

D.
- (1) 2 years experience
- (2) 3 years experience
- (3) 4 years experience
- (4) 6 years experience
- (5) 8 years experience
- (6) 10 years experience
- (7) 12 years experience
- (8) 13 years experience
- (9) 15 years experience
- (10) 9 years experience

E.
- (1) Yes, if in responsible charge of engineering teaching.
  - (1.1) If rank of assistant professor or higher.
- (2) Yes, year for year
- (3) Yes, 1 year
- (4) 2 years maximum
- (5) Yes, up to 3 years
- (6) Yes, up to 5 years
- (7) Yes, evaluated by Board
- (8) Yes, same as other experience
- (9) Yes, if of an engineering nature
- (10) Only engineering experience
- (11) No, unless planning and design involved
- (12) Experience must be documented

F. All written examinations unless otherwise noted
- (1) 4 hour sessions (see qualifications)
- (2) Two 4 hour Fundamentals; Two 4 hour PE (branch)
- (3) 8 hour Fundamentals and 8 hour Principles & Practice
- (4) Three parts, 3 days
- (5) 8 hour land surveying
- (6) Oral examination, no time limit
- (7) 4 hour basic, 4 hour general, 4 hour PE
- (8) 3 hour basic, 3 hour specific
- (9) According to qualification requirement
- (10) Fundamental 8 hour general (closed book) Principles & Practice 8 hour (open book)
- (11) Two 4 hour sessions on engineering fundamentals; 4 hours application of engineering principles; 1 hour misc. and 3 hours design problems in specialization
- (12) Treatise or course on arctic engineering

G.
- (1) 70 average all parts, 60 minimum any part
- (2) Based on examination load
- (3) Pass or fail
- (4) 70 full field, 75 limited field
- (5) 75 average all parts, 60 minimum any part
- (6) 75 on each part
- (7) 70 on each 8 hour part

H.
- (1) Supporting but not in itself qualifying (independent determination made in all cases)
- (2) Limited (accepted as evidence of qualification)
- (3) Only for verification of record
- (4) If registration is by written examination
- (5) When registration conforms to State (territory) law
- (6) Yes, provided it is up-to-date

I.
- (1) Not permitted under State Law
- (2) Rarely, must have national recognition
- (3) Not eminence but under exemption
- (4) Experience must be of high quality
- (5) Only on basis of outstanding professional accomplishment
- (6) 15 years lawful practice
- (7) Recognized standing, 12 years experience
- (8) 12 years progressive experience
- (9) 15 years progressive experience
- (10) 20 years progressive experience
- (11) 24 years progressive experience
- (12) Age 35, 12 years experience, 5 years responsible charge
- (13) Age 35, 15 years experience
- (14) Age 35, 16 years experience, 4 of which may be ECPD school
- (15) Age 40, 15 years experience
- (16) Age 40, 20 years experience
- (17) Age 45, 12 years responsible charge
- (18) Age 45, 15 years experience
- (19) Age 45, 20 years experience, 10 years responsible charge of outstanding engineering work
- (20) Age 50, 20 years experience
- (21) Age 50, 25 years experience, 15 years responsible charge
- (22) At least 10 years in responsible charge of important engineering work
- (23) Age 40, ECPD Graduate, 8 hour examination
- (24) Registered over 30 years and 15 years responsible charge

J.
- (1) Comity, not reciprocity
- (2) Recognition, not reciprocity
- (3) Recognize all states having equivalent requirements
- (4) On equal basis (States which have agreement)
- (5) If first registration based on qualification by written examination
- (6) Must meet requirement of State (territory) Law
- (7) 45 minute to 3 hour examination
- (8) Fundamental part of examination may be waived
- (9) Evaluation by Board
- (10) Eligibility for N. Y. exam and already passed full equivalent

K.
- (1) Examination fee of $15.00 (for each examination)
- (2) Includes $5.00 application fee and $5.00 Certificate fee
- (3) Student (B. S. Candidate) $8.00; others $20.00

Renewal fee: *Biennial
\# up to $10.00 allowed

NEC - Committee on National Engineering Certification

Compiled by: NATIONAL COUNCIL OF ENGINEERING EXAMINERS
Uniform Laws and Procedures Committee

Table 14.1 Continued.

MINIMUM REQUIREMENTS FOR ENGINEERING REGISTRATION

STATE	PROFESSIONS REGISTERED	E.I.T. QUALIFICATIONS	PROFESSIONAL ENGINEER QUALIFICATIONS	CREDIT FOR EXPERIENCE TEACHING	MILITARY	CONTRACTING	LENGTH & TYPE ENGRS. EXAM.	PASSING GRADE	NEC RECOGNITION	POLICY ON EMINENCE	RECIPROCITY PRACTICES	REGISTRATION FEES PROF. ENGR. APPL.	CERT.	E. I. T. APPL.	CERT.	RECIPROCITY	RENEWAL
ALA.	ENGINEERS & L. S.	B(1)+C(2) or D(3)+ C(2)	B(1)+D(3)+C(6) or D(5)+C(6)	E(1)+ E(1.1)	E(9)	No	F(3)	G(7)	Yes	I(1)	J(6)	15.00	10.00	10.00	-----	25.00	10.00
ALAS.	ARCHITECTS, ENGINEERS & L. S.	N/A	B(1)+D(3)+C(6) or D(5)+C(6) or C(2)+ I(15)+F(12)	E(1)	E(9)	No	F(3)	G(7)	Yes	I(1)	J(3)+F(12)	40.00	10.00	40.00	10.00	25.00+10.00	15.00
ARIZ.	ENGINEERS, L. S., ETC.	B(1)+C(2) or D(3)+ C(2)	B(1)+D(3)+C(2) or D(5)+C(6) or D(5)+ C(8)	E(6)	E(9)	E(11)	F(3)	G(7)	H(3)	I(11)+C(8)	J(3)	15.00	10.00	10.00	-----	50.00	15.00
ARK.	ENGINEERS, L. S.	B(1)+C(2) or D(3)+ C(2)	B(1)+D(3)+C(6) or D(5)+C(6)	E(8)	E(10)	E(11)	F(3)	G(7)	Yes	I(5)	J(6)	30.00	15.00	10.00	-----	45.00	P.E. 10.00 EIT 4.00
CAL.	ENGINEERS, L. S.	C(2)	B(1)+D(1)+A(2)+C(2) or B(4)+D(3)+A(2) +C(2) or D(4)+A(2)+C(2)	E(3)	E(10)	E(10)	F(2)	G(7)	Yes		J(1)+J(5)	60.00	-----	40.00	-----	60.00	20.00
C. Z.	ENGINEERS	B(3)+C(2)	B(2)+D(3)+C(7) or B(6)+D(7)+C(6)		E(10)	E(10)	F(2)	G(6)	Yes	I(19)+(5)	J(3)	25.00	-----	10.00	-----	25.00	5.00
COLO.	ENGINEERS, L. S.	B(3)+C(2) or D(3)+ C(2)	B(3)+C(6)+D(3) or D(5)+C(6) or D(9)+ C(2)	E(1)	E(9)	E(7)	F(3)	G(7)	Yes	I(1)	J(6)	30.00	10.00	10.00	-----	25.00	6.00
CONN.	ENGINEERS, L. S.	B(1)+C(2) or C(2)+ D(4)	B(1)+D(3)+C(6) or D(5)+I(23) or D(6) +C(6) or I(20)+C(9)	E(7)	E(7)	E(7)	F(3)	G(6)	H(3)	I(20)	J(6)	50.00	-----	25.00	-----	50.00	35.00
DELA.	ENGINEERS, L. S.	B(1)+C(2)	B(1)+D(3)+C(6) or B(3)+D(5)+C(6) or I(16)	E(7)	E(7)	E(10)	F(3)	G(7)	Yes	I(16)	J(1)	30.00	-----	10.00	-----	30.00	6.00
D. C.	ENGINEERS	B(2)+C(2) or D(5)+ C(2)	B(2)+D(3)+C(6) or D(7)+C(6) or D(7)+ C(8)	E(7)	E(7)	E(7)	F(3)	G(3)	Yes	I(7)	J(6)	40.00	10.00	15.00	5.00	50.00	7.00
FLA.	ENGINEERS, L. S.	B(1)+C(2) or B(3)+ D(1)+C(2)	B(1)+D(3)+C(6) or D(6)+C(6)	E(2)	E(9)	E(10)	F(3)	G(7)	H(3)	No	J(1)+J(6)&(9)	35.00	-----	10.00	-----	35.00	15.00
GA.	ENGINEERS, L. S.	B(1)+C(2) or D(5)+ C(2)	B(1)+D(3)+C(6) or D(7)+C(6) or B(1)+ D(9)+C(2) or D(9)+C(2)	E(2)	E(9)	E(10)	F(3)	G(7)	H(2)	I(1)	J(3)	15.00	-----	-----	5.00	15.00	10.00
GUAM	ENGINEERS, L. S.,ARCH.	B(3)+C(2) or D(3)+ C(2)	B(3)+D(3)+C(6) or D(5)+C(6)	E(7)	E(10)	E(10)	F(3)	G(7)	Yes	I(10)+(22) +C(8)	J(6)	25.00	-----	10.00	-----	25.00	5.00
HAW.	ENGINEERS, L. S.,ARCH.	B(4)+C(2) or D(10) +C(2)	B(4)+D(2)+C(6) or D(7)+C(6)	E(3)	E(9)	E(11)	F(2)	G(7)	Yes	I(1)	J(3)+C(9)	30.00	15.00	5.00	-----	45.00	15.00
IDAHO	ENGINEERS, L. S.	B(4)+C(2) or D(3)+ C(2)	B(1)+D(3)+C(6) or D(5)+C(6)	Yes	E(9)	E(9)	F(3)	G(7)	H(1)	No	J(3)+J(6)	40.00	10.00	10.00	5.00	50.00	13.00
ILL. P. E.	ENGINEERS	B(2)+C(2) or D(3)+ C(2)	B(2)+D(3)+C(6) or D(5)+C(6)	E(2)	E(10)	E(11)	F(3)	75	Yes	I(5)	J(6)	30.00	-----	15.00	-----	30.00	10.00*
ILL. STRUCT.	STRUCTURAL		B(2)+D(3)+C(6) or D(4)+C(6) or D(7)+ F(6)	E(7)	E(7)	E(7)	C(6)	G(5)	H(1)	No	J(3)	30.00	-----	-----	-----	30.00	20.00*
IND.	ENGINEERS, L. S.	B(2)+C(2) or B(5)+ C(2)+D(1)	B(1)+D(3)+C(6) or D(7)+C(6)	E(1)	E(9)	No	F(11)	G(7)	H(5)+H(6)	I(24)	J(1), J(3)	5.00+ K(1)	7.50	-----	10.00	27.50+K(2)	15.00*
IOWA	ENGINEERS, L. S.	B(1)+C(2) or D(5)+ C(2)	B(1)+D(3)+C(6) or D(7)+C(6) or I(5)	E(7)	E(7)	E(7)	F(3)	G(7)	Yes	No	J(6)	25.00	10.00	10.00	-----	25.00	10.00
KANS.	ENGINEERS, L. S.	B(2)+C(2) or D(3)+ C(2)	B(2)+D(3)+C(6) or D(5)+C(6) or B(2)+ D(7)+C(2)	E(1)	E(10)	E(7)	F(2)	G(7)	Yes	No	J(6)	25.00	10.00-student 15.00 non-student		-----	25.00	10.00

Table 14.1 Continued.

MINIMUM REQUIREMENTS FOR ENGINEERING REGISTRATION

STATE	PROFESSIONS REGISTERED	E.I.T. QUALIFICATIONS	PROFESSIONAL ENGINEER QUALIFICATIONS	CREDIT FOR EXPERIENCE: TEACHING	CREDIT FOR EXPERIENCE: MILITARY	CREDIT FOR EXPERIENCE: CONTRACTING	LENGTH & TYPE ENGRS. EXAM.	PASSING GRADE	NEC RECOGNITION	POLICY ON EMINENCE	RECIPROCITY PRACTICES	REGISTRATION FEES: PROF. ENGR. APPL.	PROF. ENGR. CERT.	E. I. T. APPL.	E. I. T. CERT.	RECIPROCITY	RENEWAL
KY.	ENGINEERS, L. S.	B(1)+C(2) or C(2)+ D(3)	B(1)+C(6)+D(3) or B(1)+C(2)+C(9)+ D(3) or C(6)+D(5)	E(9)	E(10)	E(10)	F(2)	G(7)	Yes	I(1)	J(3)	25.00	15.00	15.00	-----	35.00	10.00
LA.	ENGINEERS, L. S.	B(1)+C(2) or D(4)+ C(2)	B(1)+D(3)+C(2) or D(5)+C(6) or I(10)	E(2)	E(10)	E(10)	F(3)	G(3)	Yes	I(10)+I(17)	J(6)	25.00	-----	10.00	-----	25.00	7.50
MAINE	ENGINEERS	B(1)+C(2) or D(5)+ C(2)	B(1)+D(3)+C(2) or D(7)+C(6) or I(6)+ C(8)	E(2)	E(9)	E(9)	F(2)	G(7)	Yes	60 + Emin.	J(3)	10.00	10.00	10.00	-----	20.00	3.00
MD.	ENGINEERS, L. S.	B(1)+C(2) or B(4)+ D(3)+C(2)	B(1)+D(3)+C(6) or B(4)+D(5)+C(6) or D(7)+C(2)	E(1)	E(9)	E(11)	F(3)	G(7)	H(2)	I(7)	J(6)	20.00	15.00	10.00	-----	35.00	10.00
MASS.	ENGINEERS, L. S.	B(1) or B(2)+D(3) or B(6)+D(7)	B(1)+D(3)+C(6) or B(2)/(3)+D(5)+ C(6) or D(7)+C(6)	E(1)	E(10)	No	C(6)	G(3)	Yes	I(10)+ C(8)	J(5)	40.00+ 20.00 each examination	-----	20.00	-----	40.00+	15.00*
MICH.	ENGINEERS, L.S.& ARCH.	B(1)/(2)+C(2) or D(3)+C(2)	B(1)/(2)+D(3)+C(6) or D(5)+C(6)	Yes	E(8)	No	F(2)	G(7)	H(4)	No	J(3)	30.00	40.00	30.00	-----	70.00	15.00
MINN.	ENGINEERS, L.S.& ARCH.	B(1)+B(5)+C(2)	B(1)+D(3)+C(6) or B(1)+I(16)+C(2)	E(1)	E(10)	E(11)	F(3)	G(7)	Yes	No	J(3)	100.00	15.00	30.00	-----	100.00	15.00
MISS.	ENGINEERS, L. S.	B(2)+C(2) or D(3)+ C(2)	B(1)+D(3)+C(6) or D(5)+C(6)	Yes	E(10)	E(10)	F(2)	G(7)	Yes	No	J(1) if reg. 10 yrs	25.00	-----	-----	10.00	25.00	8.00
MO.	ENGINEERS, L.S.& ARCH.	B(3)+C(2) or D(3)+C(2)	D(5)+C(6) or B(3)+C(6)+D(3) or B(3)+ I(10)+C(8)	Yes	E(9)	E(9)	F(3)	G(7)	H(2)	I(5)	J(1), J(3)	35.00	-----	10.00	-----	35.00	10.00
MONT.	ENGINEERS, L. S.	B(1)+C(2) or D(3)+ C(2)	B(1)+D(3)+C(6) or D(5)+C(6) or I(12)	Yes	E(10)	E(10)	F(3)	G(7)	H(3)	J(3)	J(3) , J(5)	20.00	-----	20.00	-----	20.00	10.00
NEBR.	ENGINEERS, ARCH.	B(1)+C(2) or D(3)+ C(2)	B(1)+D(3)+C(6) or D(5)+C(6) or I(11)+ C(8)	Yes	E(9)+ E(12)	E(9)	F(3)	G(7)	H(3)	I(11)+ C(8)	J(3)	50.00	25.00	25.00	-----	75.00	10.00
NEV.	ENGINEERS, L. S.	B(2)+C(2) or D(3)+ C(2)	B(2)+D(3)+C(6) or D(5)+C(6) or I(5)	Yes	E(9)	E(11)	F(3)	G(7)	H(1)	I(5)	J(6)+C(9)	35.00	-----	10.00	-----	35.00	10.00
N. H.	ENGINEERS	B(3)+C(2) or D(3)+ C(2)	B(2)+D(3)+C(2) or D(5)+C(2) or I(15)	Yes	E(9)	No	F(3)	G(1)	Yes	I(15)+ C(7)	J(6)	20.00	20.00	10.00	7.50	40.00	5.00
N. J.	ENGINEERS, L. S.	B(1)+C(2) or D(3)+ C(2)	B(1)+D(3)+C(6) or D(5)+C(6) or I(13)+ C(8)	Yes	E(9)	No	F(3)	G(7)	H(1)	I(9)+ C(8)	J(3)	40.00	-----	10.00	-----	40.00	5.00
N. M.	ENGINEERS, L. S.	B(2)+C(2) or D(4)+ C(2)	B(2)+D(3)+C(6) or D(5)+C(6)	Yes	E(7)	E(7)	F(3)	G(7)	Yes	I(2)+C(9)	J(6)	25.00	-----	15.00	-----	25.00	14.00
N. Y.	ENGINEERS, L. S.	B(1)+C(2) or B(6)+ D(5)+C(2)	B(1)+D(3)+C(6) or B(6)+D(7)+C(6) or B(1)+I(5)+I(6)	E(10)	E(10)	E(10)	F(3)	75	H(1)	B(1)+I(5)+ I(6)	J(10	40.00	-----	20.00	-----	40.00 P.E. 20.00 EIT	15.00*

Table 14.1 Continued.

MINIMUM REQUIREMENTS FOR ENGINEERING REGISTRATION

STATE	PROFESSIONS REGISTERED	E.I.T. QUALIFICATIONS	PROFESSIONAL ENGINEER QUALIFICATIONS	CREDIT FOR EXPERIENCE			LENGTH & TYPE ENGRS. EXAM.	PASSING GRADE	NEC RECOGNITION	POLICY ON EMINENCE	RECIPROCITY PRACTICES	REGISTRATION FEES					
				TEACHING	MILITARY	CONTRACTING						PROF. ENGR. APPL.	PROF. ENGR. CERT.	E. I. T. APPL.	E. I. T. CERT.	RECIPROCITY	RENEWAL
N. C.	ENGINEERS, L. S.	B(1)+C(2) or B(6)+ D(4)+C(2)	B(1)+D(3)+C(6) or B(6)+D(6)+C(6)	E(1)	E(9)	E(4)	F(9)+C(6)+ C(9)	G(7)	H(6)	No	J(3)	45.00	-----	10.00	-----	45.00	10.00
N. D.	ENGINEERS, L. S.	B(1)+C(2) or B(4)+ C(2)	B(1)+D(3)+C(6)	D(1)	E(9)	E(9)	F(3)	G(7)	Yes	I(1)	J(6)	25.00	-----	10.00	-----	25.00	10.00
OHIO	ENGINEERS, L. S.	B(1)+C(2) or D(3)+ C(2)	B(1)+D(3)+C(6) or D(5)+C(6)	E(1)	E(10)	E(10)	F(3)	G(7)	Yes	I(19)+C(8) +B(2)	J(6)+J(1)	15.00	15.00	15.00	-----	30.00	5.00
OKLA.	ENGINEERS, L. S.	B(1)+C(2) or B(4)+ C(2)+D(1)	B(3)+D(3)+C(6) or D(5)+C(6) or I(16)	E(1)	E(9)	E(9)	F(3)	G(7)	Yes	I(16)	J(6)	25.00	-----	10.00	-----	25.00	5.00
ORE.	ENGINEERS, L. S.	B(3)+C(2) or D(3)+ C(2)	B(3)+D(3)+C(6) or D(5)+C(6)	E(7)	E(10)	No	F(3)	G(3)	Yes	I(1)	J(9)	10.00	10.00	5.00	5.00	25.00	10.00
PA.	ENGINEERS, L. S.	B(3)+C(2) or D(3)+ C(2)	B(1)+D(3)+C(6) or D(7)+C(6) or I(21)	Yes	E(10)	No	F(3)	G(7)	H(1)	I(21)	J(6)	25.00	-----	15.00	-----	25.00	10.00*
P. R.	ENGINEERS, L.S.& ARCH.	B(1)+C(6)	B(1)+D(3)+C(6)	Yes	E(9)	No	F(2)	G(7)	Yes	I(5)	B(1)+D(3)+C(8)	15.00	20.00	25.00	-----	35.00	-----
R. I.	ENGINEERS, L. S.	B(3)+C(2) or B(6)+ D(5)+C(2)	B(3)+D(3)+C(2) or B(3)+D(7)+C(2) or B(6)+D(7)+C(2) or I(10)+C(2)	E(1)	Some E(12)	No	F(3)	G(7)	Yes	I(10)+C(2)	J(6)	30.00	-----	10.00	-----	30.00	5.00
S. C.	ENGINEERS, L. S.	B(3)+C(2)	B(3)+D(3)+C(6) or D(7)+C(6) or I(12)+ C(2)	E(9)	E(9)	E(11)	F(3)	G(7)	Yes	I(12)+C(2)	J(6)	30.00	5.00	-----	10.00	25.00	10.00#
S. D.	ENGINEERS, L.S.& ARCH.	B(1)+C(2) or D(3)+ C(2)	B(1)+D(3)+C(6) or D(5)+C(6)	E(2)	Some	No	F(3)	G(7)	Yes	No	J(5)+J(6)	30.00	5.00	5.00	-----	30.00	10.00
TENN.	ENGINEERS, ARCH.	B(3)+C(2) or D(4)+ C(2)	B(3)+D(3)+C(6) or D(5)+C(6) or I(5)/ (10)	Yes	E(9)	E(10)	F(3)	G(7)	Yes	I(5)+I(10)	J(9)	35.00	-----	10.00	-----	35.00	10.00
TEXAS	ENGINEERS	B(1)+C(2) or B(3)+ C(2) or D(5)+C(2)	B(1)+D(3) or B(3)+D(3) or D(5)+C(6)	E(1)	Some E(12)	E(10)	F(3)	G(7)	Yes	I(1)	J(6)	15.00	10.00	10.00	-----	10.00	10.00
UTAH	ENGINEERS, L. S.	B(1)+C(2) or D(3)+ C(2)	B(1)+D(3)+C(6) or D(5)+C(6)	E(3)	E(7)	E(11)	F(3)	G(7)	J(5)	I(12)	J(6)	30.00	-----	15.00	-----	30.00	3.00
VT.	ENGINEERS	B(2)+C(2) or B(6)+ D(5)+C(2)	B(2)+D(3)+C(6) or D(7)+C(6) or I(10)/ (22)+C(7)	E(2)	E(9)	E(9)	F(3)	G(7)	H(5)+H(6)	I(5),(20) &(22)	J(5),(6) & (8)	45.00	-----	15.00	-----	30.00	5.00
VA.	ENGINEERS, L.S.& ARCH.	B(1)+C(2) or D(4)+ C(2)	B(1)+D(3)+C(6) or D(6)+C(6) or I(20)+ F(6) or D(9)+C(2)	E(7)	E(7)	E(7)	F(3)	G(7)	H(3)	I(20)+F(6)	J(6)	30.00	-----	15.00	-----	30.00	10.00*
V. I. (U.S.)	ENGINEERS, L.S.& ARCH.	B(2) or D(3)+C(2)	B(1)+D(1)+C(6) & Resident of V. I.(US) or D(5)+C(6) & Resident of V.I. (US)	Yes	E(9)	E(9)	F(3)	G(7)	Yes	I(12)	J(6)+resident of V. I. (U.S.)	50.00	-----	25.00	-----	50.00	25.00
WASH.	ENGINEERS, L. S.	B(1)+C(2) or D(3)+ C(2)	B(1)+D(3)+C(6) or D(5)+C(6) or B(1)+ I(11)+C(2)	E(4)	E(9)	No	F(3)	G(7)	H(3)	No	J(6)	15.00	10.00	10.00	-----	15.00	7.50
W. V.	ENGINEERS	B(2)+C(2) or D(5)+ C(2)	B(2)+D(3)+C(6) or D(5)+C(6)	Yes	E(9)	No	F(3)	G(7)	Yes	I(1)	J(3)+J(6)	10.00	15.00	10.00	-----	25.00	7.50
WISC.	ENGINEERS, L.S.,ARCH.	B(1)+C(2) or D(3)+ C(2)	B(1)+D(3)+C(6) or D(5)+C(6) or I(12)	E(1)	E(9)	E(9)	F(3)	G(7)	H(3)	I(12)	J(6)	12.50	12.50	12.50	-----	75.00	25.00*
WYO.	ENGINEERS, L. S.	B(1)+C(2) or D(3)+ C(2)	B(1)+D(3)+C(6) or D(5)+C(6) or I(5)/ (10)	E(4)	E(9)	E(9)	F(2)+F(5)	G(7)	Yes	I(5)+I(10)	J(6)	15.00	-----	5.00	-----	15.00	8.00*

3. To assist the young engineer in obtaining immediate professional affiliation, guidance, and protection and thus earmark him or her for the profession.
4. Through increased professional consciousness, better knowledge of the registration procedures, and reduced fear of the final professional examination, to encourage the young engineer to acquire professional status by registration as soon as he or she is legally qualified.

Specific engineer-in-training qualifications in the various states are summarized in Table 14.1. Full registration as a professional engineer generally requires:

1. Graduation from an Engineers' Council for Professional Development accredited engineering program.
2. A number of years of professional experience.
3. Successful completion of either or both written and oral examinations.

Constance (1977) provides a guide to P.E. refresher courses.

Not infrequently, an engineer's work may be conducted in more than one state and for this reason and to accommodate engineers relocating from one state to another, there is generally reciprocity among states for facilitating the registration of an engineer already registered in one or more states. Again, the specific reciprocity practices are summarized in Table 14.1.

Clincial engineering certification

Another important measure of professional achievement for clinical engineers is clinical engineering certification. Certification is the process by which a practicing clinical engineer meets a set of prerequisite requirements and is formally recognized for his or her knowledge of clinical engineering principles. Candidates successfully completing a required examination are then "certified" as competent clinical engineers. Certification is not to be confused with and directly compared with professional engineering registration. Registration requirements and procedures are regulated by law in every state. Although an engineer working in the biomedical area may be able to claim a biomedical specialty for practice, examinations for registration do not currently include specific optional questions on clinical engineering subjects and thus, these examinations for those employed in clinical engineering establish a general proficiency in the engineering sciences but not specifically in clinical engineering as a discipline.

Certification, however, is aimed specifically at clinical engineering as a professional specialization. Procedures and requirements for certification are currently established both by the Board of Examiners for Clinical Engineering Certification and by the American Board of Clinical Engineering. Addresses and procedures for certification by these two groups are outlined in Table 14.2. Note that in both cases, eligibility for certification includes appropriate formal training, experience, and successful completion of an examination.

At this point a clinical engineer might ask: "As a practicing professional, should I become registered, certified, or both?" The answer to this question in a particular case depends on local and state laws, employer's guidelines and personal professional objectives. As a general rule, a practicing clinical engineer would be best prepared to defend his or her professional competency if necessary if he or she were both registered as a professional engineer and certified as a clinical engineer. In any case, certification should not generally be viewed as a substitute for registration. Instead, it is appropriate to view certification as supplemental to registration in the sense that registration establishes overall competence as an engineer while certification establishes special training and proficiency in clinical engineering as a specialized discipline. Such a view is similar to the long-standing practice in medicine where a physician may be licensed to practice medicine with or without being "Board Certified" in a particular clinical specialty. Although the requirements for engineering registration are not as uniformly adhered to as the licensing of physicians in many states, you are advised to investigate local and state laws before pursuing registration. In addition, as indicated above, additional guidelines are set by employers. With the establish-

Table 14.2. Requirements and Procedures for Clinical Engineering Certification

Certifying agency	Board of Examiners for Clinical Engineering Certification	American Board of Clinical Engineering
Address	Suite 602, 1901 N. Ft. Myer Drive, Arlington, Virginia 22209	Box 2166 Rapid City, SD 57709
Eligibility		
(a) Education	Bachelor's degree or higher in engineering or a related field is required. A waiver of this requirement may be requested when applicant has three or more years experience in the clinical engineering field.	Bachelor's degree or equivalent in engineering or physical sciences.
(b) Experience	Minimum of three (3) years clinical engineering experience, of which at least one (1) year has been in	Minimum of five (5) years experience in clinical engineering which may include two (2) years of relevant educational

	a position of significant management responsibility. Patient contact and interface with physicians and nursing staff are required; competence to establish and maintain a clinical engineering program in a health-care facility is required	experience. Persons who are qualified for certification in all respects except length of satisfactory experience will receive a certificate of candidacy.
Examination(s)		
(a) Content	Anatomy and physiology, safety, equipment application, teaching responsibilities, equipment specifications, consulting and advising, Physical Plant, evaluation of equipment, test equipment, engineering principles, current issues.	Fundamentals of engineering, fundamentals of biological and medical sciences, application of engineering to clinical medicine, moral and ethical aspects of the delivery of health care.
(b) Format	Closed-book, oral examination consisting of ten (10) questions, five (5) of which are to be chosen by the examinee. Examinee is given ninety (90) minutes to prepare answers before appearing before two members of the Board of Examiners. The oral review lasts a maximum of two (2) hours.	Eight (8) hour written examination in four (4) parts (see above); fundamentals portions are multiple choice and true false (four (4) hours, closed book). Applications portions are problem-discussion type questions. Up to one (1) cubic foot of commercially available reference material, including a calculator may be used. Five (5) questions must be selected from a list of engineering application questions (three (3) hours) and two (2) questions from a list of questions on moral and ethical issues (one (1) hour). A waiver of any of the four parts of the examination may be granted upon the presentation of evidence of competence established by equivalent prior examination.
(c) Location	Selected sites	Regional centers and selected national conferences.
Summary outline of certification procedures	Achieve educational requirement Accumulate appropriate experience. Submit application. Take examination.	

ment of certification procedures for clinical engineers, it seems probable that certification will become increasingly important in the practice of clinical engineering. A newly graduated clinical engineer should keep informed of developments related to certification procedures and requirements.

Professional and technical societies and meetings.

All professions are identified with one or more professional societies. Societies originate as a group of professionals with similar training, interests, and work experience joined together to promote professional activities in order to advance their respective professions. Societies of interest may be grouped as professional or technical in emphasis. Professional societies, such as the National Society of Professional Engineers, sponsor meetings and other activities focused on issues of general concern such as promoting public understanding of engineering, establishing standards of ethical conduct and economic status of members. Technical societies such as the Biomedical Engineering Society are focused in their activities on a major engineering technical area. For a listing of the major engineering professional and technical societies see Beakley and Leach (1972).

Most professional and technical societies are organized with a member-elected group of officers typically including a president, vice president, secretary, treasurer and a board of directors. In addition, various committees or divisions may be organized to focus on one or more specialized activities such as bylaw revision or current trends in industrial demand for engineers in a particular discipline.

Membership in one or more professional and technical societies is important for all practicing engineers including clinical engineers for a number of reasons including the following:

1. Each engineer can identify with professional groups committed to advancement of his or her profession.
2. Professional activities of personal interest (such as training programs) may be sponsored in part by a professional group.
3. Annual and additional meetings sponsored by the society provide the opportunity for the formal exchange of recently accumulated information and aids in continuing education.
4. Society newsletters and journals provide additional media for communication of technical findings and current professional issues.

Membership in the various professional and technical societies virtually always requires submission of a completed application for membership and payment of dues. Additional requirements may include training and experience in a particular discipline, proven professional accomplishment and letters of recommendation for acceptance as a society member by professional colleagues.

Note that the large majority of professional and technical societies provide for student membership status, and in view of the possibilities for professional development, students should pursue membership in one or more societies.

Clinical engineering societies. A relatively large number of societies in the biological and medical sciences and biomedical engineering areas have been established. [See for example listings of member societies of the Alliance for Engineering in Medicine and Biology (Sec. 15.8)]. Pertinent information on a selected set of organizations having a significant concentration in the clinical engineering area are listed in Table 14.3 for convenient reference.

Table 14.3. Selected Technical Societies and Professional Groups with Clinical Engineering Involvement

Society or Group	*Activities*
(Addresses in Chap. 15) Alliance for Engineering in Medicine and Biology	Comprised of approximately 23 member societies. Sponsors Annual Conference on Engineering In Medicine and Biology (ACEMB), usually in fall of each year, publication of meeting proceedings, many instruments on displays at meetings; sponsors various programs for information dissemination and continuing education.
Institute of Electrical and Electronics Engineers Inc. (Engineering in Medicine and Biology Society)	Annual meeting at ACEMB featuring sessions on clinical engineering topics; journal: *IEEE Transactions on Biomedical Engineering*
Association for the Advancement of Medical Instrumentation	Annual meeting, usually in spring, focused on instrumentation and measurement in clinical medicine; many instrumentation displays at meetings; meeting proceedings published; journal: *Medical Instrumentation.*
Regional meetings	(1) New England Bioengineering Conference (2) Rocky Mountain Bioengineering Symposium (3) San Diego Biomedical Symposium

In addition to the publications cited in Table 14.3, it is particularly noteworthy that there is now being regularly published the *Journal of Clinical Engineering* (Sec. 15.6). This journal which began publication in October 1976, is devoted to the dissemination of information dealing specifically with topics germane to clinical engineering. Clinical engineering practitioners should therefore subscribe to this new and important journal.

14.5 Ethical Issues

Nature and relevance of ethics to clinical engineering practice

Ethics may simply be defined as the study of standards of conduct and moral judgment. Most simply stated, ethics deals with human behavior and seeks to establish the principles defining the right or correct course of action in a given situation.

Clearly, any practicing clinical engineer will have to deal with many issues requiring professional discretion in decisions affecting those for whom and with whom he is working. More specifically, a clinical engineer may deal with issues related to such sensitive topics as professional fees, salaries of coworkers, patent rights, and professional competency. In addition to such issues of relevance to all engineering disciplines, the clinical engineer will likely have to deal with additional problems related to biomedical ethical issues such as informed consent and euthanasia. The need to be able to make the best decisions in the many situations that may arise in practice requires a knowledge of general guidelines for decisions requiring ethical considerations.

Criteria for making ethical decisions

The principal method of ethics in practice is decision making based on criteria established for deciding what makes an act right or wrong, or less definitively, better or worse in a given situation. It should be noted that ethics can be both practical and theoretical since it can be applied to both actual and theoretical issues related to professional practice.

For any practicing professional, the available criteria for deciding the ethical status of a given situation may be grouped simply as follows:

1. Codes of behavior established by law.
2. Codes of ethics for professional groups (e.g., engineers, physicians, and attorneys, among others).
3. Personal moral code.

Codes of ethics established by law may be included in registration legislation and these deal with the general aspects of engineering practice. For example, the canons listed in Table 14.4 excerpted from the Texas Engineering Practice Act, are explicit statements of principles of professional responsibility. In the various state laws, ethical considerations such as those listed in Table 14.4 are elaborated in relative detail to provide fairly explicit statements of professionally appropriate or inappropriate actions.

Jendrucko and Favoriti

The codes of ethics associated with state laws are not voluntary in the sense that the laws specify disciplinary rules and actions for violations of acceptable codes of behavior. Registered professional engineers should review carefully a copy of the registration law for the state in which they will practice to determine the precise nature of the ethical guidelines. Not surprisingly, the cannons of professional ethics do not vary greatly in content from one state to another.

In addition to codes of ethics specified in state law, professional societies may specify their own codes of ethical conduct for members. Such codes originate with society members who seek to maintain the status and reputation of the profession by establishing a set of explicit guidelines for ethical professional practice. For the practice of engineering, the code of ethics adopted by the National Society of Professional Engineers (NSPE) outlined in Table 14.5 is representative. Clearly the statements comprising the various sections of the NSPE ethics code have many implications. NSPE publication #1102 (1974) provides a more detailed discussion of specific ethical guidelines for engineers.

In addition to providing a general code of ethics as presented in Table 14.5, the NSPE has a Board of Ethical Review which reviews specific cases to provide examples of ethical evaluation of issues arising in actual work situations. The accumulated findings of the

Table 14.4. Code of Responsibility for Professional Engineers (excerpted from the Texas Engineering Practice Act)

Canon I:	The engineer's paramount professional responsibility should be the safety, health, and welfare of the general public.
Canon II:	The engineer should exercise independent professional judgments, decisions, and practices on behalf of clients and employers.
Canon III:	The engineer should act competently and use proper care in performing engineering services for clients or employers.
Canon IV:	The engineer should safeguard and preserve the confidences and private information of clients and employers.
Canon V:	The engineer should endeavor to build his practice and professional reputation solely on the merits of his services.
Canon VI:	The engineer should contribute to the maintenance, integrity, independence, and competency of the engineering profession.
Canon VII:	The engineer should asssit in preventing the unauthorized practice of engineering.

Table 14.5. Code of Ethics for Engineers (adopted by the National Society of Professional Engineers; excerpted from NSPE publication #1102)

Preamble: The Engineer, to uphold and advance the honor and dignity of the engineering profession and in keeping with high standards of ethical conduct:

> Will be honest and impartial, and will serve with devotion his employer, his clients and the public;
>
> Will strive to increase the competence and prestige of the engineering profession;
>
> Will use his knowledge and skill for the advancement of human welfare.

Section 1: The Engineer will be guided in all his professional relations by the highest standards of integrity, and will act in professional matters for each client or employer as a faithful agent or trustee.

Section 2: The Engineer will have proper regard for the safety, health, and welfare of the public in the performance of his professional duties. If his engineering judgment is overruled by nontechnical authority, he will clearly point out the consequences. He will notify the proper authority of any observed conditions which endanger public safety and health.

Section 3: The Engineer will avoid all conduct or practice likely to discredit or unfavorably reflect upon the dignity or honor of the profession.

Section 4: The Engineer will endeavor to extend public knowledge and appreciation of engineering and its achievements and to protect the engineering profession from misrepresentation and misunderstanding.

Section 5: The Engineer will express an opinion of an engineering subject only when founded on adequate knowledge and honest conviction.

Section 6: The Engineer will undertake engineering assignments for which he will be responsible only when qualified by training or experience; and he will engage, or advise engaging, experts and specialists whenever the client's or employer's interests are best served by such service.

Section 7: The Engineer will not disclose confidential information concerning the business affairs or technical processes of any present or former client or employer without his consent.

Section 8: The Engineer will endeavor to avoid a conflict of interest with his employer or client, but when unavoidable, the Engineer shall fully disclose the circumstances to his employer or client.

Section 9: The Engineer will uphold the principle of appropriate and adequate compensation for those engaged in engineering work.

Section 10: The Engineer will not accept compensation, financial, or otherwise, from more than one interested party for the same service, or for services pertaining to the same work, unless there is full disclosure to and consent of all interested parties.

Section 11: The Engineer will not compete unfairly with another engineer by attempting to obtain employment or advancement or professional engagements by competitive bidding, by taking advantage of a salaried position, by criticizing other engineers, or by other improper or questionable methods.

Section 12: The Engineer will not attempt to injure, maliciously or falsely, directly or indirectly, the professional reputation, prospects, practice or employment of another engineer, nor will he indiscriminately criticize another engineer's work. If he believes that another engineer is guilty of unethical or illegal practice, he shall present such information to the proper authority for action.

Table 14.5. (Continued)

Section 13:	The Engineer will not associate with or allow the use of his name by an enterprise of questionable character, nor will he become professionally associated with engineers who do not conform to ethical practices, or with persons not legally qualified to render the professional services for which the association is intended.
Section 14:	The Engineer will give credit for engineering work to those to whom credit is due, and will recognize the proprietary interests of others.
Section 15:	The Engineer will cooperate in extending the effectiveness of the profession by interchanging information and experience with other engineers and students, and will endeavor to provide opportunity for the professional development and advancement of engineers under his supervision.

NSPE Board of Ethical Review appear in NSPE publications 1106, 1106A, -B, and -C which provide examples of professional judgments related to specific ethical concerns.

Table 14.6 provides a code of ethics that was drafted by the Ethics Committee of the IEEE Group on Engineering in Medicine and Biology (EMB) in September 1975. It was subsequently adopted by the administrative committee of EMB.

The last category of ethical guidelines dealing with your personal moral code is perhaps the most important since personal ethics are usually the overriding factor in instances of difficult decisions. Fortunately, many of the ethical issues arising in engineering practice are dealt with explicitly in state registration law or canons of ethics adopted by professional societies and personal ethical evaluation is often relatively straightforward. However, in the practice of clinical engineering additional issues and concerns, not dealt with in published guidelines may arise. In particular, clinical engineering usually involves either directly or indirectly clients who are hospital or clinic patients. Attendant to the medical services rendered in a clinical setting are many biomedical issues of concern to clinical engineers participating in team medical care. As an example, consider a hypothetical case where a group of clinicians in a medical center invite a clinical engineer to participate in the design and construction of a new device for use in abortions. If the engineer had some personal reservations (not related to law or a published code of ethics) concerning the conduct of abortions on humans, he or she might decline participation on the project. It is an unfortunate fact that there are accumulating many such issues related to clinical medical practice and it is the nature of the clinical engineer's work, dealing as it frequently does directly with human health care, which results in his or her occasionally having to make personal decisions on ethical issues. In order that the clinical engineer may be better prepared to deal with some of the problems that may arise, a selected set of current biomedical ethical issues of relevance to clinical engineering practice is presented below.

Table 14.6. Code of Ethics for Engineers in Medicine and Biology (EMB)

Preamble: The following principles are established to aid individuals practicing engineering in medicine and biology to determine the propriety of their actions with relation to patients, health-care personnel, sponsors, students, clients, customers, and the public. These standards are intended to promote and maintain a high level of ethical conduct.

A. With respect to *patients* engineers shall:
 1. Use their knowledge and skills for restoring health and well-being;
 2. Promote peace of mind, bodily comfort and personal dignity, without infringing on the roles of nurses and physicians;
 3. Adhere to instructions of responsible physicians;
 4. Answer questions clearly and sympathetically about the purpose and function of equipment for which they are responsible but refer questions about diagnosis, therapy and prognosis to the responsible physician;
 5. Refrain from incurring unnecessary costs.

B. With respect to *health care personnel* engineers shall:
 1. Represent honestly their own capabilities and specify accurately the performance and characteristics of equipment under their control;
 2. Promote interdisciplinary communication and understanding while recognizing and maintaining respect for differing cultural values;
 3. Eschew professional jealousy and destructive competition;
 4. Inform the responsible person of observations of deficiencies that may influence diagnosis, therapy, or prognosis, including specifically performance, safety or misuse of equipment, and where conflict arises, to do so formally in writing.

C. With respect to *students* engineers shall:
 1. Provide guidance in the study of significant problems that offer opportunities for professional maturation and employment;
 2. Refrain from exploiting positions of authority over students' entries upon professional careers.

D. With respect to *clients and customers* engineers shall:
 1. Provide prompt and efficient service after sales;
 2. Assure avoidance of false or misleading reports or advertising.

E. With respect to *sponsors* engineers shall:
 1. Assure accuracy and completeness of proposals and reports;
 2. Refrain from incurring unnecessary costs.

F. With respect to *public interest* engineers shall:
 1. Avoid waste of human, animal, material and financial resources;
 2. Apply their expert knowledge to inform and educate the public, including the correction of false or misleading reports and advertising;
 3. Comply with voluntary standards and recommended practices.

G. With respect to *experimental animals* engineers shall:
 1. Advocate protocols that use no more animals than are necessary to achieve significant results;
 2. Minimize pain and suffering;
 3. Adhere to regulations concerning animal care.

Issues in biomedical ethics have been receiving progressively more attention in recent years owing primarily to the great strides that are being made in the medical sciences and engineering. While there is no comprehensive list of such issues available, there have been recent attempts to elucidate some of the current ethical concerns (Bronzino, 1977; Kieffer, 1975). The clinical engineer may find himself involved with such issues either directly or indirectly depending on his specific job function and the extent to which he functions as the part of a medical team in patient care. Since patient involvement is likely for most clinical engineers, we now present a selection of "typical" ethical issues of concern.

Patient selection. In a busy hospital or clinic, the patient load may be such that many serious medical problems cannot be treated on a "first-come, first-served" basis. In such cases, specialized personnel (e.g., a neurosurgeon) or facilities (e.g., a hemodialysis system) may be limited and this may necessitate selecting patients for care from a group needing special care. In such cases, those not receiving immediate care may suffer increased risks of morbidity and mortality. Fortunately for clinical engineers, a physician is almost always solely responsible for patient scheduling for treatment. However, with the rise of more complex and specialized equipment for diagnosis and treatment, physicians will be increasingly reliant on the opinion of clinical engineers for an assessment of suitability of prospective patients (in terms of such aspects as measurement system limitations) for specialized instrumental analysis or treatment. The advice of the clinical engineer may especially be sought when new instrumentation and methods are being selected or evaluated. Thus in his role, the clinical engineer either directly through communication with physicians or indirectly through his choice of equipment characteristics may limit the suitability of certain patients for diagnostic examination or treatment. No general guidelines exist for assessing the overall impact of engineering decisions on the patient selection process but in general it appears that clinical engineers will become more influential in matching patients to hospital systems and in so doing, they will assume part of the overall responsibility for the patient selection process.

Confidentiality of patient records. A clinical engineer in his work will frequently have access to or may in fact be instrumental in the procurement of various patient data. Certain transcriptions of these data or discussions with colleagues will be necessary during the course of patient care. However, since personal and medical records frequently contain sensitive information, discretion must be used in the use and publication of such data. Professionals have traditionally protected the confidentiality of client files and this tradition applies equally appropriately to clinical engineering practice. Again, in most hospitals and clinics, codes of conduct for employees may specify limitations on release of patient data. However, instances that are unique may arise, and in cases of doubt, a clinical engineer should assume personal and medical data are confidential until the appropriateness and justification for release are established.

Benefit-risk analysis for diagnostic and treatment procedures. Many medical procedures instituted for the ultimate benefit of the patient are associated with significant risks. Such risks are relatively well documented for procedures such as radiotherapy and catheterization. As new systems for patient instrumentation are developed and evaluated it will frequently fall on the clinical engineer to objectively assess system capabilities, expected reliability and in particular any significant attendant risks. Although the desire for professional recognition suggests early introduction of new systems for patient use, the clinical engineer must consider the ethical implications associated with risks to patients arising from use of an instrumentation or therapy system. Ultimately, the overall appropriateness of continued use of a new diagnostic or treatment system must be determined by a frequently qualitative "cost-benefit analysis." Simply, the clinical team must decide in each case, or for a class of patient disorders, whether the risks of contributing to the development or advance of pathological conditions are outweighed by the benefits anticipated.

Experimental instrumentation and procedures: informed consent. Not infrequently in large medical centers, medical school faculty or other university faculty will participate in training and research programs as well as patient care. In such institutions, many new procedures occasionally utilizing new instrumentation systems are developed and used. The final stage of evaluation of such new procedures involves what is essentially experimentation with human subjects. When human subjects are used in research, "informed consent" for their participation should be obtained to afford legal protection with respect to patient rights as well as from an ethical point of view of giving patients enough information to make a rational decision whether or not to participate. In order to be eligible for federal funds for research involving human subjects, each university or research institution must have a committee on the protection of human subjects (Code of federal regulations, 1975). Virtually all universities have such committees, which have their own informed consent requirements and procedures.

Informed consent in brief consists of adequate written and oral explanation to the would-be subject of the objectives of the procedures to be used. In particular, any likelihood of pain or discomfort or risk to health must be specified. Alternative procedures must be discussed and it must be made clear that a subject may withdraw from an experimental program at any time.

The implications of informed consent for clinical engineers should be quite apparent. If an engineer will personally administer or supervise the administration of a new or experimental diagnostic or treatment procedure, it may be that informed consent of a patient is required. Such situations may be complicated if the subject is not conscious or is mentally deficient and a new procedure is suggested. In the area of clinical engineering, instrumentation will be the most likely association of the engineer with clinical research. Thus, in any case of patient instrumentation, with new or relatively untried devices, care should be taken to see that the patient has been informed of the experimental nature of the procedure and any significant risks that are involved.

Dying, death, and euthanasia. Hospitals by their nature are centers where individuals with various degrees of pathology are concentrated, and among the most critically ill patients (e.g., coronary and intensive care) a significant fraction die while under evaluation or treatment. Thus, the role of the clinical engineer requires dealing with the many aspects of dying and death.

The major consequences of an association with dying patients relates to (1) the psychological well-being of the patient, (2) personal response of the engineer, and (3) psychological status of the family and friends of the patient. Although a clinical engineer may not always be able to avoid discussions of death, especially in the case of the patient, engineers are not generally trained in death psychology, and therefore discussion should probably be limited in most cases to mild personalized reassurance. In any case, the patient's family and frequently clergy will usually be available for personal attention in cases of terminal disease.

The effect of patient death on the function of the clinical engineer is yet another matter and some of the personal adjustment will occur only after enough experience is gained to provide a set of personal perspectives on death.

In many instances (e.g., coma) it is possible to extend life indefinitely with the use of available life-support systems (e.g., ventilators). It has been suggested in a number of medical circles that in cases where pathological damage (and particularly brain damage) is irreversible, it may be that such treatment is futile and probably should be discontinued. Obviously in many such cases discontinuance of life support results in the biological death of the patient. The legality of such "passive" actions is being tested in the court systems. However, until some precedents are set, decisions whether or not to continue a treatment will be difficult to make. Again, the clinical engineer, with a major role in instrumentation and data acquisition may be called upon to provide information that will at least in part affect such decisions.

A separate but related issue of current concern is euthanasia or "mercy killing" of hopelessly ill (and often aged) patients. This subject differs from the former in that at issue is whether or not it is humane and legal to end the life of a suffering patient by active means. At the present time, while it is under discussion by many medical and nonmedical groups, euthanasia is viewed as murder and as such is illegal. However, as a possible means to justify euthanasia, the Euthanasia Society, 250 West 57th St., New York, NY 10019, has suggested the option for the establishment of a "living will" in which a person while in good health prepares a written and legally binding will directing that if he is taken critically ill and death appears certain, that extraordinary means of life support not be used (Kieffer, 1975).

Individuals may also carry a card in their wallet which is sometimes called a *Uniform Donor Card.* Such cards are available from Living Bank International, Box 6725, Houston, TX 77025. Completion of this card—or in some states, an indication on the driver's license—permits removal immediately after death of needed organs such as kidneys, hearts, corneas, etc. for transplantation. The clinical engineer should consider executing such a card himself and informing others of this possibility, so that the critical shortage of organs for transplantation might be reduced.

14.6 Study Questions

14.1 List the items that fall within the concept of a *product.*

14.2 What are the two more prominent ways that a defect may show up?

14.3 Name the three legal theories of product liability recognized in a court of law.

14.4 Name two legal defenses that may bar recovery in a product liability case.

14.5 Give four required characteristics of an engineering professional.

14.6 Name and describe the two stages in the process of engineering registration.

14.7 Distinguish between registration and certification.

14.8 Describe the procedure required for certification.

14.9 Name two societies that have clinical engineering involvement.

14.10 Define ethics.

14.11 List three groupings of criteria for deciding the ethical status of a given situation.

14.12 List four ethical factors required by the NSPE code of ethics.

14.13 List three typical ethical issues of concern to the clinical engineer.

14.7 References

Beakley, G. C., and H. W. Leach, (1972). Engineering: an introduction to a creative profession, pp. 113-4. New York: McGraw-Hill.

Bronzino, J. D. (1977). Technology for patient care. St. Louis: C. V. Mosby.

Code of federal regulations (1975). Protection of human subjects. Title 46, Section 46, November 6, 1975.

Constance, J. D. (1968). How to become a professional engineer. Chapter 5. New York: McGraw-Hill.

Constance, J. (1977). A guide to P.E. refresher courses. *New Engineer*, 6(9) October:28-37.

Ellis, P. G. (1972). A case for professional registration. Publication 2210. Washington, DC: National Society of Professional Engineers.

Fine, M. S. (1977). Engineering registration and the law. *IEEE Spectrum*, 14(10):60-64.

Jarzembski, W. B., and B. A. Rowley, (1976). Professional engineering registration and clinical engineering. *J. Clin. Eng.*, 1:56-62.

Kieffer, G. H. (1975). Ethical issues and the life science. American Association for the Advancement of Science Study Guide on Contemporary Problems, test edition.

National Society of Professional Engineers. (1974). Code of ethics for engineers. Publication 1102. Washington, DC.

Reich, W. T. (ed.) (1978). Encyclopedia of bioethics. New York: Free Press.

15
Resources

ROBERT A. PEURA and JOHN G. WEBSTER

15.1 Learning Objectives

Upon completing this chapter you should be able to:

1. Find information on equipment function and operating principles.
2. Locate product listings and addresses of equipment manufacturers.
3. Perform a literature search using indexes, abstracts and computerized retrieval systems.
4. List two each of clinical engineering (1) periodicals, (2) newsletters, (3) conferences, and (4) standards agencies.

15.2 Introduction

As a clinical engineer, you need information about medical equipment. You, your department, or your library should purchase a number of the information sources described in this chapter. Books provide information on equipment function and operating principles. Product listings lead you to the companies that supply medical equipment. You can contact the companies for detailed product specifications, application notes, and prices. Indexes available in libraries organize your search for specific information. We provide a list of the most important periodicals, which may be found in either the medical or engineering libraries. Newsletters provide up-to-date information on new techniques, devices, standards, regulations, and training. You can obtain information at conferences, either by attending or by reading the proceedings. We provide addresses for standards agencies, so you can obtain the latest standards.

Frequently someone else knows the information that you need. Ask engineering and medical coworkers, clinical engineers at other hospitals, equipment salesmen, and others you think may know the answer.

15.3 Books

The first place to start a search for information is with books. A book, such as the one you are reading, organizes information on a subject in one convenient place. It is a good source of information on equipment function and operating principles. Advantages are: (1) it can be found through library card catalogs, (2) it can be obtained from most libraries, and (3) it has its own index to find information within it. Disadvantages are: (1) it is normally one to two years out-of-date when published and (2) it continues to become more out-of-date as time progresses. For information that changes slowly, this is not a great disadvantage. Following are a few of the most important books that should be valuable to you.

Ackermann, P. G. (1972). Electronic instrumentation in the clinical laboratory. Boston: Little, Brown.

Becker, R. F., et al. (1971). The anatomical basis of medical practice. Baltimore: Williams & Wilkins.

Bray, W. E. (1968). Clinical laboratory methods. St. Louis: C. V. Mosby.

Caceres, C. A. (ed.) (1977). The practice of clinical engineering. New York: Academic.

Cobbold, R. S. C. (1974). Transducers for biomedical measurements. New York: Wiley.

Cromwell, L., F. J. Weibell, E. A. Pfeiffer, and L. B. Usselman (1973). Biomedical instrumentation and measurements. Englewood Cliffs, NJ: Prentice-Hall.

Dorland's illustrated medical dictionary, 25th ed. (1974). Philadelphia: W. B. Saunders.

Geddes, L. A., and L. E. Baker (1975). Principles of applied biomedical instrumentation, 2nd edition. New York: Wiley.

Graf, R. F., and G. J. Whalen (1976). The Reston encyclopedia of biomedical engineering terms. Reston, VA: Reston.

Jacobson, B., and J. G. Webster (1977). Medicine and clinical engineering. Englewood Cliffs, NJ: Prentice-Hall.

Jung, W. G. (1974). IC op-amp cookbook. Indianapolis, IN: Bobbs-Merrill.

National Fire Protection Association (1973). Essential electrical systems for health care facilities. NFPA 76A. Boston.

National Fire Protection Association (1973). Electricity in patient care facilities. NFPA 76B. Boston.

National Fire Protection Association. (1973). Electricity in patient care facilities. NFPA 76B. Boston.

National Fire Protection Association (1977). National electric code 1978. NFPA 70. Boston.

Ray, C. D. (ed.) (1974). Medical engineering. Chicago: Year Book.

Selkurt, E. E. (ed.) (1976). Physiology. Boston: Little, Brown.

Simmons, D. A. (1972). Medical and hospital control systems. Boston: Little, Brown.

Simmons, D. A., and J. O. Wear, (1977). Introduction to clinical engineering management. North Little Rock, AR: Scientific Enterprises.

Strong, P. (1970). Biophysical measurements. Beaverton, OR: Tektronix.

Texas Hospital Association (1975). Electrical safety guide. Austin, TX: Health Services Corp.

Thomas, H. E. (1974). Handbook of biomedical instrumentation and measurement. Reston, VA: Reston.

Weast, R. C. (ed) (1976). Handbook of chemistry and physics, 1976–1977. Cleveland: CRC Press.

Webster, J. G. (ed.) (1978). Medical Instrumentation: application and design. Boston: Houghton Mifflin.

15.4 Product Listings.

While books provide information about equipment function and operating principles, they rarely list equipment manufacturers. You may need information to help select an instrument from those manufactured by different companies. The following sources provide this information plus manufacturer's addresses.

Annual guide to scientific instruments (1978). In annual (November) issue of *Science*. American Association for the Advancement of Science.

Dictionary Buyer's Guide (1978). Annual. Medical Electronics & Equipment News. Radner, PA: Chilton.

Electronic Design's gold book. Annual (1978). Rochelle Park, NJ: Hayden.

Electronic engineers master (1978). Garden City, NY: United Technical Publications.

Medical Electronics Buyers Guide (1978). Appears serially in the six annual issues of *Medical Electronics*. Pittsburgh, PA: Measurements & Data Corp.

15.5 Indexes and Abstracts

If you need more detailed or more current information than that from books it is probably available in journals. It is difficult to find information on a particular subject because journals publish articles dealing with subjects at random. Journals have annual subject indexes, but these help little if your subject is covered in dozens of journals. To solve this problem there are index publications that cover large numbers of journals. A good summary of the procedures for using printed and computerized literature searches is found in Beatty, W. K. (1973), Searching the literature comes before writing the literature (*Ann. Int. Med.* 79:917–924). The most helpful indexes follow.

Index Medicus is available in most medical libraries and indexes 2200 medical journals. The January issue lists "Medical Subject Headings." You find the term that most closely resembles your topic. Then check each annual index to find references to journal articles on your topic. Each issue also contains a "Bibliography of medical reviews." You should be able to find most journals in your library. The librarian can obtain others through interlibrary loan.

Medline is an on-line computerized information retrieval system. If your library has a terminal, you can list keywords taken from *Index Medicus* and tell the librarian what you seek. The system scans the past three years of 1100 medical journals to pick out articles containing combinations of keywords. You may interactively constrain the search to yield a reasonable number of references. Then the terminal prints out the references. There is frequently a charge of several dollars for each search.

Excerpta Medica is an index that provides abstracts for each article. Of the 41 separate sections, you will most frequently utilize section 27: Biophysics, Bio-Engineering and Medical Instrumentation. There are annual indexes that summarize the monthly issues.

Science Citation Index indexes 2000 journals and patents from all fields of science. It compiles cited references. Thus if you find a particularly useful article in 1970 you may find if any author has used this article as a reference after 1970. For example, you look up the 1970 article in the 1975 Science Citation Index and find six articles in 1975 that referred to the 1970 article. By checking these six articles you may find later improvements or modifications in the 1970 technique.

15.6 Periodicals

The foregoing methods of searching for information are efficient when you know what you are looking for. In addition, you should spend a portion of your time acquiring new information by non-directed reading. This enables you to keep up with new developments that affect your work. You should read the contents pages of the periodicals that follow. If you find an interesting title, you can read the abstract within the periodical. If the abstract is interesting you can read the whole article.

Biomedical Engineering. Monthly. 42–43 Gerrard St., London W1V 7LP: United Trade Press.

Computers in Biology and Medicine. Quarterly. New York: Pergamon Press.

CRC Critical Reviews in Bioengineering. Quarterly. Cleveland: CRC Press.

Electronic Design. Biweekly. 50 Essex St., Rochelle Park, NJ: Hayden.

Electronics. Biweekly. New York: McGraw-Hill.

Health Devices. Monthly. 5200 Butler Pike, Plymouth Meeting, PA 19462: Emergency Care Research Institute.

Hospitals. Semimonthly. 840 N. Lake Shore Dr., Chicago, IL 60611: American Hospital Association.

IEEE Transactions on Biomedical Engineering. Monthly. 345 East 47th St., New York, NY 10017: The Institute of Electrical and Electronics Engineers.

Journal of Applied Physiology. Monthly. 9650 Rockville Pike, Bethesda, MD: American Physiological Society.

Journal of Clinical Engineering. Quarterly. P.O. Box 4141, Diamond Bar, CA 91765: Quest Publishing Co.

Measuring for Medicine & the Life Sciences. 175 Wyman St., Waltham, MA 02154: Hewlett-Packard Co.

Medical & Biological Engineering & Computing. Bimonthly. Station House, Nightingale Rd., Hitchin Herts SG5 1RJ, England: Peter Perigrinus.

Medical Electronics. Bimonthly. 2994 West Liberty Ave., Pittsburgh, PA 15216: Measurements and Data Corp.

Medical Instrumentation. Bimonthly. 1901 N. Ft. Myer Dr., Arlington, VA 22209: Association for the Advancement of Medical Instrumentation.

Medical Electronics and Equipment News. Monthly. Chilton Way, Radner, PA 19089: Chilton Co.

15.7 Newsletters and Information Services

The most up-to-date information is found in newsletters. These cover new techniques, devices, standards, regulations, training, etc. for medical equipment. The most important clinical engineering newsletters follow.

AAMI News. Bimonthly. 1901 N. Ft. Myer Dr., Arlington, VA 22209: Association for the Advancement of Medical Instrumentation ($35/yr).

Biomedical Electronics. Semimonthly. 399 Howard Blvd., Mt. Arlington, NJ 07856: Girard Associates. ($60/Yr).

Biomedical Safety & Standards. Monthly. P. O. Box 4141, Diamond Bar, CA 91765: Quest Publishing Co. ($60/yr).

Biomedical Technology & Human Factors Engineering. Weekly. National Technical Information Service, Springfield, VA 22161: U.S. Department of Commerce ($45/yr).

Biomedical Technology Information Service. Monthly. P.O. Box 4141, Diamond Bar, CA 91765: Quest Publishing Co. ($80/yr).

BMET News. Bimonthly. 1901 N. Ft. Myers Dr., Arlington, VA 22209: Association for the Advancement of Medical Instrumentation ($10/yr).

Clinical Engineering Information Service. 5104 Randolph Rd., N. Little Rock, AR 72116: Scientific Enterprises ($65/yr).

Clinical Engineering. Bimonthly. 1901 N. Ft. Myers Dr., Arlington, VA 22209: Association for the Advancement of Medical Instrumentation ($16/yr).

Computer Medicine. Monthly. 399 Howard Blvd., Mt. Arlington, NJ 07856: Girard Associates. ($40/yr).

Devices & Diagnostics Letter. Weekly. 1080 National Press Bldg., Washington, DC 20045: Washington Business Information, Inc. ($205/yr).

Medical Devices & Instrumentation Reports. Weekly. One National Press Bldg., Washington, DC 20045: Drug Research Reports, Inc. ($180/yr).

15.8 Conference Proceedings

There are several conferences that usually have sessions on clinical engineering. If possible you should attend at least one each year. Since it is impractical to attend all of them, you can examine abstracts of clinical engineering sessions in the following proceedings.

Biomedical Sciences Instrumentation. (Proceedings of the Rocky Mountain Bioengineering Symposium). Annual. Pittsburgh, PA: Instrument Society of America.

Digest of the International Conference on Medical and Biological Engineering. Triannual. International Federation for Medical and Biological Engineering. Secretariat, National Research Council, Ottawa K1A OR8, Canada.

Proceedings of the AAMI Annual Meeting. Association for the Advancement of Medical Instrumentation. 1901 N. Ft. Myers Dr., Arlington, VA 22209.

Proceedings of the Annual Conference on Engineering in Medicine and Biology. Annual. Alliance for Engineering in Medicine and Biology, 4405 East-West Highway, Bethesda, MD 20014.

Proceedings of the New England (Northeast) Bioengineering Conference. Annual. New York: Pergamon Press.

Proceedings of the San Diego Biomedical Symposium. Annual. New York: Academic Press.

15.9 Regulatory and Standards Agencies

Regulations and standards are continually being updated. Chapter 4 describes many of the standards related to medical equipment. You can write to the sources of these standards to obtain the latest editions.

American National Standards Institute (ANSI). 1430 Broadway, New York, NY 10018.

American Society for Testing and Materials (ASTM). 1916 Race St., Philadelphia, PA 19103.

Association for the Advancement of Medical Instrumentation (AAMI). 1901 N. Ft. Myer Dr., Arlington, VA 22209.

Canadian Standards Association (CSA). 178 Rexdale Blvd., Rexdale, Ontario, Canada M9W 1R3.

Compressed Gas Association (CGA). 500 Fifth Ave., New York, N.Y. 10036.

Food and Drug Administration (FDA). Bureau of Medical Devices, Division of Medical Device Standards and Research (HFK-300), 8757 Georgia Ave., Silver Spring, MD 20910.

General Services Administration (GSA). 18th & F St., Washington, DC 20405.

International Electrotechnical Commission (IEC). One Rue de Varembe, 1121 Geneva 20, Switzerland.

International Organization for Standardization. (ISO). One Rue de Varembe, Case Postale 56, CH-1121 Geneva 20, Switzerland.

Joint Commission on Accreditation of Hospitals. (JCAH). 875 N. Michigan Ave., Chicago, IL 60611.

MET Electrical Testing Company (MET). 916 W. Patapsco Ave., Baltimore, MD 21230.

National Eelectrical Manufacturers Association (NEMA). 155 East 44th St., New York, N.Y. 10017.

Resources

National Fire Protection Association (NFPA). 470 Atlantic Ave., Boston, MA 02210.

National Safety Council (NSC). 425 N. Michigan Ave., Chicago, IL 60611.

Occupational Safety and Health Administration (OSHA). 200 Constitution Ave. NW, Washington, DC 20210.

Occupational Safety and Health Review Commission (OSHRC). 1825 K St., N.W., Washington, DC 20006

Underwriters Laboratories (UL). 207 E. Ohio St., Chicago, IL 60611.

Utah Biomedical Test Laboratory (UBTL). 520 Wakara Way, Salt Lake City, UT 84108.

Veterans Administration (VA). 810 Vermont Ave., N.W., Washington, DC 20420.

15.10 Study Questions

15.1 List four sources of information about medical devices.

15.2 List two advantages and one disadvantage of a book as a source of information.

15.3 Why would a clinical engineer need to refer to product listings?

15.4 List the steps involved in making a computerized literature search.

15.5 How can you find work that has been done after original publication of a technique?

15.6 Name a journal that would be useful to a clinical engineer.

15.7 Name a newsletter that would be useful a clinical engineer.

15.8 Name a conference that a clinical engineer should consider attending.

15.9 Name a regulatory agency that is important to clinical engineering.

Index

®NEC and National Electrical Code are the registered trademarks of the National Fire Protection Association